THE FREE LIFE OF A RANGER

To Bill
Hope you enjoy this
bit of Forest History.

Archie Murchie, 1990

THE FREE LIFE OF A RANGER

Archie Murchie in the U.S. Forest Service, 1929-1965

R. T. King

*University of Nevada
Oral History Program*

Publication of *The Free Life of a Ranger*
was made possible in part by grants from
the USDA Forest Service,
Toiyabe and Humboldt National Forests

University of Nevada Oral History Program
Reno, Nevada 89557

© 1991 by the University of Nevada Oral History Program
All rights reserved. Published 1991
Printed in the United States of America

Publication Staff:

Production Manager: Helen M. Blue
Senior Text Processor: Linda J. Sommer
Text Processors:
Kay M. Stone, Verne W. Foster and Ann E. Dalbec

Library of Congress #91-30443
Cataloging in Publication Data Available

ISBN #1-56475-000-0

Contents

TABLE OF ILLUSTRATIONS	xi
INTRODUCTION	xiii
LIFE BEFORE THE FOREST SERVICE	1
FIGHTING FIRE ON THE KOOTENAI FOREST, 1929 　　Locating trail. The Granite Creek fire. Fire at Geiger Lake: dynamite and death. Inspecting shelterbelt.	13
HELENA FOREST, 1931-1933: ARSON, TRESPASSING AND WHISKEY 　　"I'm no buckaroo." Postgraduate study. Incendiary Wobblies. Making money by trespassing. Forest stills. Acting ranger.	31
BURNING BUGS AND CRUISING TIMBER ON THE WYOMING FOREST, 1933-1934 　　A permanent appointment. Firing bug trees. A Forest courtship. Cruising timber on the Green River. More bugs and fire.	51

SCALING IN A TIE CAMP ON THE WASATCH FOREST, 1934-1935 71
Tie timber. Hewing and hauling. "They took everything."
Among the tie hacks. Scaling ties. Okies at Archie Creek.
Winter at Hewinta. A wild ride.

KAMAS RANGER AND HORSE CREEK TIE SCALER, 1935-1936 103
A brief appointment. Kamas. "We've never paid for mine prop." "It's the fault of the damned haulers." CCC trouble. WPA: "A rough, ringy bunch." An extra-period fire. Just a punk ranger. Preparing to leave. Tie jam at Horse Creek. Sawed ties.

ON THE CHALLIS FOREST: RAPID RIVER AND LOON CREEK, 1936-1942 131
Interim appointments. Seafoam: the end of the road. Dutch Charlie's vegetarian fur farm. Working in to the Loon Creek station. Tracking a dude. A ranger is always prepared. Fire up Pioneer Creek. We decide to winter over. "Worlds of time" to build a cabin. Mules, horses, and treacherous ice. Snowshoeing over the summit. Cabin fever. A serious injury.

ON THE CHALLIS FOREST: WILDHORSE, 1943-1947 173
An easy district to administer. A dispute and a killing. Roundup. Trespass is trespass. Wartime use of the Wildhorse. Poison larkspur and warbles. Frostbitten on a snow course. Not an administrator.

RAISING A FAMILY ON REMOTE RANGER STATIONS 193

NEVADA AND HUMBOLDT FORESTS: ELY DISTRICT, 1947-1959 207
Bad relations with permittees. Rosendo, the sheepherder. Politics on the range. Deception on Success Summit. A mile-wide sheep trail. Balancing sheep with cattle. Reducing the deer population. Elk poachers. Controlling cougar and bobcats. Dynamiting beaver dams. Dealing with the mustang problem. Winter disaster: the Forest Service responds. The Humboldt and the Nevada Forests are combined. We thwart a national park proposal. "I'd never refused a transfer."

Contents

STAFF OFFICER ON THE TOIYABE FOREST, 1959-1965   247
    Productive supervision. The Dog Valley fire: watershed rehab. Meadow gullies. Reseeding range land. Trespass: fugitives, gardeners and non-shady characters. "Getting to the age . . . . "

UNEXPECTED ADVENTURES: ENCOUNTERS WITH ANIMALS   265
    Mule in torrent, with pack. Running naked in the night. Huckleberry bears. "Instead of a Hereford, it was a grizzly!" She-bear and cub. "You could get hurt." A bear defeats a trap. Cougar kill at Hospital Cove. How many feeds in a deer? A killing frenzy. Porcupine: pretty tasty, salt hungry and in the sourdough. Reading tracks in the snow. Photographing a moose calf. Apache, my mustang colt.

RIDDING THE RANGE OF UNDESIRABLES   291
    Coyotes get the blame and the poison. Cyanide guns. Larkspur non-eradication. Sagebrush removal. "They poisoned an awful lot of rabbits." Tame horses gone wild. Capturing mustangs. Bad mustangers. Hunting mustangs. "An island surrounded by mustangs."

MANAGING THE RANGE   305
    Tepeeing and salting out. Close herding ruins the range. Transient bands. A counting game. Salt in the saddles. Sacrifice areas. Improving the range. Erosion of the watershed. Controlling erosion by trenching. Snow surveys. Reseeding. Equity of grazing fees. Right or privilege? "We should have started cutting numbers."

FIGHTING FIRE   333
    The best fire detection system. Smokechaser fires. Maintaining telephone lines. CCC: "They were ready to go." Fire plows and Pacific Marines. When a fire was crowning. Fire in Hells Canyon. Hot shots. Let some fires burn?

AN EVOLVING MISSION: TRAINING AND POLICY   359
　　The Buckskin Bibles. A forestry education. Blasting: training and regulations. Property transfer. "Rangers had to furnish everything." "Better not be any squawking." Administrative authority. "You had to live a Forest Service life." Self-supporting districts. Armed rangers. Contracting out. Summer homes on Forests. Regional differences: trespassing, fire and budgeting. "I was never directed to preserve anything." Public relations, public service. Vandalism.

GLOSSARY   385

INDEX   401

PHOTO CREDITS   407

Illustrations

Frontispiece	v
Tree-top lookout	17
Burning snags	25
Map 1	33
Treating beetle-infested trees	57
Map 2	58
Peeling and counting ties	77
Splash dam construction	78
Scaling logs	91
Dutch Charlie	135
Map 3	136
"The end of the road"	195
Murchie family	196
Archie Murchie on the Ely district	223
Map 4	224
Map 5	249
Archie Murchie with pack string	267
Sheep wagon	313
Counting sheep	314
Snow course/contour trenching	325
Maintaining telephone lines	341

Introduction

ORGANIZED, DESTRUCTIVE AND unrestrained exploitation of forestlands was an unfortunate feature of economic growth and western expansion after the Civil War. In 1891 Congress responded to this ruinous looting of a national heritage by authorizing the withdrawal of forest reserves from the public domain. To the Department of Interior went the responsibility of administering these national assets; the first to be established was the Yellowstone Timberland Reserve. When Congress transferred the forest reserves to the Department of Agriculture in 1905, they became national forests, and the United States Forest Service was created to manage them. Today the National Forest System encompasses almost one hundred and ninety million acres of land, the vast majority of it in western states. In the five states where Archie Murchie was posted between 1929 and 1965 (Montana, Wyoming, Utah, Idaho, and Nevada) the Service controls almost sixty million acres, upon which logging, grazing, mining, hunting, fishing, and many other activities are permitted and regulated. National forests may not be inhabited, but their Western extent is so vast, and the use and exploitation of them so varied, that they and the people who have managed them must be considered important factors in the twentieth century social and economic history of the region.

While the legislative, regulatory and statistical dimensions of Forest Service administration and management are well documented, there is scant record of how the Service has actually operated in the national forests during the eighty-six years of its existence. In *The Free Life of a Ranger* I set out to capture something of the fundamentally human quality of national forest problems and Forest Service solutions through the memory of someone who had been charged with taking direct action at the most basic level–the ranger district. Rangers operate at the point where policy encounters the reality of nature and the forces of human enterprise, human politics, and the human social dynamic. Results are often unpredictable. In the period spanned by Archie Murchie's career, rangers had considerable latitude to interpret policy, and thus, perhaps, were more personally influential in determining how a district was managed than today's rangers can be. As Archie might say, sometimes that was a good thing, and sometimes it was not.

The reader will learn some things about Archie Murchie in the main body of the book, but he deserves an introduction. By all accounts Archie was an extraordinary forest ranger. Certainly, he was recognized as such by his peers; and while preparing the interviews from which this book is derived, I learned that his name lives on in Region Four, a quarter century after his retirement. Although the specifics of his renown are now blurred in the institutional memory, various documents support Archie's enduring reputation. A 1949 confidential communication to the regional personnel office, written by the Forest Supervisor of the Nevada National Forest, specifically addresses his role in fighting the treacherous Hells Canyon fire: "Mr. Archie Murchie . . . was assigned to the Hells Canyon fire as a Division Boss–a position of extremely great responsibility in view of the hazards which existed on this extremely stubborn and dangerous fire. He was regarded as an excellent fire man and turned out a tremendous amount of work in his division. He has the ability of sizing up a situation readily. He knows what to do and how to do it. He is steady and positive in emergencies. He is a good organizer and drives forward with everything he has. We feel that Archie could handle a big fire situation almost anywhere in the region." This could easily serve as an official summary of the qualities that Archie regularly demonstrated throughout his career.

Although he has been retired from the Forest Service for many years, Archie's body still carries the marks of his chosen profession. Thousands of hours in the saddle left him noticeably bow-legged, his face is deeply weathered, and for some time now the doctors have been removing keratoses from parts of his fair skin that were routinely overexposed to the sun. The man is also smaller than he used to be, as one would expect in a person of his years whose life's work was so physically demanding: he was about six feet tall, and throughout his career his weight averaged around one hundred and eighty pounds; now he is not so tall or muscular. However, he remains lean, fit, active, and self-sufficient.

With his wife Jane, Archie lives in Jacks Valley, south of Carson City, Nevada, in a comfortable house that he built with his own hands and considerable skill. The house and its lawn occupy a quarter of a one-acre plot, the remainder of which is cultivated. In the rain shadow of the Sierra Nevada, Jacks Valley is naturally a sagebrush desert, so when he was building his house Archie sank a deep well from which he irrigates his land. Behind the house he maintains extensive vegetable and flower gardens and numerous fruit trees, all enclosed by borders of pines and ornamental trees, and he also keeps laying hens and a hive of bees. Following each growing season the Murchies put up the produce of the gardens and fruit trees, and when the hunting season arrives Archie consistently gets his deer, usually on familiar hunting grounds in eastern Nevada. As the frontispiece photo reveals, in his early eighties Archie Murchie is still a tough, alert, confident, self-reliant fellow—the consummate ideal of a Western forest ranger living in exemplary retirement. It is just about too splendid to be credible. There is no grain in the image; it is a picture taken with a fast shutter through a small aperture.

Archie Murchie was and is as described above; but clearly this cannot be all that he was and is. The outline is based on what I know of him from Forest Service files, conversations with some of his acquaintances, and my own limited time with him. Although Archie certainly possesses a full human complement of complexities and contradictions of character and personality, they are only obliquely glimpsed in *The Free Life of a Ranger*. This is because the work was not intended to be biographical, and the preparation and interviewing that I did were directed less toward learning about Murchie the man than toward other goals: I wanted to record in some detail the life of a ranger in the decades before the position became bureaucratic and office-centered; and to reveal, as much as is possible through the experiences of a single ranger, the philosophy and methods of the United States Forest Service in its management of the national forests. As the months of interviewing and review unfolded, however, I found myself following as often as leading, moving in unexpected directions that added texture and depth and fuller development to the record. This always happens in the best, most satisfying oral history encounters, and it is part of the allure of the method.

Archie may have been a terrific ranger and he is a hell of a guy, but obviously these virtues alone would not make him a suitable subject for the project I had in mind. Fortunately, he brought considerably more to the exercise. I have been employing oral history methodology for nineteen years, and I have never had an interviewee who embodied so many of the qualities that we look for in the model. Archie is bright, articulate, and thoughtful, and he has remarkably strong and precise powers of recollection. (I was able to check a variety of things against the documentary record, and I found that he rarely erred; when he did, as is so typical of human memory, the error was generally one of quantity or number, such as an incorrect date.) He had clearly been passionate about his profession, and had paid close attention to what was happening around him during his thirty-six years in the Forest Service. Because his mind is inquisitive, he had always sought to understand what was going on, and he was able to provide not only detailed description, but also analysis that he could support from direct observation.

My approach to the interviews was calculatedly historiographic. By this I mean that since our interviews would produce a collaborative, creative interpretation of the past, I wanted that interpretation to be informed by the historical imagination. Once I was satisfied that Archie had the memories, I set about trying to draw them out of him in an organized fashion consistent with what I believe to be the strengths of the historical process. The structure provided by my questions was intended to reveal change over time, causality, and sequential development (one thing leading to another) where they existed. The product of all of this was fifty-one tape-recorded hours which resulted in sixteen hundred pages of transcription—and very little discernible organization of the contents. Only in the particular, only within its discrete elements, is structure apparent in the original transcription; and lack of structure is not the only thing that inhibits accessibility to the information embedded in the transcript. Such is the dynamic of oral history interviewing that in a work of this size, the verbatim transcription, taken as a whole, speaks with a voice that is almost unintelligible. Oral communication, when represented in print–stripped of gesture, inflection, tone, and other components that go unrecorded on tape, or for which there are no symbols on the keyboard–is encumbered by fractured syntax, false starts, repetition, and numerous other impediments to clear understanding of its intended meaning. Disorganized, incoherent, and longer than the telephone book–clearly, the number of people interested in reading that sort of thing would be few, and even they would have to be interested principally in the way the interviews developed to attempt it. I have tried to make that experience unnecessary.

Although it has been composed to read as a first-person account by Archie Murchie, *The Free Life of a Ranger* is not an oral history transcript: it is instead my *interpretation* of the work that Archie and I created over all those months of interviewing, reviewing, and documentary research. This constructed narrative, as I call it, is an effort to make the fruits of oral history methodology coherent and accessible to the average reader. These are not Archie Murchie's words precisely as he spoke them, in the order in which he spoke them, but I have recreated his speech as faithfully as possible consistent with the aim of composing a readable volume from the elements of the interviews. In addition, my questions, which established the structure and elicited the detail of the work, have been subsumed into the narrative, and I have imposed a measure of chronological and topical order on the whole that was not always evident in the interviewing. Nonetheless, it was not my intent to polish Archie Murchie's story to a high shine, and *The Free Life of a Ranger* reveals some of the problems inherent in working with what is available when it does not precisely fit the chosen form.

This constructed narrative to a certain extent reflects the episodic quality of the oral history exercise from which it was drawn. That quality is apparent not only in the way in which chapters are internally organized, but also in the way in which they all fit together. Chapters One through Eleven provide the reader with a fairly straightforward chronological reading of Archie's career, and they introduce (and frequently fully develop) most of the topics and characters that are central to understanding the themes of the book. The final five chapters provide analysis, discussion, and more detail about previously introduced subjects; they are not chronological in their sequence, but I have tried to give them some internal order. Perhaps someone more adept than I at working with this sort of material could have brought it to a graceful conclusion, but I could not. The book ends abruptly.

Archie Murchie has read the finished manuscript in page proof form, and affirmed in writing that it accurately interprets the content of the interviews upon which it is based. Still, I hope that there will be some readers who are interested in examining the unaltered record. Copies of the tape recordings of the interviews, their transcriptions (with corrections noted), and extensive collateral material are in the archives of the Oral History Program of the University of Nevada, Reno. As with all such efforts, while I can vouch for the authenticity of *The Free Life of a Ranger*, I do not claim that it is entirely free of error. It should be approached with the same caution that the prudent reader exercises when consulting government records, newspaper accounts, diaries, and other sources of historical information.

The reader is advised that some of the sentence construction and punctuation that will be encountered in this book are departures from standard practice: they derive from my efforts to preserve the flavor of Archie's speech. (For example, ellipses are used not to indicate that material is missing, but to represent pauses or unfinished thoughts.) Archie also used quite a few colloquialisms, inside terms, and expressions that have fallen out of use. These appear in the text without clarification, but an extensive glossary is provided at the back of the book. Some capitalization may also appear odd, such as Forest when "national forest" is meant but not said, but I have tried to keep this to a minimum, and it should be easy to catch on to the usage. Finally, although I have attempted to match the illustrations as closely as possible to the passages that they illuminate, perfect conformance was not always attainable. For a precise accounting of dates and locations, please consult the photo credits.

A work like this is necessarily something of a collective enterprise, beginning with the participation and cooperation of the interviewee. Archie Murchie was a willing and enthusiastic subject, and one who patiently endured questioning on a number of topics that he would not personally have placed on the agenda. My naivete about many things having to do with the Forest Service and national forests must have amused him, but he never laughed openly–at least not while I was around. Working with him was truly a pleasure, and I hope and believe that we have established a lasting friendship. All of the interviewing was done in the Murchie house, where Jane Murchie was an unfailingly cheerful hostess. She also contributed a good deal of information about raising a family in the Forest Service, and sat in on some of the interviewing on that subject.

I am indebted to Tom Wagner for putting me in touch with Archie. Tom was then (1989) on the staff of the Carson Ranger District of the Toiyabe National Forest. Guy Pence, the Carson ranger, was instrumental in procuring Forest Service funding in partial support of the publication of this book. He was also of great assistance in helping me locate collateral documentation and photographs, and in serving as liaison with the other offices in Region Four that provided photos, documents and the correct spelling of the names of the many Service employees who are mentioned in the narrative. I am grateful to Guy and his staff for their considerable efforts on behalf of the project.

The camera-ready master of *The Free Life of a Ranger* (including its dust jacket and photo pages) was typeset and formatted on personal computers, and printed on a laser printer in the Oral History Program office. This considerable achievement can be credited entirely to the energy, intelligence, imagination and perseverance of Helen Blue and her assistant, Linda Sommer. The program had never before attempted such a polished publication; its computers are obsolescent and temperamental; but the finished appearance is, I think, extraordinarily professional. Helen and Linda were helped within the program by Verne Foster and Ann Dalbec. Nick Cady, Cam Sutherland and Tom Radko of the University of Nevada Press were generous with their advice and suggestions, as was Bob Blesse, who teaches printing at the university. Kris Pizarro created the clear, clean, computer-generated maps, and Steve Davis made the copy prints of the photos that appear in the work as halftones. Fred Radford, Skip Broten and Larry Tuteur accepted the various technical problems we had with our computers and printer as personal challenges, and they were always successful in overcoming them so we could stay on schedule.

Working on this book has been a rewarding experience, due in no small part to the efforts of the many others who have been involved in its production. But it has also been satisfying because I found the subject fascinating. I hope that it can be read with profit by those with a specific interest in the history of our national forests, and I believe that the human themes that are developed within are sufficiently universal that it can also be read as an exploration of twentieth century America's ambivalence about nature and man's place in it.

ROBERT THOMAS KING
Reno, Nevada
August, 1991

1

LIFE BEFORE THE FOREST SERVICE

THE BEST JOB in the Forest Service was a ranger's job. A ranger was his own boss and most of his work was out-of-doors; and if it were possible, I would go back, start all over and live my career through again—I enjoyed it that much.

I always was an outdoor person, beginning in my youth on the plains of North Dakota. Grandfather Peter Murchie was born and raised in Quebec, but in 1881 he took off for North Dakota with his brother-in-law, Dan Shank, their two horses, a wagon, a plow, and a primitive reaper. They couldn't drive a team from Quebec to North Dakota, so they loaded their stock and equipment on the Canadian Pacific Railroad, which hauled them west to Crystal City. There they unloaded their belongings and drove the wagon down to where Grandfather was going to homestead. He took up 160 acres right on the Canadian border, but some years later the border was resurveyed and it was found that part of his land was actually *in* Canada, so the government gave him additional acres on the United States side to compensate for the part that was lost.*

Dan Shank helped my grandfather build a dwelling shack prior to putting a crop in. There were quite a few cottonwoods growing along the coulees, so they hewed cottonwood logs into framing timbers, and they hauled lumber for the shack from Crystal City, about fourteen or fifteen miles away. Since they didn't have a seed drill, they planted their grain by hand, and that fall they cut the grain with their reaper, which was a forerunner of the binder. A binder would cut grain and tie it into bundles, but this reaper just laid it out on the ground, so they had to gather bundles of grain by hand; they also thrashed the grain by hand.

That fall of 1881 both men went back to Quebec, but the next spring Granddad and his wife Mary moved their family of three children to the one-room shack on the homestead. My dad was four years old. Only my Uncle Will, who was the oldest, Dad, and the oldest girl, Mamie, were born in Canada. Two other children, Susie and John, were born later in North Dakota, and another child, maybe more, died in infancy. At that time it wasn't uncommon at all for children to die fairly quickly after they were born, for lack of a doctor and proper care.

Granddad was of Scottish ancestry, and Grandma was born in Inverness, Scotland. Five or six Scottish friends came from Quebec to North Dakota with them in that spring of 1882, and they also took up homesteads in the area. (Dan Shank took up a homestead just east of Granddad's, so close that you could see his buildings from Granddad's farm.) They formed a Scottish community, isolated from anybody else, and they called it Woodbridge Settlement. Later on they had a post office and also a school at Woodbridge.

Some time later the Great Northern built a railroad through North Dakota, more or less parallel to the Canadian line, but about sixty miles to the south. They ran a number of spurs up toward the Canadian boundary, most of them ending around three miles from the line or even closer than that. Along these spurs townsites were located about seven

---

* The Homestead Act of 1862 offered to any U.S. citizen (or intending citizen) who was the head of a family and over twenty-one years of age 160 acres of surveyed public domain after five years of continuous residence on it and payment of a registration fee ranging from twenty-six to thirty-four dollars. As an alternative, after six months of residence the land could be purchased for $1.25 an acre.

to nine miles apart. When those townsites were set up the city of Sarles was the last town at the end of one spur, and about fourteen miles east of Sarles was Hannah, which was the last town on its spur. Grandad's place was closer to Hannah, so he traded mostly there.

Granddad had taken up 160 acres, which was a regular claim. Later on, he staked a 160-acre timber claim. Back in those days—with lightning and the Indian fires, and with all that tall native grass—the land was burned off periodically and there was very little timber. In places where there was a live stream, you had cottonwoods, box elders, willows and trees like that growing, but out away from water there were practically no trees. Any homesteader that would plant so many acres of trees could take up another 160 acres as a tree claim.* Then there were what they called preemption claims. I don't know where the "preemption" came from or just what it means, but a homesteader could buy an additional 160 acres for $1.25 an acre.† So Granddad wound up with 480 acres of land. That constituted the farm, the home place. There he mostly raised wheat, oats and barley, and later on some flax and rye.

My dad, James Murchie, was nineteen when he decided to homestead. He borrowed a horse from his father and he borrowed money to buy a second horse, so that he had a team. He took a plow and the same old reaper, and he moved about fourteen or fifteen miles west of Granddad's place and took up a 160-acre homestead right next to the Canadian boundary near Sarles. To prove up on a claim you had to build a house on it, plow the land and put in a crop. After a certain number of years the government would give you title to the land. There also was a policy that if you had not been on the land long enough to prove up on it, but you wanted to leave, you could sell your interest in the 160 acres for what you could get out of it, and the buyer could finish proving up on the land and receive full title. The young fellow who had homesteaded just west of Dad wasn't much of a farmer, so he sold his interest to Dad.

---

* An 1878 revision to the Timber Culture Act of 1873 authorized any person who planted and kept ten acres of trees in good condition to acquire 160 acres thereof.

† The Distribution-Preemption Act of 1841 authorized those who actually settled public lands to stake claim to and purchase up to 160 acres at $1.25 an acre.

Dad finished proving up on it, and this gave him two quarter-sections or 320 acres.

Later on, when North Dakota became a state, the federal government gave the state sections sixteen and thirty-six, and they called them school sections. They were strictly for the support of schools. The states could sell these sections to raise money for building schools, and Dad bought the west half or the two west quarters of section thirty-six, which was just south of his land. That gave him a full section of land. Years later, shortly before Dad died, he gave me the two school quarters, and my youngest brother got the two quarters along the Canadian line. Jack still owns his two quarters, and I still own mine. I rent mine out to the same farmer that's renting Jack's two quarters.

My mother was born in Bergen, Norway, March 19, 1882. With her parents, Lars and Olive Peterson, she came to the United States in 1884. They settled first in LaCrosse, Wisconsin. I don't know what kind of work Granddad was in, but in 1898 they moved to Edinburgh, North Dakota, and in 1905 they moved to Sarles.

Some time before 1905, the grain separator had been invented and came into use. The early ones were forty-eight inch separators, *big* separators, and they required anywhere from twelve to fourteen bundle teams to haul bundles in to the separator. The thrashing outfits would move around the country to one farm and then another, thrashing grain. Their separators were powered by big self-propelled steam tractors, which also pulled a big bunkhouse on four iron wheels, where the men slept in double bunks; and then they had a cook house about the same size, and pulled the same way, where the men ate. When my mother and her folks were still living in Edinburgh, she and her sister got a job on one of these thrashing outfits as cooks and helpers. This outfit happened to thrash my dad's grain, and that's when my dad first met my mother. They carried on the relationship, and in 1905 they were married. From this union three boys and one girl were born, and I was the second boy.

About the time Mom and Dad were married, Grandma and Grandpa Peterson moved to Sarles, and there Grandpa worked on the railroad. I remember that he had two big oxen that he kept at his place, and we kids were scared to death of them. Grandma was a registered nurse and she helped deliver all us children. She helped different doctors deliver a lot of babies, and she was well on in years before she had to stop. Grandma Peterson was a very happy, congenial individual right up until

Life Before the Forest Service

she had a stroke before she died; she was just bubbly all over, and us kids really got along with her. She was very agile for an old woman: any trick that we could do as a kid, she could do . . . and we were very close to her. Now, Grandpa was very different, and I don't have too many happy thoughts of my Grandfather Peterson.

On the Murchie side of the family, we kids were a lot closer to our grandpa than we were to my grandmother. Grandma Murchie was very strict, and you toed the mark when she was around. But Grandpa Murchie was just the other way–he was a lot like my grandmother on my mother's side. He was a happy, enjoyable man, and he had a good memory and a real good mind. I think he was ninety when he finally just died in his sleep, but right up until the last his mind was just as good as could be.

One windy, dusty day in the spring of 1918 Dad was seeding a field. He had hauled the seed grain in a wagon out to where he was working, and on one trip back to it to re-fill the drill he found some stray horses eating the grain. Dad chased the horses away, but in the dust he couldn't see well, and when he walked around the end of the wagon one horse that had not fled kicked him in the face and broke his jaw and his eye socket. They thought for sure he was going to lose his eye.

Dad was laid up for quite a while with that, but he finally got over the injury so he could go back to work. Then in the late summer of 1919 he got hurt again on one of the quarters in section thirty-six that we called the Pasture Quarter. It was kind of rocky, and we had fenced it off for pasture for the milk cows and beef cattle. Every evening one of us boys went up to the pasture and got the cows and drove them the half mile to the farm where we'd milk them; then we'd keep them there at the farm and milk them in the morning, then take them back up to the pasture. Late one afternoon my older brother jumped on a saddle horse, no saddle or anything, and went up there to get the milk cows. My dad also raised Angus beef cattle, and he had just bought an Angus bull. The bull was as gentle as could be . . . never had a bit of trouble around the barn. We'd turn him loose to water and he'd come to his stall, and we never had a bit of trouble. But when Pete rounded up the milk cows this afternoon and was bringing them down to the gate, the Angus bull was off to the side snorting and bawling and pawing the ground. Pete could see that he was getting mean, so he bailed off his horse and crawled under the fence and ran home.

Dad was mowing hay to the south of there a little ways, and later on he came by, heading home, and he saw the saddle horse out there in the pasture. Dad figured maybe the horse had stumbled and thrown Pete or something, and he looked around, but he couldn't see Pete lying on the ground any place; so he climbed on the horse and rounded up the milk cows and started toward the gate. The old bull started acting up again, but Dad said he never thought anything of it. He was riding right alongside the three-wire fence, and all of a sudden, there the bull came! He hit the horse and tossed him over the fence, and when the horse went over the fence, he turned over and came down on his back with Dad underneath. The horse lit across Dad's chest and broke a whole bunch of ribs and drove them into his lungs . . . but the horse rolled off of him, and Dad said he got up and didn't think he was hurt. He got up and the horse was just laying there, so he reached down, took ahold of the reins to see if he could get the horse up, and the minute he pulled, he just passed out. Apparently the horse wasn't hurt, because he got up and came home. When we saw him there we couldn't figure out how he got out of the pasture, but we thought nothing of it.

Later on, getting pretty late in the evening, Dad regained consciousness. He crawled over to the mower, climbed up into the seat and tied himself to it with the reins and got the team started, and then he passed out again. By then we were all outside, wondering where Dad was, because he should have been home a long time before that. Finally we see the mower coming, but nobody's with it, and the horses are coming up the lane going to the house. When they got pretty close, we could see Dad slumped over with his head between his legs. As soon as the team got to the house, Pete stopped them and we got Dad untied and laid him on the ground alongside the house. He was out like a light.

There was a farmer that lived a mile east of us by the name of Martin Estinson who had a new baby-grand Chevy, and he came up and got Dad and took him into Sarles to the doctor. They took him away from there to a hospital, and it was quite a while before he came home. The doctors said that he'd never work again. Boy, you wouldn't even know it was the same man when he came home, he'd lost so much weight; and pale . . . .

Us kids ran the farm while Dad was gone. I was eleven and Pete was thirteen, but we'd been doing farm work for some time. We could drive teams . . . we didn't have any motorized vehicles, but there was nothing on the farm we couldn't do, and during thrashing neighbors came over

Life Before the Forest Service

and helped. But Dad decided that if he couldn't farm any more, we'd move into Sarles, which we did in 1920. And then he rented the farm out.

I had started school in the one-room Dash School, where one teacher taught all the grades from the first through the eighth. The Dash School was the closest to our farm, but later they said we had to go to the school in our own township, and we transferred to the Bryan School. In fair weather we got to school in a buggy with one horse, but when the snow came we rode on a sleigh that was used to haul manure out of the barn. The manure sleigh was just two four-inch by six-inch runners with planks nailed across them. During the coldest part of the winter we didn't go to school at all, but made up for it in the summer when we could.

I was in the sixth grade when we moved into town and I transferred to the Sarles grade school and went on to graduate from high school in 1926. In high school I engaged in sports quite a bit. I was pretty good in track (running the mile and the half-mile) and fair in basketball, but we had such a small school that we didn't have a football team. We did have a band, and I played the trumpet, but I never stayed with it. My youngest brother played a clarinet, and he got pretty proficient at it. All three of us boys played an instrument, but he was the only one that stayed with it.

My dad was the head push around our house [laughter]–main guy in the family–and he was very strict on us getting our homework done. Boy, that homework was done before anything else was done! I don't know just how far Dad had gone in grade school, but I doubt it was beyond the fourth or fifth grade, and my mother probably didn't go much further in school than the fifth grade, either. Comparing theirs to education now, my parents really didn't have that much, but we had lots of books in the house, and both of them were well-educated, self-taught people. My dad was just as good a mathematician as you'd want to come by, and he was, you could say, a jack of all trades; he was a good carpenter, and he could do most anything. Quite a lot of my early knowledge of building I got from him, and he played an important part in what I turned out to be.

We subscribed to a very good monthly magazine that had a lot of outdoor stuff–I believe it was called *Boys Life*. I may have read a story in it and gotten the idea that I wanted to be a forest ranger. The free life of a ranger–that's what appealed to me. I liked just getting out and prowling around, and in a way I was kind of a loner. I used to take my .22 rifle and go out and hunt rabbits by myself . . . or my youngest brother'd go with me. By the time I was ten years old I had my mind made up to become a forest ranger, and I never did deviate from that goal.

Dad was a stout-built man, a lot stouter than any of us kids were, and he eventually recovered fully from his injuries. Even in advance of complete recovery, he took up operating a dray line in Sarles. I was twelve years old when Dad got the dray line, and we helped him as much as we could. On the dray line he hauled the freight that came in by train from the depot to the different stores or people who had ordered it. He also hauled all the coal that came in. Everybody burned coal then, especially in the wintertime, and there were carloads and carloads of coal that we had to haul and unload. Boy, I've shoveled a good many tons of coal [laughter], and I wasn't too old when I was shoveling it, either. There was sand that was shipped in, and carloads of cement and all different kinds of machinery that had to be unloaded and hauled.

Dad was Scotch and he saved money; he laid quite a bit away, but he was still in the dray business when the banks closed in the Depression, and he had money in both the Sarles banks. He had to carry big accounts because he paid for the freight and then collected from the stores, so his checks always had to be good. When the banks closed he lost quite a little bit of money.

During the summer months, if Dad didn't need me, I'd usually hire out to shock grain on different farms around there. I would also do other kinds of labor. Besides my dad, the man who probably had the greatest influence on me was an old English stonemason by the name of Elva Phillabaum. I was about fourteen years old when I hired out to him, and he did everything. He laid sidewalks, put basements underneath houses, laid brick, did a certain amount of building . . . and in those days you never built a concrete wall or a concrete floor unless you broke the rock first. Every rock that he put in a floor or in a wall had to have a broken face, because he said (and in a way he was right) that the concrete wouldn't hold to it good unless it had a raw, broken face. My first job was breaking rocks for the old guy. I also mixed concrete for him, and I got so I could lay up a wall and do everything from laying brick to any kind of concrete work. He was an awfully good teacher, who was real quiet and never bawled me out or really criticized me.

I was lucky to acquire these skills, because when I got into the Forest Service.... Back in those days, a ranger did everything. If he had concrete work to be done, he did it, unless it was a big job like a basement underneath a house or something. But if it was ordinary concrete work, or laying brick, or minor carpentry work, or wiring, the ranger did it. I did a lot of that sort of thing as a ranger, and I am thankful to have had a chance to work with old Elva Phillabaum, because he taught me an awful lot.

I graduated from high school in 1926. Although I wanted to go on and become a forest ranger, it was not entirely clear to me what the best preparation would be, and money was a problem. I had written to the North Dakota Agricultural College in Fargo and learned that they offered forestry-related courses. Fargo was not too far from home, and in-state tuition was quite low, so in the fall I enrolled there. However, I soon learned that though the college offered forestry courses, it did not offer a *degree* in forestry, and I had come to realize that I would be better off with a forestry degree. I resolved to transfer to a university with a forestry school as soon as I could manage it.

One of my aunts had taught at the Great Falls, Montana, High School for many years, and she'd been over most of the state and knew a lot of the forestry people. She encouraged me to get a forestry degree from the University of Montana, so I wrote to their forestry school and got all the literature and a list of costs. It was affordable and it looked like just what I wanted. In the fall of 1928 I packed up my bags and headed for Missoula, where I enrolled in the University of Montana School of Forestry.

I transferred from the Agriculture College of North Dakota with two full years of college, so I'd already taken a number of subjects that were required for a degree from Montana. Montana offered both forest management and range management. I decided to take range management, but by not having to repeat the courses that I'd already taken at North Dakota, I had room for working in a lot of forestry courses. So when I got through, I had taken most of the forestry courses except logging engineering.

The forestry school basically trained their people for two careers: to go with the Forest Service or to work for the logging industry. Those that could pass the civil service exam went to the Forest Service; those that couldn't, went to private industry. One big appeal of the Forest Service–I'm just speaking for myself now–was that it paid a fair salary. It didn't pay as much as the logging industry, but it paid a fair salary. You had good security, far better than any other job probably in the range or forest industry, and you had what was a good retirement program for the time. Now, you take a young fellow planning on maybe getting married and having children . . . that steady wage coming in month after month, the security, the good retirement and good sick leave, all meant a lot.

There were far more opportunities in the logging (or forest management) end of the Forest Service than in the range end of it, and probably twice as many students took forestry as took range. We went on a number of field trips to see how things were done in the logging industry. About 1930 I went on a forestry field trip to a Great Northern or Northern Pacific operation up north of Missoula that was treating railroad ties with creosote. Practically all the ties they were treating were hewn by hand out of round timber, but a few sawed and squared ties had just started to come in. They had a big cylindrical tank at least eight feet in diameter and probably sixty or seventy feet long. One end could be opened, and little railroad tracks ran into the tank. Ties would be loaded on small flat cars and run into the cylinder until it had taken all the cars it could, then the door was closed and creosote was pumped in until around two hundred pounds pressure built up in the tank. I don't remember how many hours they held that pressure before they drained the creosote out, but they would get very good saturation of the ties. They also had the ability to create a vacuum after draining the tank, and this vacuum would suck all the surplus creosote off of the ties, so when they came out they were practically dry. You could touch them and you wouldn't get any creosote on your hand. With squared ties, they could do even better. They would run them through a machine that had spiked rollers that put holes about a half inch deep in all four sides, so when the pressure came on it forced the creosote to penetrate the ties a lot better.

We also spent a lot of time out in the field along the Blackfoot River in western Montana, going through those Anaconda sawmills and logging camps, and we saw the whole operation. All the logging was done in the wintertime, and the company had a Shay engine and train that ran up the river and hauled the logs down to the mill at its mouth. (The Shay was a narrow-gauge steam engine. It was different from other railroad engines in that it was quite small and power was applied to the wheels on only one side of the engine. Shays were used a lot on logging operations in the West at that time.) The actual logging was all horse-and-chute logging, done in the winter when the chutes would be slippery from snow and ice. It was strictly the easiest logging, and it was better in a lot of ways than logging today, because there was very little of the surrounding forest torn up. They could build these chutes in the summertime in any direction or place they wanted, so that in the winter they only had to skid logs a very short distance with their teams to get them into a chute. To make a chute, first they would lay what they called bed logs—logs about fourteen to eighteen inches in diameter and probably five feet long, bedded in the ground. Then they'd take two logs and face them with a broad axe, and set them so the faces were at about a forty-five degree angle to create a kind of cradle for logs to slide down in. These faced timbers were separated by about four inches at the bottom so any snow or debris could fall through and wouldn't slow the logs down.

Anaconda built log chutes that ran way back up the side of the mountain, and they would skid the logs to a chute, dump them into the chute, and they would come down by gravity. At some places they would run too fast, and there would be a guy with a fire burning next to a little bank of dirt to keep it thawed, and he would be tossing dirt into the chute to slow the logs down. Further on down, maybe they wouldn't run fast enough, and a man would be shoveling snow in the chute to keep them going. Then there were other places where they had to go over a little rise, and there would be a fellow stationed there with a team and a light chain and a set of what they called snatch tongs. He would wait until three, four or five logs had piled up, and he would snatch these tongs into the tail log and drag the logs over the high place, and on down they'd go.

The chutes would run down to the bottom of a canyon, where they had a number of landings. At each landing there was a pole about six or seven inches in diameter and sixteen to eighteen feet long, with a pin in either end, and on either side of the chute they had holes bored. They would set the pole at an angle across the chute and drop these pins in the holes to lock it. The logs would come running down, hit the pole and roll out onto a landing. If the log didn't roll straight, or was twisted, men down there with peaveys would kick it around so that it rolled straight onto the bed of a waiting logging car. Then when this landing filled up, they would move their whippoorwill to another landing and roll off another bunch. There was very little actual hand work.

Even though forest management was emphasized in the forestry school at Missoula, I went there with the intention of going on into range management after graduation. The main reason was that I already had a background in range management: I was born and raised on a farm; I knew livestock from one end to the other; and I'd already taken two years of schooling in an agricultural college. At that time in the Forest Service you had forest districts and you had range districts. There was opportunity in Region One for somebody who was in range management, but the percentage of positions available wasn't near as high as it was for forestry. Now, for a person that went to a college down in Region Four, the opportunity would have been greater for a range man than it would be for a timber man. In Nevada, except on the Sierra front, most of the districts were what they would call range districts, and many of them have never sold even a foot of saw timber.

Missoula is in Region One in what they used to call the Inland Empire. That included Montana, Idaho, eastern Washington and eastern Oregon . . . that was about the size of it. There were probably eight to ten national forests in the area, and the various Forests tended to hire people from the forestry colleges that were located within the region. This practice may have had something to do with summer employment: forestry students weren't required, but they were strongly advised to put in their summers working for the Forest Service. In doing that, students got acquainted with a lot of rangers and supervisors, and occasionally even with people from the regional office. If they were good men and had worked on a Forest for three summers, that Forest would often try to pick them up after graduation. (And if the Forest couldn't pick them up, the regional office might.) I believe the Depression changed that custom.

2

Fighting Fire on the Kootenai Forest, 1929

TRAIL LOCATION ON THE Kootenai in 1929 was my first job with the Forest Service. A buddy and I wrote around to most of the Forests in Region One, asking for a summer job. From the Kootenai National Forest we got our only offer, so we took it. My buddy was chief of party, and I was his assistant; I got $105 a month, and he got $115.

Locating trail Most of your main streams on the Kootenai had a good Forest Service trail running up them, but all the country in between was trail-less. In places the timber and brush was so dense that you were lucky if you could make a mile an hour, and sometimes you wouldn't even make that. Our job was to locate trails up through this growth to the top of a ridge and follow the ridge a ways–ten or twelve miles–and then drop back into a trail on the creek. It was just opening up country, but we had standards that we had to go by. Our ruling grade for a trail was 15 percent, and then up to a certain distance we could go about 30 to 35 percent, and for an extremely short distance you could go 45 percent. It was a very interesting job.

When locating trail we didn't carry a tent. We had something called a bed tarp, which was seven feet wide and fourteen feet long, and we'd either just sleep under that or we would make an A shelter with it and sleep inside. That was before the days of sleeping bags, and we were checked out two Army wool blankets. Then we had a little mess kit for two men, which included three galvanized pots, each of them with a lid. The pots fit one inside another, and the lids of the two smaller ones were our plates. Nested within were two cups and knives and forks, and the lid of the big pot was what we used for a frying pan.

Everything was to be packed from place to place on our backs, but neither one of us had done any amount of backpacking; so before departing for the Kootenai we went to one of our profs who was a real outdoorsman—he'd hiked all over the country—and he showed us how to make a pack harness. In addition to shoulder straps he used what was called a tup-line—a strong leather band that comes around your forehead. (It was about two and a half inches wide, with a buckle at one end and holes at the other.) You'd get your pack adjusted to your shoulders, and you'd put this tup-line over your forehead, run it down along the sides of your pack, and tighten it up underneath so that when you leaned over you took most of the pack's weight off your shoulders. And if you leaned your head back, your shoulders would take the weight.

When I went in to the Kootenai I took a fifteen-inch collar, but when I came out I couldn't even button a sixteen-inch collar! [laughter] Packing with a tup-line developed your neck that much. At first it was difficult, but your neck muscles finally got so stout that you very much preferred to pack with a tup-line, which was far safer and more comfortable than packing with shoulder straps alone. Crossing a stream on a foot log, you keep your balance better with a tup-line. Also, packing with shoulder straps alone bothered your arms, which would eventually get kind of numb from the pressure. Leaning into the tup-line relieved the pressure and left your hands free, and in some of that country you needed both hands to go where you wanted to go. We got so we'd pack mostly with the tup-line.

The Granite Creek fire

Although my partner and I were on trail location, when the fire season got bad they called us in to the Turner guard station, which was about twenty miles from the end of the road, and they also brought in a ten-man fire crew. It was heavy-timbered country, and they didn't have very much horse pasture near the station, so when the fire crew wasn't fighting fire they were clearing timber off for a pasture. There was always plenty of work to do.

Fighting Fire on the Kootenai

One day in late July or early August, shortly after my partner and I were detailed to the Turner guard station, the fire crew had gone on the trail and the dispatcher called up and told me that a fire had been reported on Granite Creek, and I was to take fire tools and go to it. This was my first small fire–the first I would be fighting alone. The packer had pulled in just a little before that with a pack string, so he told me that he'd take me as far as we could go by trail. We took off, and we finally came to where I had to leave the trail. When I got ready to go on alone, he asked me if I had enough tobacco. I told him no. I said, "I don't smoke."

He said, "You can't fight fire without chewing tobacco. You'll die of thirst."

And I said, "No. I'll get by."

But he insisted, and he pulled out his plug of tobacco and cut off a chunk and gave it to me. Well, he was right. Boy, at times you get thirsty when you're fighting a fire. So, when I got to the fire I cut off a little bit, and I chewed it, and it tasted pretty good. That started me chewing, and I chewed for about three years before I finally quit.

This was a lightning strike fire. Nearly all of your lightning strikes are single strikes. Sometimes you have two strikes, but this was three strikes–came down one right after the other–and it was in a spruce bottom. There wasn't any water running, but I dug a hole where I could get enough water to drink.

In those days, if you were on a fire for three days and you didn't come in, they'd send a replacement. Three days was the limit. (Of course, if you got on the fire and they had given you time to control the fire, and the fire didn't diminish, you had help right away.) Finally, at the end of the third day they sent a replacement in. I had the fire pretty well controlled; it wasn't out, just stopped and controlled, and I showed him what I'd done and told him where the possible hot spots were. Then I took off and went back to the station.

The lookout on Turner Mountain had a little farm down on Pipe Creek, and he had a little bit of hay on it, and his wife couldn't take care of it. He wanted a couple of days leave to go down and put up his hay, so the fire dispatcher told me to go on up and pinch hit for him while he was gone. His lookout was a tree-top platform.* The practice of placing a lookout platform atop a tall tree was eventually phased out, but it was common in Region One at that time, perhaps because of the terrain. This is interesting country–the Kootenai River and all that country to the north. It is old, old country. That area was glaciated thousands of years ago, so the valley bottoms are real wide, and the ridge tops are gently rolling, not coming up to any sharp point. There was good soil, clear over the top of the ridges, and about anyplace you wanted to go. This was all virgin timber country, and there was a very heavy stand; in fact, it was the heaviest timber that I've seen anyplace I've ever been. Even on top of a ridge, you had tall trees all around you. There was no way you could see out unless you climbed high up into a tree, which I've done lots of times.

To make a permanent lookout they would top a tree and perch a platform on it, high enough so you could look out over the top of all the timber and see in whatever direction you wanted to watch. At the Turner Mountain lookout there was a tent with a dirt floor, near a tree about thirty inches in diameter at its base. Up where the tree trunk was about fourteen inches across they'd cut the top off and built a lookout platform. This platform was braced down to the side of the tree, so the platform was sturdy enough–it was solid. Then they had a handrail about three and a half feet high around the outside of the platform, and there was a hole in the floor, through which you climbed up onto the platform. I didn't question how sturdy it was; the thing that bothered me was the swaying. When the wind was blowing the whole thing swayed back and forth pretty good. Even a tree that size, you'd be surprised how much it would sway back and forth. The first time I went up there, boy, I sure hung onto that rail for a while! You got used to it, so it didn't really bother you, but I wouldn't want to be up there in a lightning storm, because there was no protection.

About my third day there I saw the lookout coming back up the trail, so I thought, "I'll scan my country once more." You always started at a certain point, and you scanned the country slowly around, and you always came back to the same point. The point I started from was where

---

* In Forest Service parlance, "lookout" could mean either the person who watched for fires or the vantage point from which this was done.

Lookouts were often constructed on topped trees in heavily timbered areas. "The thing that bothered me was the swaying."

Fighting Fire on the Kootenai

I'd been on the little fire up on Granite Creek, and I looked there. I hadn't seen any smoke since I went up to the lookout, but I scanned around and finally came back to where my fire had been . . . and a puff of smoke came up! Boy, in just a little while there was a lot of smoke, and I hurried down the tree, which had spikes set in it like they put in telephone poles to climb on. One of these iron mine telephones was on a post at the base, and I rang the dispatcher and told him that the Granite Creek fire had blown up. He wouldn't believe me. He said, "That's impossible. That thing's been dead for days."

I said, "Well, she's blown up. What do you want me to do?"

And he said, "You get back to the guard station. If the fire crew has already left and there are horses still there, pack all the grub you can get on a horse and take it to the fire for the fire crew. We'll have a pack string coming in to replace the grub at the station, and then we will also be bringing fire equipment and grub and supplies in to the fire. So as soon as you leave the trail, you start blazing a trail up to the fire so that the packer has something to follow to get there."

As soon as the lookout got to me I was all ready to go. I told him what had happened, and I took off. When I got to the station, I found that everybody had gone ahead to the fire, so I packed all the grub I could get on the one horse that was left, and I started out afoot leading her. By the time I got to the mouth of Granite Creek where I left the trail it was getting pretty close to dark, but I started blazing a trail. When I got about halfway between the end of the trail and where I remembered the fire to be, I saw a light ahead. And I thought, "What in the world? That fire can't be down this far." I kept on going, and here was the fire crew sitting around. It kind of teed me off, because I'd worked my fool head off getting down from that lookout, and getting that horse and grub and blazing that trail. I said, "What are you guys doing down here?"

They said, "Well, the fire is just up the canyon a short ways." I could see the light of the fire, and I asked, "How did it start? How did it move so fast?"

It turned out that the fellow they'd sent in to replace me was a jungle bum. (They called them jungle bums because they lived down by the railroad tracks, in what they themselves called the hobo jungle. These were unemployed, homeless men.) Now, back in those days, the Forest Service hired a lot of jungle bums as fire fighters out of Spokane, and they were dispatched all through Washington, Oregon, Idaho and

Montana. Many were experienced men who had fought a lot of fires. Most of them were damned good workers.

The fellow who had taken my place was named Finnegan . . . I'll never forget his last name. He told me what he'd already told the crew: that after I left there were very few smokes showing up, and as soon as one showed, he would put the fire out. He sat at the upper end of the burned area and watched, and whenever a smoke showed, he'd go down and dig the fire out and bury it. The fire appeared to him to be completely under control . . . but down in that spruce bottom there was a lot of moss that hung from the trees, long stringers of it. This moss would hang right to the ground, and in the middle of the summer it was as dry as tinder. If you touched a match to it, it would burn just like a torch. Any tree that had been inside the burn area before the fire had been brought under control had this all scorched off, but outside the fire there was still a lot of moss hanging. On an afternoon when Finnegan hadn't seen a smoke for a while, the wind came up. Apparently there had been fire in a root that went down underneath the fire line, and it had been smoldering, and with the wind it burned out under the trench and finally came to the surface. He said it was just a little flame when it first flared up, but it caught in the moss hanging from one of these spruce trees. That moss was like kerosene: it just exploded when it was hit. In a matter of seconds, the whole stand was engulfed in fire.

Finnegan had his bedroll on down a little ways, and he didn't have enough time to rescue that, but he gathered his tools and started down the canyon. The canyon ran east and west, and the wind was from the west, coming down the canyon, and the fire was right behind him, and he started dropping tools. He didn't know whether he was going to make it or not, but finally he outran it. When the fire crew found him, he'd had it–he was all in.

The next morning Finnegan's feet were so sore from running down the canyon in his cheap, worn shoes that he couldn't work, so we left him in camp. The rest of us went out and fought the fire. We made progress on our side, but it was spreading to the west. Late in the afternoon a lone aeroplane–it looked like a World War I biplane–flew over the fire up high, circled around for a while and left. He was trying to learn the extent of the fire.

Since Finnegan had to stay in camp, he was going to be the cook and have supper ready for us in the evening when we came back. Well, he was a jungle bum, and in the hobo jungle, as I understand it, they had a custom of having a big pot and dumping in it anything that anyone could scavenge. I'd brought a fairly good-sized kettle in from the guard station, and when we got back from fighting fire that first day we discovered that Finnegan had dumped into the kettle about everything we had except our fruit. He had a grin from ear to ear, and said, "Boy, do I have a mulligan for you guys!" It wasn't too bad; it tasted pretty good, but to have it for breakfast and for supper was too much. Then the second day it soured, and we were in there without any grub.

The packhorse I had brought in was an old mare. There was very little feed for her around the fire camp, but there were ferns that grew as high as your head, and I turned her loose in them. She fed on these ferns, and talk about halitosis! . . . I never knew a horse to smell the way she did. [laughter] In the evening two days after we started fighting the fire we came back to camp and discovered that the old mare had pulled out. I was sent after her. If I was closer to the guard station than to camp when I found her, I'd keep on going to the station. She had grazed her way down the creek, and it was after dark when I caught up with her, so I went on down to the station. As soon as I got there I called the dispatcher. He said, "Where in the hell have you guys been?" And I told him. He said, "Didn't someone go around to the fire and tell you fellows all to come out?"

I said, "No. We haven't seen a soul all the time we were up there."

He swore; he said, "The fire's so big that we've decided to put a camp on the west side of it. Go back up and get everybody and pull them out." He said, "Turner guard station has been left with no one there, and if we get another fire we have nobody to put on it." So I got a good night's sleep and took off the next morning and went up and pulled the crew out.

Fire at Geiger Lake: dynamite and death

Well, the Granite Creek fire kept on burning; it burned until snow finally put it out in the fall. It burned over into Idaho; it burned even up into Canada, I was told–a tremendous fire. But our crew wasn't again involved in fighting it. We worked out of the Turner guard station for a couple of weeks, and then on a Saturday toward the middle of August we were pulled out and taken in to the supervisor's office in Libby, Montana, for two days off. I hadn't had a lick of ice cream all summer, and I was hungry for it. My buddy, Bill Brown, and I went down to the drugstore and ordered us each a dish of ice cream. We finished it and I ordered a second, but Bill returned to the supervisor's office. Pretty soon he came rushing back with news that a fire had started up at Geiger Lake, and we were being put on a crew that would fight it.

The Geiger Lake fire was more than thirty miles northwest of Libby, with no road up to it. Going in we hiked clear into the night of the first day, and then stopped and built a campfire. There were about forty of us, and I heard more bear stories that night than I had ever heard in my life! [laughter] Next morning real early we took off, and it was well after noon before we got to the fire. We had brought in fire packs on our backs, and we had a mule string come along behind with more equipment. There was already another camp in there, but we were welcome reinforcements.

The Geiger Lake fire was not nearly so large as the one that had started up Granite Creek, but because of special circumstances it proved difficult to fight. The summer of 1910 had been one of the worst fire seasons that the United States probably has ever known. The 1910 fire burned through Idaho and Montana and some of it might have got into Washington, and there was, I guess, millions of acres of timber burned. Fires burned together, so that they didn't know where the fire lines were. A small part of the Kootenai Forest had been swept by the 1910 fire, but the rest of it was mostly virgin country, and boy, some of the prettiest timber that you ever laid eyes on grew in that Forest! It didn't make any difference whether it was cedar or white pine or tamarack or hemlock or what it was, it was exceptionally good timber, and it was worth preserving; it was worth protecting. But before we could get to it, the Geiger Lake fire spread into a valley bottom where there were old cedars that had burned in that 1910 fire.

Some of these old cedars were seven, eight, even nine feet in diameter, and they were swell-butted. All of them were rotten in the center, and over the years the tops had broken off of a lot of them. When fire burned through a shell into the soft center, the trunk could act just like a flue, building up intense heat. This heartwood was dry; and I won't say it was weightless, but it didn't weigh very much. Sometimes when a burning snag like that got a good draft it would breathe out pieces of flaming heartwood debris. If you had any wind, those big sparks would carry ... oh, I saw spot-fires three-quarters of a mile ahead of the fire. They could spot clear out in this virgin timber. Those trees had to come down, but some of the trunks were too big to cut with saws (the saws weren't long enough); and those that were afire that you *could* cut, as soon as your saw went through the shell and got into that inferno, it lost all its temper and its blade turned blue. We either had to fell the burning cedars with hand axes or blast the biggest of them apart with dynamite. It was the only time in my life that I ever saw dynamite used on a fire.

With us we had an old man who had done a lot of mining and knew powder. He was told to pick somebody out of the crew to help him . . . to be his powder monkey. For some reason he picked me, and I was later glad that he did. He took me out, and in the first two hours he taught me more about the use of dynamite than you could ever get out of reading a book. At first we just sat on a log and the powderman showed me how to put the dynamite in holes, and how to load the cap in the stick of dynamite, and how to cut the fuse and everything. Then I went with him and did all the digging to place the dynamite under the roots of the trees. He'd load a hole and I'd watch him; and he'd go to another place and he'd load a hole and I'd watch him; and then he'd have me load a hole. Finally about the second day he turned me loose putting the dynamite in the holes, but it was quite a while before he'd let me load the charge–put the cap in the last stick to ignite the rest of the dynamite.

If the powderman loaded six holes he'd cut all his fuses the same length . . . they were quite long. Then he'd walk slowly around the tree to all loads and time himself to check just how long it took. Then he would trim the fuses, leaving the first fuse the longest. The second fuse was cut a little bit shorter, and so on around, each one a little shorter. His idea was that when he lit the fuses all the charges should go off at the same time.

To light the fuses he would make what they called a spitter. He would take a piece of fuse about two feet long, and bend it so he was holding both ends of it together. Then he'd cut slits half an inch or less apart all the way around. (You want to be sure you cut clear to the powder.) When he was ready, he would light his spitter, and fire would shoot out of each cut as the powder burned to it. With this he would go around and light the longest fuse first, and so on around to the shortest fuse; and when they were all lit he'd throw the spitter back at the base of the tree and we'd run like everything for cover, because there would be dirt and rocks and wood flying all over. [laughter] That old guy was good enough at it that you'd swear there was a single explosion!

You had to load a tree heavy enough that you knocked it down the first time. If you didn't, there was no use putting any more dynamite under it because the ground was so shook up that your dynamite wouldn't do any good. So you always loaded them plenty heavy. With some of those great big trees we'd have to put four or five (or one time even six) charges all around the base. A few of those trees actually lifted straight up out of the ground for several feet when the charges blew. The smaller ones would lift up a few feet, and then they'd come toppling down.

There's a condition that you have in snags, especially when they're burning on the inside: way up towards the top–maybe where a branch has broken off, or where there is a woodpecker hole–some opening will let the fire come out, and it will eventually burn the top of the snag off, and down it comes! We call these *widowmakers*. They're correctly named, because sometimes they let loose without any sound or warning, and the first thing you know, here they come toppling down. At the Geiger Lake fire there were six of us in a crew with a foreman, and the foreman had given us very definite instructions that when anybody yelled, "Widowmaker!" we were to drop everything and get out; not to look one way or the other, just get out of there. And everybody knew it–every one of us knew it!

We were felling many of these burning cedars by hand axe, with two of us chopping at one time. I was ambidextrous–I could chop left-handed or right-handed–so one fellow chopped on one side and I chopped on the other. The fire was burning all around the trees. We had to put green boughs down to stand on so we wouldn't burn our shoes, and the smoke was so heavy that you could only chop for a couple of minutes or so and then you had to get out of there.

The foreman called to us to drop our axes and step out, which we did, and this young fellow stepped in and took my axe, and another went around and took my partner's axe. I had so much smoke in my eyes that I couldn't see anything, but I stumbled out of there. Suddenly somebody yelled, "Widowmaker!" One fellow ducked his head and ran, but the man who had taken my axe just stood and looked up, and the end of that widowmaker caught him and caved his forehead in. It didn't knock him out–in fact, at no time was he unconscious–but it knocked him down,

Smaller burning snags were felled with hand axes or saws. The largest of them had to be blasted apart with dynamite.

and he started crawling, and he was getting into the fire, and a couple of guys ran in and grabbed him and got him out. They sat him up against a log, and he just sat there. You could see a little bit of his brain in the hole, but he didn't bleed very much for some reason. It surprised me that he didn't bleed very much. He just sat there. He'd start laughing, and he'd laugh like everything; then he'd stop; then he'd start crying, and then he'd sit there and just stare straight ahead, and pretty soon he'd start laughing again. We cut two poles and ran them through the inside-out sleeves of three jackets to make a stretcher, and we got him back to the fire camp. By saddle horse it was a long trip from the camp to the closest road, so some of the guys made a sling between two horses and covered it with a canvas and laid him on top of it and started out with him. I don't think they were over seven or eight miles from camp before he died.

We lost another man on that fire. The easiest way to travel was down the fire line, because you had your trench to walk in, just like walking on a trail. But it could be dangerous, and everybody knew to run for cover or get behind a tree when somebody yelled. The next day a crew was coming into camp, the foreman leading them. Somebody yelled and everybody went for cover, but the foreman looked back and thought he could outrun the falling tree. Well, he missed outrunning it by about two feet, and this snag hit him on the head and it killed him dead. (I was on another fire, the 1949 Hells Canyon Fire, where a man was burned to death. I fought quite a few fires in my Forest Service career, but those three fatalities were all that occurred during fires I was on.)

The men who were killed on that Geiger Lake fire might not have died if they were wearing helmets, but we didn't have helmets or protective clothing or protective gear. Back then they fed you good, but you furnished all your own clothing and gloves and stuff like that, and there was no protection of any kind furnished you. We didn't even have headlights; we had what they called a Stonebridge lantern.

Stonebridge lanterns were made out of zinc-coated metal, and there were four windows in them. The lantern was open at the bottom; it had a kind of frame that you shoved a candle up through. At first we used beef tallow candles, which you never dared to leave out anyplace, because pack rats and mice would eat them. They'd eat them just as fast as they could get to them! [laughter] Finally they came out with wax candles, which the rodents didn't bother.

The windows in Stonebridge lanterns were made of isinglass, which didn't let all that much light through in the first place, and would soot up and become practically useless. A lot of us preferred what we called Palouser lanterns . . . perhaps named after the Palouse country of eastern Oregon. Lard could be bought in five-pound and ten-pound buckets, wide at the mouth and tapered down to the bottom, and the inside of them was highly tinned. You'd take the bail off a lard bucket, punch a hole under the rim at the top and punch another hole at the bottom, and reattach the bail lengthwise for a handle. On the side opposite the handle, at the bottom end, you'd take your pocketknife and cut an X in the tin, and bend the four points up. Then you'd shove a candle up, and these four points would hold it. After the candle was lit, it had to be pushed further in periodically as it burned down.

The Palousers were good; you couldn't beat them. In lots of ways these tapered lard-bucket lanterns were better than headlights. The light didn't shine out very far ahead, but it gave you a wide body of light, and going through country without any trails, they were good. Of course, you had to carry a Palouser in your hand, and that left only one hand free, so if you were going through brush and timber, it was kind of inconvenient. Nineteen twenty-nine was the only time that I really depended on Palousers for light. I made them afterwards, but then you depended on them; that was your only light, period!

Inspecting shelterbelt   In the fall I returned to forestry school. One day about two or three weeks before school let out in the spring of 1930, the dean called me into his office and asked, "Do you have a job lined up for the summer?"

I said, "Yes, I'm going back to the Kootenai on trail survey."

He says, "Well, I got a request here from the Northern Great Plains Experiment Station in Mandan, North Dakota, for a man to do shelterbelt work for the summer."

On the Great Plains at that time there were very few trees. Homesteaders had planted some, and farmers had planted trees around their places, but there weren't that many, and when the Department of Agriculture put in the experiment station at Mandan they grew trees that through experiment they found suitable for those cold, blizzardy, Dakota winters. These trees were made available to farmers who applied for them, with certain conditions. The station would send a man out to look the ground over and to decide whether the location was suitable, how wide a shelterbelt was needed, and what species of trees should be planted. Trees would be furnished to the farmer for free the next spring if he would sign an agreement that he would do certain things: he would prepare the soil the way the station wanted it prepared; he would cultivate the trees for so many years; and he would keep the shelterbelt fenced so livestock wouldn't get in and destroy the trees. If these were all agreed to, next spring he would be sent the trees with planting instructions.

The summer following the planting, the experiment station would send a man out to inspect the planting and determine if it was satisfactory. Then five years after the shelterbelt was planted, it was inspected. They were inspected again in ten years, and a few after twenty years. My job in the summer of 1930 was doing the inspecting. They gave me a brand new Ford roadster and all the folders for the shelterbelts on the different farms and ranches. There were photos of each farm or ranch, and a map on how to get there. When you got into Montana, some of those ranchers were way out in the boondocks, and I had an awful time finding some of them. But it was an interesting job. Some of the farmers abided by their planting agreement, and, of course, some of them didn't.

3

Helena Forest, 1931-1933:
Arson, Trespassing and Whiskey

I RETURNED TO SCHOOL in the fall of 1930 for my senior year. By 1931, what with the Depression, jobs were really getting scarce, but I got together with another young fellow in forestry school who was taking range management at the same time as me, and we wrote to every single national forest in Region One, looking for work. We never got an answer from any of them. I didn't know what to do, but finally, less than a month before graduation, there came a letter from the supervisor of the Helena Forest, asking me if I was interested in a job on the Lincoln district. It was temporary, only for the summer, but I was to function as an assistant ranger. Boy, I didn't even write; I fired him a wire! He wrote me back to be there as soon as I could after school let out.

"I'm no buckaroo" When I reported to the Helena Forest supervisor he told me that I needed two horses, which I would have to buy.
I had brought my saddle along because I knew I'd be riding horses on the district, but I didn't think I would have to *buy* any. This put me in a quandary. I didn't have much money to start with, and I didn't know what to do. But there was a long-geared ranger sitting there in the office who said, "Say, I got two horses I'll sell you cheap. I'll sell you the saddle horse for seventy-five dollars, and the packhorse for fifty dollars."

I thought, "Well, that's the best I can do."

He says, "Throw your saddle in my pickup, and I'll take you out to the station." His station was on the east side of the Missouri River, and I learned that the station I was going to was west of the river, clear over the top of the Continental Divide. It was located near Lincoln, Montana, a small town of about forty people. He says, "I'll take you out to my station and you stay overnight. Then from there it's a good, long, two-day ride to Lincoln.

We went out to his station; his wife was still in town and his kids were still in school, and he was out there all by himself. We had an early breakfast the next morning, and as I was doing the dishes, he says, "I'll go out and run the horses in." So he went out and ran in a good-looking white and a black that seemed like he was plumb gentle but looked pretty old. He put a halter on both of them and he says, "Maybe I better put your saddle on the white. He isn't bad, but he may be kind of rough in the mornings." So he saddled him up and rode him around the corral for a while, got off, and I climbed on . . . no problems. He says, "Well, you are ready to go. I'll open the gate."

We'd already packed the black, and he handed me the lead rope. He'd drawn me a map on how to get to Lincoln by the shortest route across country. There wasn't much detail on the map, but it was pretty accurate and I didn't have any trouble following it. I had to cross the Missouri River on a dam, and I didn't miss it; I hit it all right. Along late in the evening, I finally made it to the road into Lincoln. There was a little store, post office and ranch on the road at a place called Canyon Creek. I rode in there and asked the rancher, "Any chance you can put me up for the night?"

"Yes, sure," he said. I noticed him eyeing my white horse, and finally he said, "You must be quite a buckaroo."

"No," I says, "I'm no buckaroo."

He says, "Well, unless I'm mistaken, they've bucked that horse all over the country," and he walked around him. "Yes," he said, "That's the horse. Where'd you get him?" And I told him. "Yes," he said, "that's the horse, all right."

So I got off and we fed the horses and gave them a good feed of oats. They had a little bunkhouse there, and I went to bed, but for a while I didn't go to sleep. I was thinking about that darn horse, and that when I left Canyon Creek I had to head across country again, and I thought, "Boy, if he ties into me out there someplace and I get hurt, that's it." But

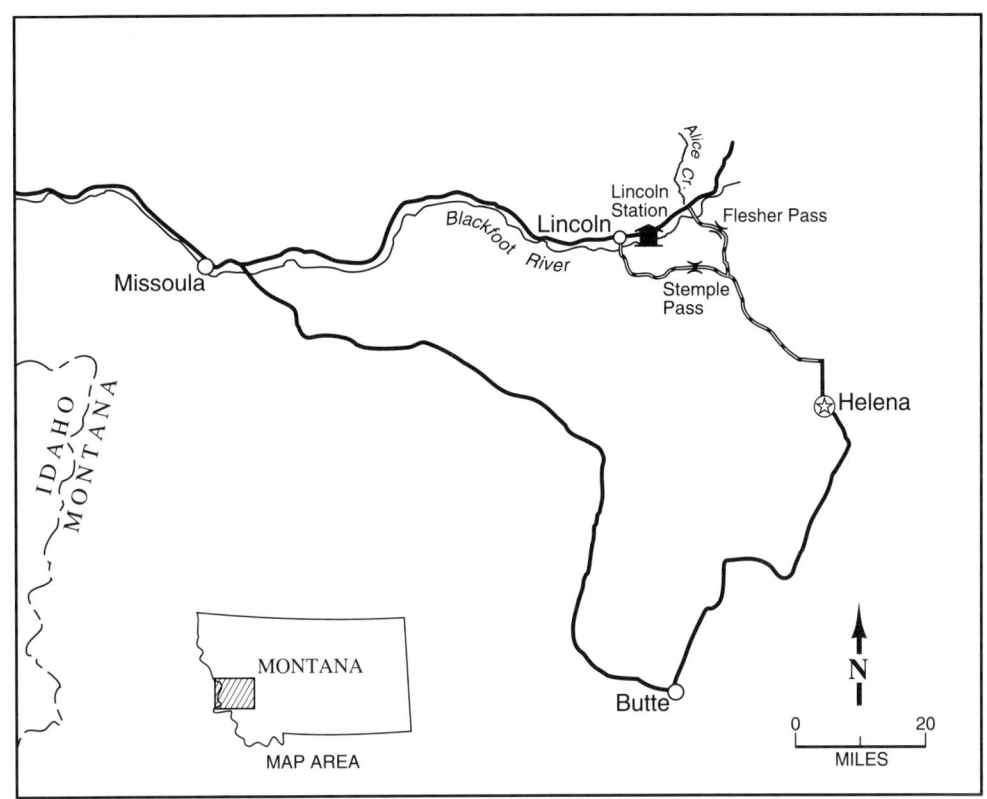

Map 1. Archie Murchie served on the Lincoln district of the Helena National Forest in Montana, 1931-1933.

in the morning I got on him in the corral and rode him around a while and he was all right. Then the rancher handed me the halter rope of my packhorse and I took off. I didn't have any trouble with the white, but I never got off him until I got to the ranger station; when I had to urinate, I urinated right from the saddle! [laughter]

That white was actually a good horse. The only time he ever bucked with me on him was one time when I was fixing supper and was out of butter . . . and the station was a mile from Lincoln. I always had to keep a saddle horse up during fire season until dark, and he was up, so I just ran out and jumped on him and went down to Lincoln, tied him up to the fence and went in and got my butter. I was in a hurry to get back, and I jumped on him and gigged him in the ribs a little hard, and he bogged his head. Boy, he could really get into it! But I turned him into the fence and he stopped, and that was the only time that he ever bucked.

He was a good horse, but he was a tough son-of-a-gun. I sold him to the ranger at the end of the season, and the ranger kept him because he knew I was going to be back the next summer. I didn't lose any money on him, and I rode him again the next year.

That summer of 1931, I wouldn't say I did everything the ranger didn't want to do, but . . . . He had heart problems, and I did most of the range work and all the fire suppression and most of the timber work, and he did all the office work. We had a pretty good-size trail crew, and he did some supervision of the crew, and whenever the supervisor came out he rode with him. But mostly I did all the field work, and I enjoyed it. (I was tickled to death that he did the office work.) It was an enjoyable summer, and it gave me excellent experience that really qualified me when I took over the district on my own a couple of years later.

Postgraduate study I didn't have a permanent job, so in the fall of 1931 I went back to the forestry school just to talk things over with the profs and the dean. The dean suggested that I go on and try to gain a master's degree. Although the forestry school had never issued a master's degree before, the dean discussed it with the university president and they figured that the school was qualified to issue one. To get me through the academic year I had my summer's wages, and I got three part-time jobs: one job was assisting the prof teaching a course in range management; one was a janitor's job; and then I was also conducting a seed viability experiment for the Range Division of Region One. It kept me busy.

I asked the Forest Service if it would be OK to use the data from the experiment for my master's thesis. They said it was OK, but I didn't want to quit my other jobs to concentrate on getting the degree, because money was scarce. Finally I wrote my dad and got some money, thinking, "Well, I'll just quit everything, and then I'll really get going on writing my thesis." But I found out that you can't write a thesis in a week or a month . . . that if you're going to do it right, it is going to take a lot longer. I didn't get it done; I just didn't get it done.

This would have been the first Master's Degree that the Montana forestry school had ever issued. Too bad. But I got to thinking that it wasn't really good to have a Bachelor's Degree and a Master's Degree from the same school, anyway, so I planned on going to California to get my Master's from the University of California at Berkeley. I knew it would take much more money than I had, so I figured on staying out of school a year and working for the Forest Service and accumulating enough. Well, betwixt and between times, I got my civil service appointment and I was transferred to Pinedale, Wyoming, where I met my wife and we were married, and that pretty much ended my quest for a Master's Degree, which I don't really regret. [laughter]

Incendiary Wobblies In the summer of 1932 I was again given a job on the Lincoln district. It turned out to be an extraordinary season. Due partially to the way railroad construction had been subsidized years before, some people were confused about the location of Forest boundaries on the Helena. This confusion led to arson on Forest land in 1932 and 1933.

In the nineteenth century, when the transcontinental railroads were going through, the federal government gave them every odd section of land on both sides of the right-of-way to a considerable depth as a subsidy. Both the government and the railroads soon found it was next to impossible to manage their checkerboard lands, so an act was passed which enabled the railroads and the government to swap sections and consolidate their land in large units. This was an excellent idea, but in practice there was a lot of skulduggery on the part of the railroads in influencing the selection of the lands to be swapped.

When the Northern Pacific through Montana made application for a swap, the government sent out surveyors to look things over and ensure that it would only get land equal in value to what it surrendered. Well, the story goes—and it came from good authority, because it came from one of my profs at the forestry school—that when the government surveyors came out to the area, the railroad wined and dined them and had women for them and whatnot in Missoula, and very few of them even got out to look at the timber. [laughter] As a result, the railroad wound up with some of the best and choicest timberland in that part of the state, the timber along the Blackfoot River.

Some time after that, the Northern Pacific sold all this timberland to Anaconda Copper. At the mouth of the Blackfoot River where it flowed into the Clark's Fork of the Columbia, Anaconda put up a big sawmill and a planing mill. They cut a lot of lumber for sale and for their own use, and, as I remember, they had three logging camps in full operation at one time along the Blackfoot.

In 1932 (and in 1933 as well) one of the big problems along the Blackfoot River, which included much of the Lincoln Ranger District, was arson fires set by IWWs (also called Wobblies).* It was a time when the IWW was active in the area, and they had been having trouble with the Anaconda Copper Company . . . there was a lot of animosity between the IWW and Anaconda. That summer arsonists burned down two Anaconda logging camps when the caretakers were away. (In the summertime nobody was at the logging camps but a caretaker and a bunch of cats to keep the rats and mice down.) Arson even began occurring on sections of the Helena Forest when Wobblies came out and set fires thinking that they were setting them on Anaconda Copper land. They didn't have any idea where the boundary lines were, and we were getting a fair part of these fires on the Forest. We had *many* incendiary fires that summer of 1932, in addition to the expected lightning fires.

The trouble with incendiary fires is you never knew when they were going to start; you didn't have any idea. Now, lightning fires, you know they are going to appear after an electrical storm, and most of them start

---

* Founded in 1905, the Industrial Workers of the World (IWW) was a radical international labor union that had lost most of its vigor and membership by 1924. In the early years of the Depression the IWW enjoyed a brief and only marginally successful revival, principally in western states.

at higher elevations. But incendiary fires would nearly always start at a lower elevation, and they'd always be where there was good fuel. Actually, there was some predictability about *when* the incendiary fires would start. Arsonists set most of their fires on Saturday or Saturday evening, and they would often set two or three or four fires simultaneously. I kind of figure they thought you wouldn't have your full force of men on duty—we didn't have that many to start with—or the men would be lax in their work. Anyway, most of our incendiary fires started Saturday or Saturday evening.

There was a big dance hall at Lincoln, and a lot of people from the Helena area came to the dances that were held there every Saturday night. We were fortunate, in that to get to Lincoln from Helena, you had to cross the Continental Divide through Stemple Pass, and when you crossed over you could see a big expanse of the district before you. About a third of the way down from Stemple Pass we had an old mine telephone, an iron telephone. (They were darned good telephones, but you didn't want to be near one when there was lightning because, boy, you sure could get a shock off of it!) Travelers were very good about reporting fires on that telephone. Coming through the pass, if they saw smoke or light from a fire they would stop and call it in and we'd get men right on it, so as a rule these incendiary fires did not get to any size. We had only two that got big enough to be called project fires.

Very few of the arsonists were ever caught, although Anaconda Copper had its own fire protection organization—the Blackfoot Protective Association—which the forestry dean at Missoula headed up. One time that summer of 1932 I had been out riding, and when I pulled in to the ranger station I saw the Blackfoot Protective Association pickup truck sitting in the yard. Before I even got off my horse, the ranger came to the door of the office and told me, "As soon as you put your horse away, come to the office." So I put my horse in the corral and gave him a feed of oats and went over to the office. The forestry school dean was there, and they all had grins on their faces and Mac said, "Well, we've got a job for you."

I said, "What's that?"

He said, "You are going to do some detective work."

The deal was there were two IWW leaders who had set fire to the summer home of one of the officers of the Anaconda Copper Company there at Flathead Lake. They had also started a fire trying to burn one of the logging camps, but hadn't succeeded. The Forest Service knew they were in the area, and they wanted a Blackfoot Patrolman and me to act as detectives. We were given a little typewritten description of the two arsonists–a column about an inch long. One fellow's name was Nesbitt, and the other one I can't remember. The one who was the ringleader had been doing a lot of talking and speaking to the workers there at Anaconda Copper in Butte. He was described as dark-complected, with a foreign accent. The other one was lighter-complected, and I think he was an American–he did not speak with any dialect.

Down below Lincoln were two big valleys, and there were big cattle ranches down there. In those days the ranchers put up all their hay by hand. One of them, Roy Thompson, had right close to forty men during haying, and a lot of the others had crews pretty near as big. It was figured that the best way for these Wobbly arsonists to hide out until they got ready to do something would be to get on a hay crew, because hay crews were having men coming and going all the time. So they gave us an old beat-up pickup, and I got an old ragged pair of overalls and got them good and dirty, and not too clean a shirt, and I took a darn good hat and tore the band off of it and roughed it up, and dressed like this we started going to different ranchers in the two valleys. I knew quite a few of them, and I told the ones I knew what we were up against. They'd say, "Fine. Go down to the mess hall and have a meal and look the crew over." If anybody asked us why we were there, we were looking for a job, see? We'd just tell them that the boss might be able to put us on.

We hunted and hunted for a week and a half, I guess, until we came to one rancher who had a small outfit–he didn't hire too many men. We didn't know him, and we didn't tell him what we were doing. He wouldn't hire us at first; he wouldn't even invite us to have a meal. But we kept after him, kept after him, and finally he said, "OK. I'll put you on." He told us to hike out in the field and get to work. We did, and we worked until evening. We'd sized the crew up pretty well by then, and the IWWs weren't there. Then we made a mistake: we should have eaten first, but we didn't; we told the rancher that we were quitting. Mad . . . oh, he was mad! He didn't pay us, either. [laughter]

Finally, the last rancher that we figured we ought to go see was this guy, Roy Thompson. I knew him because I'd had his crew help out a number of times on fires, and I told him exactly what we were up against. It was just before noon, so he says, "Well, I don't believe they're here. But you go down to the mess hall, and when the crew comes in you look them over and have a meal, then come up and tell me what you found out."

We went down to the mess hall about a half-hour or so before the haying crews were to come in for the noon meal. Out in front of the building they had a long bench and a bunch of washbasins for the men, so we washed up. Sitting at the end of one of these benches was a sandy-haired guy, probably in his late thirties, who had a broken arm. We went over and got talking to him . . . wondered what happened to his arm. He'd fallen off a hay rack and broke it. Then he wanted to know why we were there, and we told him that we were looking for a job. We visited there, and finally we were pretty sure this was one of the men we were looking for. Then the crews started coming in, jumping off the hay rack. One of the men was a dark-complected, heavy-set guy about five-foot-seven . . . fit the description of the leader perfectly. The minute he got off he eyed us, and then he went over and talked to the fellow at the end of the bench.

While the crews were washing up, the patrolman and I moved into the mess hall and sat down at a table. Sure enough, when these two guys came in, one sat on one side of us and one sat on the other side. And boy, they started quizzing us! They wanted to know where we'd worked. The patrolman had good answers, because he'd worked all over the country, but I didn't have too many. I couldn't say I was in college, and I had to fake that I'd been in North Dakota working in different places. As soon as we got through eating we all went out and stood around for a while. Then they climbed on the hay racks and went back out to work. We thought we had them fooled.

When we had been sent on this expedition, my dean, who was head of the Blackfoot Protective Association, told us that if we located the suspects we were to call Anaconda. (Of course, the Blackfoot Protective Association *was* Anaconda Copper.) So we went up and told Roy Thompson that we'd found the two guys and asked him if we could use his telephone. We called Anaconda, and they were just tickled to death. They said, "We'll have somebody out there right away." Unfortunately, by the time they got out there, the two suspects had vanished. There was something about us that didn't ring true to them, so they took off. But the authorities were waiting for them, and when they pulled into Butte they were captured, and I was told they served a jail sentence.

Making money by trespassing

There were about five bands of sheep and probably around two thousand head of cattle that grazed on the Lincoln district. The grazing was primarily in timbered country, and forage was scattered. Consequently, the grazing allotments were quite large, so there was adequate forage on an allotment if a rancher or his herder wanted to use it. The range was broken down into individual allotments, and the allotments into units, but there hadn't been too much regard given to the location of allotment lines. Some permittees did a good job of staying within their allotment lines, and others didn't. Still, none of the allotments were overgrazed. That's uncommon with sheep, but they had a lot of range on the district and they were moving all the time.

We had a form that was put out by the region that had the grazing units listed on it, and we asked the ranchers to show the dates they went in so we could compute what use they got out of each individual unit. Of course, the form also recorded the permitted number of sheep in the band, the name of the permittee, and the name of the herder. Then there were four spaces at the bottom for losses. One was for strays, one for snagging, one for predators, and the last was for death from foraging poisonous vegetation. These grazing record forms were returned in the fall. I tried to collect the forms in the field, because sometimes when the herder got back to the ranch he'd give the form to the owner, and sometimes he wouldn't. We were supposed to use these in compiling our annual grazing report, but many were practically illegible.

The biggest problem with grazing on the district was that there was a lot of downed timber. Moving from place to place or going to water the sheep had to cross a lot of windfalls, and many of them got snagged. Some sheep would have small pieces of skin torn loose, and on a few of them half of a side might be ripped open. Flies would get to the wounds, and in just a few days maggots would set in. As a result ranchers probably lost more sheep from snagging than they did from anything else—even from coyotes. (Of course, coyotes could rip a sheep up, too, and the wound would easily flyblow.) They had very little medication to treat the wounds with. They used pine tar melted together with axle grease, I believe, and swabbed the torn flesh with that. I don't know how much it speeded healing, but it would help some to keep the flies off. The two main poison plants on the region were low larkspur and death camas. And then there was poison hemlock. Poison hemlock on swampy ground was deadly: livestock could eat the tops and it wouldn't hurt them, but in the swampy ground they'd grab ahold of the top, and if they pulled the plant and the white root came up, and they took one good bite out of it, that was it. That stuff was deadly.

We had one permittee named Chevalier–he was a Frenchman. Chevalier lived on the east side of the Continental Divide, but he was a local permittee. He had three bands of sheep, and in 1931 and 1932 we caught him in trespass. Now, why he wanted to trespass out of his allotment at that time we didn't know, because he had adequate range, but from what we found out afterwards, he deliberately turned the sheep in on this range off of his allotment. In 1931 we caught him in trespass, and the ranger wrote him a letter and warned him. In 1932 we again caught him in trespass. That time, the ranger got a good count on the number of sheep, and the herder told him how long he'd been in trespass in that area, so the ranger trespassed Chevalier, and charged him for the amount of forage that he'd taken.

Betwixt and between the incidents in 1931 and 1932, Chevalier had been talking, blowing his mouth off. He told other permittees that he was glad to pay trespass fees because he got far more feed off of Forest land when he was in trespass than he ever paid for–that he was actually making money by trespassing. Following the 1932 incident Lyle McKnight, the ranger, took the written charges in to the Forest supervisor and told him what the deal was and what Chevalier had been saying. The grazing fees weren't very much then, and I don't think the trespass fee amounted to much more than forty-five or fifty dollars at the most. So the Forest supervisor wasn't too happy, and he said, "What do you think we ought to do?"

Mac says, "Well, I think we ought to make it a lot stiffer than we did last time."

"OK," says the supervisor, "It's going to be five hundred dollars." (The amount beyond the trespass fee was what you'd call punitive damages. That's the same as punishment for trespassing.)

So the Forest supervisor typed up a letter of transmittal to send in to the regional office with the trespasser's money. Mac and I drove on out to Chevalier's ranch with the letter, and it was right around seven o'clock by the time we got there. When we drove into the yard, Chevalier came out. He was a big, stout, happy-go-lucky guy. We visited, and he said, "Have you had anything to eat?" And we said no.

He said, "Come on down to the cook shack." So we went down, and he had the cook fix us some nice, big lamb chops. Boy, we ate royally! We went back out, and he said, "Let's go into the office," where he turned to Mac and said, "What are you guys here for? You're not here just for a visit."

So Mac told him; he said, "We found your sheep in trespass again."

"Oh," he said, "that blankety-blank herder! You can't trust those herders for love or money." He blamed it all on the herder, so Mac told him that I had talked to the herder. (I shouldn't have done it, because it probably got the herder fired.) Mac repeated what the herder had told me–that "he went on because you had told him to go on." And then Mac said, "You've been telling other permittees and other people that you're actually making money off the arrangement when we trespass your sheep. And we don't like that."

Chevalier says, "How much this time?"

So Mac gave him the letter of transmittal, and he looked at it and said, "Five hundred dollars! Man, that's awful steep, isn't it?"

Well, the supervisor had told us before we left, "If he objects to it, you tell him the next time it's going to be a thousand dollars." So Mac told him.

Chevalier sat there a minute, and he said, "Well, there won't be a next time." And there wasn't. He never trespassed again. [laughter]

(In Region One we could do that, but when I got into Region Four–Challis and Ely and any place in Region Four–it was entirely different. I don't know whether they were afraid of lawsuits or whether it was word that had come down from Washington or the regional office or . . . that's something I don't know. For a while after I came to Region Four, it was hard for me to adjust to this. Time and time again when we were charging them for trespass, the fines didn't anywhere near cover the value of the forage that they had taken, and it was a lot more profitable for them to trespass and pay the trespass fees than it was not to trespass. Region Four had a lot of trespasses, but there was no place in Region Four that I was ever allowed to charge punitive damages. I discussed this with supervisors, but they never had a reason for it. They just said, "We can't do it.")

Forest stills Prohibition was still in effect in 1932, and there were a lot of stills in the district—a lot of them. These stills were scattered all around, and the operators would keep moving them. An operator would run off a batch of shine, and then he would move his still before the Prohis could locate it. They made varying degrees of moonshine—I've tasted some, and some of it was pretty good and some of it wasn't.

Most of the moonshiners were darned good fellows. If there was any lightning strike anywhere near them, boy, they were on it, because they did not want anybody prowling around where that still was! They were one of the best fire-detecting forces I had. They would also come in when they were going to move their still and tell us, "If you see smoke coming up from the head of such-and-such a creek, don't worry. That is my still." But there was a few that didn't notify us, and they caused problems; and once in a great, great while, a still would blow up and start a fire. The ranger and I took our trail crew once and went to a fire, and it was a still that had blown up. I don't think the owner of it was there when it happened, because if he'd been there, it never would've blown up. Of course, we put the fire out.

Prohibition officers would drop in every once in a while and ask if we knew if there were stills someplace on the Forest. We never told them, and the main reason was that as long as we were on the good side of those still operators, we never had to worry about them. But if you got on the wrong side, they could set more fire and do you more damage than you could ever think of; and they knew it and we knew it. So, as much as we could, we cooperated with them.

One time the ranger and I almost got ourselves shot by fooling around a still. There were two roads coming in to the Blackfoot River or to Lincoln. One came over Flesher Pass and the other came over Stemple Pass. The two passes were on the Continental Divide, and between them there was a distance of close to twenty miles. There were no trails in the area. Ranger McKnight decided to build a trail between the two passes, so if we had a fire in the area we would have more easy access to it. He and I were locating this trail and we'd taken a canteen of water with us, but we'd used it; and boy, we were thirsty! We were working on the east side of the Continental Divide, and saw a big spruce grove below us. Mac says, "Well, where there's spruce, there's probably water." So we dropped off and went down just a little ways and hit a trail that had been chopped out recently, and there were horse tracks on it. Mac said, "Oh, where does that go? That has never been here before." So we followed it into this spruce grove, and sure enough, there was a still.

I'd never seen a still before. I saw the remains of the one that was blown up, but I'd never seen one in operation, and it was kind of interesting. Boy, it was running the home brew off like mad! We saw that somebody'd been around there just recently, and he had to be running the stuff out, so we didn't touch anything. We walked around, looked at it, took a good drink and left.

It wasn't too awful long later that I was down to a dance there in Lincoln. They had a nice dance hall, and the moonshiners would bring their whiskey in to sell to the dancers. (One guy would bring his moonshine in and sell it to whoever wanted to buy it. He had a cabin there in Lincoln, and if anybody got too much to drink, he'd take him over and put him to bed in his cabin until he sobered up; then he could go.) I got talking to one heavy-set fellow at the dance, and he says, "You remember a while ago when you and Mac found my still over there on the other side of the Continental Divide?"

I said, "Was that your still?"

He said, "Yes. I was lying there in the timber with a .30-30 rifle on you, and if you had started tearing it apart, I would've dropped you." And I don't doubt that he would, because there was no way in the world that anybody could ever have pinned it on him . . . no way in the world.

Each moonshiner had his own system. They used a lot of sugar, and some of them would bring corn in and some of them would pack barley in . . . they prided themselves on their own formula for the whiskey they made. Some of it, I would say, was just as good as the whiskey you get now. But some of it . . . a guy would go blind, I guess, if he drank enough of it. [laughter] Of course, they had a tremendous market. They could sell every pint they made. Helena was just real close; Butte was close; Missoula was close. They didn't have any trouble selling what they made.

Acting ranger That fall, after I had completed my summer appointment on the Lincoln district, Bill Chapin, a young fellow that I had graduated with, called and asked if I was interested in going down to Denver to get a job driving a Greyhound bus. His cousin was married to the guy that was manager of the Greyhound bus depot there, and she had told Bill all he had to do was come down and a job would be waiting for him. It sounded good, and we went down. Well, the jobs were just as scarce driving buses as they were doing anything else. They had a waiting line of drivers a mile long, so we started looking for other jobs. The Depression was really tough then, and man, we sold insulation, we sold pictures, we solicited for milk for orphan children, and that's not all we did! And we were just barely breaking even. Finally Bill went broke; I went on for a little bit longer, and then I had to give it up too. I was going to hitchhike back to North Dakota, but Bill's cousin (the wife of the Greyhound man) wanted to go see her folks in Lincoln, Nebraska, and asked me to drive her car. I stayed a few days with her family, and then her mother gave me a train ticket to Sarles and a little money. (I exchanged Christmas cards with the mother for a few years, but her husband died around 1935, and she passed away soon after.)

It was good to see my folks, because I hadn't seen them since the summer of 1930. I still had two brothers living in the Sarles area. My older brother was farming and my youngest brother was helping my dad on the dray line. Later in the spring he joined the CCC program and worked most of the time in Minnesota. I laid around home for a while and helped my dad whenever he needed help on the dray line.

One day, a farmer by the name of Martz came in. I had known him most of my life. He was an old man in his late sixties and was pretty well stove-up with rheumatism. He asked my dad if I would be interested in coming out and working for him, and my dad told him, "Sure, Archie would be glad to go out." So I went out and I intended to work for him for nothing, just to have something to do. I told him that, and he said, "No. I have never had anybody work for me for nothing; I will pay you ten dollars a month." [laughter] This was the Depression—I had a college degree in forestry, but I worked for Mr. Martz for ten dollars a month. He had quite a few cattle and horses and some hogs, and that was about all I had to take care of. Hauled quite a bit of hay. But we had an early spring and I started putting in the grain crop and I got 50 percent or better of his grain planted.

One evening my dad called and said, "I've got a telegram for you."

I said, "Well, I'll jump on the saddle horse and come in." So I went in, and the telegram was from my supervisor. It said, "Archie, when can you report to work on the Helena?" So I went out and told Mr. Martz that I had to leave.

I went home, got what little stuff I had together, and the next morning I was on the train for Helena. It was late at night when I arrived, and I got a room in the hotel. In the morning I went up to the supervisor's office. I didn't have any idea what the deal was, and he told me, "Well, it looks like you are acting ranger of the Lincoln district." The ranger was Lyle McKnight. They were establishing a bunch of CCC camps over around Butte, and they wanted McKnight to be kind of a liaison officer between the CCCs and the Forest Service, so he would have to be relieved of duty on the Lincoln district.

The supervisor told me, "You are going to need a car." (There were lots of districts in the Forest Service then that didn't have government pickups for the rangers, and this was one of them.) At ten dollars a month I hadn't saved too much money, but I went down to a garage and got a Model A Ford two-door se   for $160, to be paid for on time. The Forest Service did pay mileage when we used our own vehicles; I believe it was four cents for every mile driven on Forest Service work. We could also buy tires, batteries and a few other things under a federal contract, real cheap.

I got my stuff together and drove on out to the ranger station. McKnight was already gone, but his wife was still there. They had tried to find a place to live where he was working, but it was quite a ways out, and she didn't really want to live alone in Butte; so they decided that it was cheaper for her to stay there at the station, which was all right with me because she was my telephone operator all summer. [laughter] And if I wasn't there she would take the calls and act as a dispatcher in case of a fire.

When I had first gone to the Lincoln district in the summer of 1931, I had needed two horses. When I left that fall, I didn't have any use for them and I wasn't going to pay for feed for them all winter, so the ranger bought them at the same price that I paid for them. Now in the summer of 1933 I had to have two horses again, so Mrs. McKnight said she'd rent my horses back to me for five dollars a month, which wasn't bad.

The Lincoln district was a good district–probably the best district that I've ever had as far as cooperation between the ranchers and Forest Service. Some of the ranchers hired as many as forty men during haying season, because their haying was all native hay. They extended their season over quite a period of time, and all the hay was stacked rather than baled. But we had an agreement with the ranchers that if we had a fire, regardless of what time of the day or night it was, their men would drop everything and pile into their trucks and head for the fire. The understanding was that they would stay on until we had the fire stopped and controlled; then we had to relieve them. We'd have to fill in behind them with fire fighters we could get out of Helena. Well, that gave the supervisor's office a good twelve hours or better to round up crews in Helena and get them out to us. So we had a very quick, experienced firefighting force, and if it hadn't been for that, we'd have had a lot more big fires. The ranchers may have been willing to cooperate like that partially because there was timber that came down pretty close to that valley, and they grazed their cattle back in the timber. Also, there was just a good feeling between the ranchers and the Forest Service, and they wanted to cooperate with us. We knew that we always had a first-strike force; all we had to do was call and they'd be there.

We always had a bunch of rations at the station, and we had fire outfits and tools for up to fifty men. Most of the time we wouldn't take all our men from one ranch; that wasn't quite fair. We'd maybe take men from two ranches, or if we needed forty men, we'd even take them from three ranches. But all I had for transport was my old Model A, so we'd ask one of the ranchers to send a truck by the station to load up the fire tools and the rations so the men would have something to eat.

My lookout wanted a horse to pack some supplies up to his station, so I lent him a Forest Service packhorse. (Why the Forest Service bought that horse, I'll never know. He had a tendency to bolt at the least little thing . . . if a twig snapped him, he was gone.) I'd take the supplies up to a nearby ranch, and he'd come down and get them. Once when he was coming down the packhorse bolted, and it spooked his saddle horse, which threw him. He was going under a yellow pine with a great big limb, and it threw him right up into the limb. This put an awful gash in his head and knocked him out for quite a while. To top it off, the spooked packhorse broke a front leg running down the mountain, down an old abandoned mine road that had channeled rainwater for so many years that it was nothing but boulders.

The lookout finally came to, and he caught his saddle horse and got on it. There was a black porter in the Plaza Hotel in Lincoln, whose wife and family had a summer home out of Lincoln just a short distance from where this accident happened. So the lookout made it down to this summer home. The only one there was the daughter, a girl of about seventeen. She took him in and washed his head off and patched him up as best she could, and then she got on the telephone and called me and told me that he was hurt and the packhorse had a broken leg.

I said, "Well, we'll have to take him in to the doctor." As for the horse with the broken leg, the only . . . . Well, I owned a .44 revolver—I've still got it—but I'd left it in some stuff that I had stored at Missoula. All I had was my .22, but I went up and got the horse and led him off a way so he wouldn't stink everything up (he hopped along on three legs), and I tied my handkerchief over his eyes, and I shot him right in the middle of the forehead and killed him deader than a mackerel, with just that .22.

When I got down to the summer home, the lookout was conscious, but he didn't have all his marbles. So I loaded him in the car and took him to the doctor in Helena. The Forest Service paid all his expenses, and they paid his salary while he was off, too. He was off for about ten days before he could go back on the lookout.

While I was acting ranger on the Lincoln district I had very little assistance in the office, particularly in comparison with what is normal today. I wasn't even permanent personnel, but I was the only administrative person. Two years before, when I was assistant to the ranger, there was the ranger and myself doing the same work that I was required to do alone that summer of 1933. The district had a four-man improvement crew, which built fences, trails, or anything that had to be done around the station or on the district, and during the fire season they became a fire crew. We also had a patrolman at the guard station and a second patrolman stationed on the Blackfoot River at the head of a stream called Alice Creek. I also had two lookouts, and that constituted the manpower on the district. As far as the operation of the district went, I was solely responsible, and I was lucky that the ranger's wife was there. She could take calls, and sometimes she sent the fire crew out on fires when I wasn't there. When the fire season ended, I was there all alone, and that was the size of it. Now I bet the district has double or triple the manpower that I had at that time.

4

BURNING BUGS AND CRUISING TIMBER ON THE WYOMING FOREST, 1933-1934

FIRE SEASON STARTED early in 1933—we were having fires by the middle of May—but on the twenty-third of August the season ended. We had a little bit of rain on the night of the twenty-second, and it turned cold. The morning of the twenty-third it started snowing, and snow fell through most of the day. Before the storm was over there was fifteen inches of snow at the district lookout. Of course, that ended the fire season. I turned my two lookouts loose, and all my temporary men were terminated, and I was on the district by myself. Due to the storm, some of the livestock left very shortly afterwards, and by the tenth of September there wasn't a hoof of livestock—sheep or cattle—left on the district.

A permanent appointment About the fifteenth of September I got a wire from personnel management in Region One. They asked me if I'd accept employment at the experiment station in Miles City, Montana. I'd been waiting for a permanent appointment for a long time, and was thrilled that I was finally offered one. But before I wired them back, I received a wire from personnel management in Region Four asking if I'd accept an appointment there! I didn't know what to do, so I called my supervisor, John Templer, and told him what had happened. He said, "Well, Arch, inasmuch as you graduated in range management, I suggest that you take the Region Four appointment, because there's far more grazing in Region Four than there is in Region One."

That was good enough for me, so I told him, "I'll wire Region Four and tell them that I will accept their appointment."

He said, "Well, just a minute; I can't let you go right now. There's too many things to be done on the Lincoln district, and you're the one to do them. You've been there all summer, and the ranger isn't even around."

So I says, "How long will it be? I don't want to take a chance on losing my appointment."

He said, "I think you can probably clean it up in a week. I'll wire Region Four and tell them that you're going to be delayed and the reason why."

I said, "OK, that's good."

Well, a week wasn't enough! I didn't get everything finished; I knew I wasn't through, but I went in and told the supervisor that I had better go. I told him "There's still work to do, but I got the essential things done."

He says, "Well, I can't let you go. I'll call Region Four and ask for another week's delay," which had me worried. You wait all this time–you go to school and get a degree and pass the civil service exam–and you don't want to fool around too much and take a chance of losing an appointment. But he said, "I'll straighten it out so you don't have to worry about that." This time when he called, he learned that I was to report to the supervisor on the Wyoming National Forest in Kemmerer.

I went back and worked my head off, and on a Wednesday I got through. The ranger's wife, who had been living at the ranger station all summer, didn't want to stay on by herself, and she had asked me to take her to Butte when I left. She wasn't taking very much, so we loaded up my old Model A and went into Helena, where she had some friends she could stay with overnight. I had some stuff over to Missoula that I'd stored with folks, so I made a flying trip over and got back the next day. Then we took off, and we made it to Butte. We had a hard time finding Mrs. McKnight's husband, but we finally found him, and he gave her the key, and I took her to her apartment, and I took off.

It was getting on pretty late by then, so I just pulled off to the side of the road and rolled out my bed and slept the night, got up in the morning and figured I could make it into Kemmerer. The shortest route was through Yellowstone National Park. The park was closed for the winter as far as tourists went, but you could still drive through it. Well, I'd never been to the park, and I didn't know when I might get back, so I thought I'd see a few of the things; I saw Mammoth Hot Springs and the lake, and then I went on around to see Yellowstone Falls. It was after noon when I got away from the park, and I had supper in Jackson and then took off. Before I got to Daniel, Wyoming, I got so sleepy I couldn't go any further, so I just pulled off and rolled my bed out and slept again, and took off early, and had breakfast in Big Piney.

When I got in to Kemmerer I reported to the supervisor, Mr. Favre. He asked me, "Do you have transportation? Well, I'm sending you out to a bug camp on Greys River, where they are treating mountain pine beetle. But," he said, "I'd like to have you wait until morning if you will, because there's two more men coming in, and I don't know whether either one of them has transportation. You can just roll your bed out here in the office, and you won't have to pay rent for a room."

I got up early the next morning, went down and had breakfast and was all ready to go, but the two men hadn't come in. We waited and waited, and finally at about ten o'clock one of them showed up, and, of course, he didn't have any transportation.

"Well," the supervisor said, "you might as well take off." So we did.

My passenger was a heavy tobacco chewer, who had a big chew in his mouth all the time. About every five minutes he'd roll down the window and take a big spit. When we got to the bug camp I looked, and that side of my Model A was just covered with tobacco juice. I didn't say anything, but he saw me look. Pretty soon I saw him with a bucket and a rag, and he washed it off! [laughter]

There were forty to fifty men in the bug camp on Greys River in 1933. They were local men that were picked up in Afton or one of those towns close to Greys River. It would be just seasonal labor: they were just hired for the duration of the job, and these were men who were paid very low wages. I reported in to the office tent, and, lo and behold, the fellow who ran the bug camp was Tom Matthews, who was also a graduate of the University of Montana. He had graduated before I did, but I knew Tom real well. It was nice to see him; we had a nice visit. Then they got me straightened out with a bunk in one of the tents. It was a Sunday and nobody was working, so I went to my tent and got my bed lined out and everything, and was lying there when a young fellow came over and said, "You're wanted over at the office tent."

I went over, and Tom said, "I just got a telephone call from the supervisor, who wants to know if you would accept an appointment as a junior forest technician."

That's the first time that I knew what kind of job I was going to have, and I said, "Sure. I'll accept anything, as long as it's more or less permanent."

If you gave me my druthers, I would probably have preferred something in range management, but I was not really disappointed. I'd worked in timber a lot, and I liked timber; I still like timber. [laughter] Anyway, I was happy, and I went back to my tent, and after another hour or so, a fellow came over: "You're wanted back in the office tent again." So I went over, and Tom asked me if I'd accept an appointment as a forest technician, which was a grade up. So I said, "Sure. Sure, I'll accept it."

The thing was so messed up, in fact . . . . [laughter] I was appointed for only a short time as a forest technician, and then I was made a junior forester! That was about as high as you could expect to go out of college. (If I had received a range appointment it would have been as a junior range examiner.) My appointments were so fouled up that I didn't even get a paycheck until about the middle of December.

The assistant forest technician and forest technician positions were New Deal make-work appointments—Depression-era jobs that were essentially created by the government—and they weren't permanent. I didn't know that at the time, but I found out afterwards. I think the confusion over my appointment may have been because they were trying to make room someplace for me as a permanent appointment. But Tom was also a junior forester. One afternoon the Forest supervisor summoned me, and he says, "Well, we can't have two junior foresters on the same project. What we're going to do . . . . Over on the east side of the Forest—on the Big Piney, the Horse Creek, the Fremont and the Green River districts—the rangers have been reporting fairly heavy bug invasions. The first thing we want to do is make a survey of those invasions to see if any of them justify a treating project. We are sending you over there."

Firing bug trees The mountain pine beetle is a bark beetle about a quarter-inch or maybe a little bit longer. After emerging in spring or summer, the adults swarm and invade weakened trees, excavating egg chambers between the bark and sapwood. They usually attack old, decadent trees or trees that aren't in too good health, because younger trees can produce enough pitch to cover and kill them as soon as they start their channels up underneath the bark. The female beetle bores a little hole through the bark into the cambium; then she starts digging a channel straight up the tree, and on either side of this channel she lays eggs. These eggs hatch, and the larvae work out at right angles to the channel. In other words, they will ring a tree; and if the tree is hit very hard, they will kill it. The larvae girdle the tree in the cambium layer just like if you took an axe and girdled it.

The needles on infested trees turn kind of a reddish color, so they called them sorrel tops. With binoculars you could stand on a ridge and look over the country and pick them up for quite a distance. If you just saw one sorrel top, you didn't worry too much, but if you saw a little grove of them you knew you had bug trouble. Sometimes you'd actually see acres of them where you had heavy infestation.

Beetles could cover a large area if it were old, decadent trees, and you often found them in big stands of old lodgepole. (At that time it was believed by the Forest Service, and I assume by entomologists, that they *only* attacked lodgepole. We later learned different.)* The larvae would live through the winter in the larval stage. Late in the spring they'd pupate into the beetle, and then they'd chew through the bark and you'd have another emergence. I've seen dead bug trees that had been hit so heavy that the larvae ran out of food. They had *completely* eaten all of the cambium, and you could take a piece of bark and lift it off the tree, and the dead larvae would roll out in handfuls, just like rice–they'd eaten themselves out of house and home.

---

*Forest entomology was a relatively new branch of the science in the 1930s, and there was still some confusion regarding the preferred hosts of certain pine beetles. The mountain pine beetle (then known as *Dendroctonus monticola*; now named *Dendroctonus ponderosae*) and the western pine beetle (*Dendroctonus brevicomis*) are now known to attack a variety of trees, including lodgepole. *Dendroctonus murrayanae*, the lodgepole pine beetle, principally invades lodgepole pine.

The Forest Service experimented with a number of different ways to combat the pine beetle years ago, but they found that the best way was . . . . They wanted to wait until the emergence had been completed, so that all the beetles were in the trees and had produced their larvae. Where the beetle'd go in, there'd always be a cylinder of pitch come out, and we called those "pitch tubes." Even if they didn't kill a tree and turn it into a sorrel top, you'd still see these pitch tubes on a tree and know it was infested. Of course, one that was dead, there was no question, but there was a lot of them that they wouldn't hit heavy enough to kill outright. You were always looking for pitch tubes.

To attack the pine beetle, the Forest Service used heat. We had oil of about the same flash point as diesel, which we sprayed on infested trunks with a regular old garden sprayer. It had two extensions on it, so you could reach up six or seven feet with the nozzle, and if you put a lot of pressure in the can you could send spray up at least twenty-five feet on a day that wasn't windy. You would be sure the tree was completely covered with oil, and then you'd take a match and touch it off. The heat generated from the burning oil would generally be hot enough to kill the larvae. After we'd burn a tree–after it had cooled down–we'd always go back with a hand axe and slice some of the bark off to see if the larvae were dead.

I can tell you this: firing a tree infested with pine beetle larvae was 100 percent effective . . . if you did it right. Now, there was always a chance you might miss a tree, which would leave a start for another epidemic, but if you treated a tree right, you got them. When I say you treated it right, I mean you sprayed it high enough that there wasn't a chance of any pitch tubes above where you sprayed; that you put enough oil on it; and that you got it hot enough to kill the larvae. If you did that, you got them. There was no question about that. But it was costly: it was all hand labor, and it took quite a while. Even during decent weather it took quite a little while to spray a tree and to ignite it, and then we always came back, at least on my crew. I wanted everybody to go back and check every tree that we worked on. We generally shut down an hour or so before our quitting time, and then we'd go back and check out our trees and see that we had gotten a good kill. After 1935 I didn't work in timber anymore, so what happened after that in treating bug timber, I really can't say, but I was told the Forest Service eventually went to the use of insecticides rather than burning the trees.

Treating beetle-infested trees with fire.

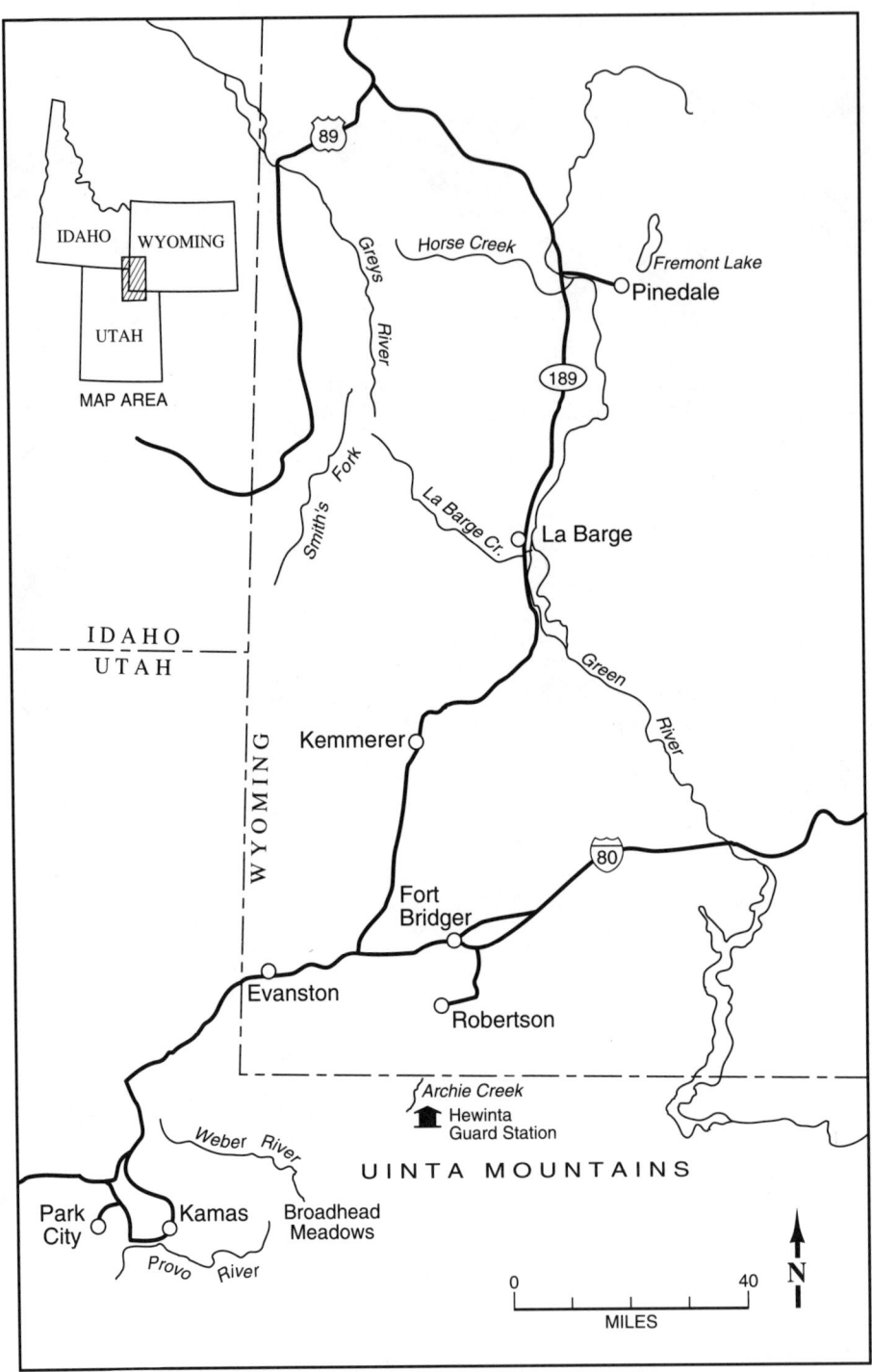

Map 2. From 1933 to 1936 Archie Murchie served on the Wyoming and Wasatch National Forests. His first appointment as a ranger was on the Kamas district in 1935-1936.

Boy, they had plenty of bug trees; they had plenty of them! I had studied some about forest insects in school–the pine beetle and others–but when Tom found out that that was going to be my job, he took me out. He gave me a full course in bug-tree identification (that's for bug-infested trees, so that you can identify them from a long ways off); how the bugs actually attacked the tree; what they did; what had to be done to control them–he gave me the whole thing. When we came in that evening I had a *full* knowledge of how to cruise for pine beetles and how to treat them. [laughter] Cruising for pine beetles was a little like cruising timber: you'd go out and tally the number you found to see if the infestation would warrant treatment. A lot of the infested places were small, and they wouldn't justify taking a crew in to treat them.

The supervisor told me to ask Tom to give me one of his best bug-trained men to go with me. Tom gave me a guy by the name of Perry Skarra, a little fellow who had been born in Russia. I don't know whether his family got into trouble, but his sister went to Canada, and he left Russia and went and stayed with her for a while. He was just a boy when this happened. Later on he came to Minnesota and went to the University of Minnesota and took forestry, but he never finished; I think he had a year left. He always said he was going to go back. He was my partner. We took off in my old Model A Ford, and the first ranger we reported to was Ed Cazier on the Big Piney district. He told us of the bug infestation areas he'd noticed on his district, and he marked the areas on a map so we could find them. We cruised all his areas, and we found patches, but they weren't big enough to treat. Then we went to the Horse Creek district, but there were very few bug-infested areas on the Horse Creek. Then we went to the Green River district; and there was one area up on the head of the Green River, above the Green River lakes, that was probably big enough to treat, but it was so far back in that they decided not to do it.

So then we went to the Fremont district out at Pinedale. Bob Dalley was the ranger there, and he had some pretty-good sized infested areas. We cruised them all out and decided on the ones that would have to be treated. To treat the biggest area we hired eight ranch boys from Farson, Wyoming. They had their own truck, and I'll never forget–they didn't have sleeping bags; they had bed rolls. We'd work ten days on and four days off, and every time they'd go home they'd bring back another blanket or two. They weren't blankets; they were quilts! By the time the last of December came, it was all two men could do to lift one of those beds into the truck. Of course, they had a mattress with it, too.

We started up the treating job, but we started too early–we started while it was still pretty dry. The ranger had said, "Well, if you don't start now, you're going to get snowed out before you get through." (Even starting as early as we did, we got snowed out!) [laughter] One of the dangers of treating bugs too early in the fall or too late in the spring is starting forest fires. This day we'd gone up the canyon; there was a fairly small area that we thought we could treat in a day with the ten of us, and we treated it, and the next day was Saturday. (Now, I don't remember whether we were still working half a day Saturday or not. For a long time, we worked half a day Saturday and got off at noon.) Anyway, after the Farson boys had gone home, Perry and I saw a smoke coming up in the canyon. We went up there and found some fires among a bunch of trees we had treated that were growing in a big bed of boulders. (This country was all glaciated.) Most of the fire was in the duff that was down in between the rocks, but we couldn't leave it because it could burn out to good stands of timber, and then it could really take off. We finally got the fires out.

Well, I called the men at Farson and told them what had happened, and told them not to come back until I notified them that conditions were right to resume firing the bug trees. There was a few places–say, along a stream or in a meadow–where you could treat a tree without any danger, so Perry and I'd just go out and treat these trees by ourselves. Then we found another little fire where we'd treated for bugs a couple of days before. So then the ranger said, "That's the end of it. No more until you get a storm."

We finally got some rain, and I got the men back to the bug job. We finished up the project over by Fayette Lake. The next job was closer to Pinedale, up on Halfmoon Mountain. When we moved in there, there was quite a bit of snow, and just spraying oil on the trees and setting it afire in *real* cold weather like that didn't get the bark hot enough to kill the larvae under it. So we shoveled the snow back, then cut brush and branches and piled them around the bottom of the tree and sprayed the oil on the brush *and* on the tree, and set it all afire. That way we could generate enough heat to kill the larvae.

A Forest courtship

When we were cruising bugs on the Big Piney district I did not get any mileage: all the Forest Service was paying for was gas and oil for my Model A. We'd done a lot of running around in the Forest with the old Model A, and I'd done a lot of running around up there in Montana with it, and my tires were getting pretty thin. Perry and I would work right on through the weekends because there was no use hanging around camp. What could you do . . . ? Oh, maybe wash some clothes or something . . . so we just worked right on through. We'd worked several weeks, and in the middle of a week I was low on gas, and we were low on grub, and we were on our way in to Big Piney to get some supplies.

The assistant supervisor on the Forest was a fellow by the name of Van Meter. He was kind of a rough character, and you either liked Van real well or you disliked him . . . and there were a lot of people who didn't care too much for Van. He was a very outspoken person, and he stammered. We were going in to Big Piney and this Forest Service car passed us. I looked back in the rear-view mirror, and I saw the dust flying, and old Van had slammed on the brakes, and back he came. Of course, when we saw him stop, we stopped. He wondered what we were doing that time of day, and why we weren't out working. So we told him. "*Well*," he says, "you don't have to take all day about it."

We told him we wouldn't, and we got to talking, and I told him, "Van, I haven't had a paycheck yet, and I'm practically broke. I need tires, and there's no way I can buy tires."

"OK," he said, "you get your grub. Then you go over to the CCC camp at Fremont Lake and tell them to put four new tires on."

We took off and went over there, and with all those CCC boys it didn't take them long to put the tires on, new tubes and everything. There was also a broken bumper on the car that needed mending. We went to the blacksmith shop and asked the fellow if he could do that little bit of welding. He said, "Sure." So we got to talking, and he introduced himself, and his name was Reuben Boulter. I did not know it, but I had just met my future father-in-law.

Before we moved out to the bug-treating job, when Perry and I were still conducting our bug cruise on the Fremont district, the CCCs had built a brand-new ranger station in Pinedale. The vacated old ranger station was located just back from the shore of Fremont Lake, and the ranger said while we were cruising for bugs we could stay in it. We had a Coleman lantern. In moving around, you would always break mantles off your Coleman, and then you'd have to put on another set of mantles. Well, one day we discovered that we were out of mantles, and I went over to the CCC camp to borrow some from Reub. His daughter, who had a job in Washington, had come for a visit. Reub asked me to step in—they were just eating supper—and I saw Jane sitting across the table. I got the mantles, and the next night Jane and her sister came over to the old ranger station. A lot of bedding had been left in the old house, and they asked if it would be all right if they took one of the pillows. Of course, we didn't have any jurisdiction, but we said, "Sure." So we talked, and I talked to Jane a little. That weekend there was a dance downtown, and Perry and I went to the dance, and Jane and a girlfriend of hers were at the dance. Jane was the only girl there I knew, and we danced quite a few times. [laughter] That was the start of a romance that has lasted over fifty-seven years!

Shortly after that the ranger invited Jane and me over to the station for Thanksgiving dinner. We started going together, and before long Jane and I decided that we would get married in June of 1934. That winter I was still working out to the bug camp, and driving back to see Jane at night and on weekends. And betwixt and between times, some way or another, apparently a piece of felt or something had gotten in my gas tank. Felt, of course, was made out of hair of some kind, and it slowly disintegrated and plugged the jet to my carburetor. I could figure about every twenty-five or thirty miles I'd have to get out and undo the gas line. The only way you could get the hair out would be to suck the gas out and then spit it on the ground. And, boy, it was getting *cold*, and having to stop every so often to suck gas out of that darned carburetor didn't do me too much good. So finally my paychecks came, and we decided . . . well, we both had our minds made up, so why wait until June? So we decided to get married on the twenty-third of December, 1933. It was a short courtship, but the marriage has lasted for over half a century, so I guess we can't kick. [laughter]

Jane and I got married on the twenty-third, a Saturday; the twenty-fourth was a Sunday, and the twenty-fifth was Christmas. The next day, Tuesday, I got started on bug work again. The crew was staying at the CCC camp then, because it was just too cold to try to camp out. The CCC would drive the men out every morning and bring them back in the evening, but I always took my car in case somebody got hurt or something happened. We went out on the east side of Halfmoon Lake the Tuesday morning after Christmas, and I don't know how cold it was, but it was way below zero. We were firing bug trees, and it was so cold that we had to actually build a fire at the base of each one to get the temperature high enough to kill the larvae. While cutting wood to make a fire, I swung at a deadfall limb with my axe. It was a pretty-good sized limb, but it had been wet when it froze: when the axe blade hit it, it just popped off, and the axe came around and hit my shin. It went right in to the bone; in fact, it even chipped the bone.

I pulled my pants down and looked at the wound, and at first it didn't bleed. You could just look in, and you could see the bone. Then it started bleeding pretty heavily, so I tied my bandana tightly around it, and the bleeding slowed. I figured I could make it back to town on my own all right, and I didn't want to take a man off the crew to drive me in, so I went down and got in my car and drove back to Pinedale. The only doctor in town was an old, retired mine doctor who had worked in the coal mines over at Rock Springs for years and years. He was a rough old character. He washed the wound off with some disinfectant, sewed it up, put a bandage on it, and told me not to bother with it for three days; then come back, and he'd change the dressing on it.

During the night of the second day my leg started hurting, and I put my hand down, and there was a big lump where the cut was. So at seven o'clock in the morning I went over and woke him up, and I told him, "Something's radically wrong."

I pulled up my pants leg, and he said, "There *is* something wrong." He took the bandage off, and the whole thing was infected. The infection was due to out-and-out carelessness, because there was no excuse for him not being able to properly disinfect that wound and the thread and needle that he was sewing with. Even *that* was infected: there was pus around each of the stitches and down the middle where it was cut, and it looked awful. So he cut the stitches, and the thing just kind of opened up, and he cleaned it out. But he couldn't sew it back up, so I had to go over every day. He had a solution–something mixed with water–and he'd bathe it. It was a month and a half, I guess, before it finally healed up. It was difficult for me to walk on it, too.

The men finished up the bug job without me, and that was it for the year. They were glad to get out of there! [laughter] And I was glad to see them get out, because another storm and they'd've had to get out, anyway. When the job was over and the men released, I still had a bum leg. The Forest Service said they had work for one man in the supervisor's office, and they asked me whether I wanted to move in to Kemmerer and take this job . . . which I didn't really care to do–Kemmerer was a mining town, and the rent was pretty high. I was given a choice: either take the job in the supervisor's office or take three months furlough, and Perry could go in and take the job. Jane and I, of course, were just married, and we had a nice two-room log cabin for the winter. "Well," I says, "OK. Let Perry take it, and I'll take the three months furlough." Jane and I were glad that it worked out that way.

Cruising timber on the Green River   Next spring when I went back to work, they told me that I was going to head up a crew surveying timber on the upper Green River. As soon as the snow left so that we could get around, we would get started on this job . . . but I was getting camp fever; when they started the road crew out and somebody in jest suggested that I go out and cook for them, I thought, "Well, why not?" So I went out and cooked for the road crew for about three weeks. Then the snow left and the ground dried up enough so that we could start our timber survey job.

I was chief of party, and Perry Skarra was still with me. A young fellow by the name of Lowell Woods, a recent graduate of Utah State, was on the crew. There was also a young man from Washington and another young fellow from Ames, Iowa. These were all forestry men, so I had a darned good crew. Then, too, I had a cook/packer who had cooked for fishing trips up the Green River and for hunting parties. He could really cook, and he was a darned good packer, too–he had his own stock and everything. I never ate so good in a camp in all my life as I ate in that timber survey camp.

This country was supposed to have been surveyed way back in the early 1900s–some of the township plats even went back a long time before that. But in all those early surveys, they surveyed the townships first, and then they went in with another contract and surveyed out the sections within the township, and set all the section and quarter corners. Well, the township survey was a real good survey. They set their township corners and their quarter corners and their section corners, and even that long after, we didn't have too much trouble finding those corners. But whoever got the contract to survey the interior of the townships did a poor job. There were townships in which *nothing* was ever done. They'd never even set a corner, and yet their plats showed that the township had been surveyed, the date when it was concluded, the head of the survey crew, and all that. And not a lick of work was ever done on it! That wasn't just true on the Green River; up in Montana there was a lot of places where the same thing was done. They just beat the government out of a lot of money.

The Green River survey I was on was what they called a one-and-a-half-percent survey, as I remember. One and a half percent of the timber was cruised. You see, it was a very, very small sample, and the survey wasn't for the sale of timber: it was to determine what timber was on the district–it was more or less a resource survey so the Forest Service would know what they had. Before this they didn't have *any* idea what amount of timber they had . . . or the species or anything. At some later date–based, perhaps, on our survey–they figured that they had enough timber there to justify an intensive survey and a sale, and they went ahead and did it.

In our survey the crew spent all their time in the field, but most of my time was spent in the office because I had to transfer all the information that they brought in on their work sheets. They ran four strips through a section on their survey, and along each line, at ten-chain intervals, plots were established. In these plots, all merchantable trees were tallied on a cruise sheet as to species, diameter and number of logs in the tree. I had to transpose all of the data from the cruise sheets to a thirty-six section township plat. This took quite a bit of office time. There are thirty-six sections in a township, and in transposing the section maps I'd take a piece of carbon paper and a cruise sheet and center it exactly over the section on the township, then trace it all and ink it in with India ink. Afterwards, I entered the streams in blue, and there were different symbols for the type lines and all that.

Every so often I would go out with the crew. We would mark where a cruiser started with a stake, and we tried as much as we could to tie into quarter corners and section corners. Where it was surveyed right, the cruiser in his pacings could tell when he'd come to a section line. Then he'd pace off either way, and either tie into a section corner or tie into a quarter corner. But on that poorly-surveyed Green River country they couldn't do that.

On the baseline, going through pretty heavy timber, the head chainman would carry a mirror, and he'd flash it to reveal his position if we couldn't see him. If I'd want him to go one way or the other, I'd just yell left or right. Our cruisers would start from the baseline (which, of course, was accurate), but by the time even the best cruiser had run a line out for three miles or more, his pacing or alignment could be off a little. So if a cruiser couldn't periodically check his position against a known section or quarter corner, the further he got from his baseline, the more inaccurate his map would become. There was no way we could correct that. We'd just have to take that into consideration, and leave it as it was. But by running a compass line and pacing, some of those guys could come sometimes within just a few feet of a section corner or a quarter corner.

To prepare for this work, when we first went out to the camp all we did was run compass lines and pace, just for training. Every man's stride, of course, is different, and each man had to learn the length of his pace. Then I had them pace uphill and downhill. Your measurement was supposed to be two-dimensional; it was not to include the contours of the terrain, but to be between two points as if the surface were flat. So going up and over a hill or down a draw, they had to compensate, but they even got pretty good at that with practice. The two experienced men, Perry and Woody, had to have very little training–just kind of a refresher course. But the other two boys took quite a lot of training, and they also had to have training in tree measurement. From its diameter and the number of sixteen-foot logs it could be cut into, the volume, or board feet, of lumber in a tree could be determined. When they first started I required them to measure every tree with a diameter tape and measure the height of every tree with the Abney level. After a couple of days, they could go to estimating. It was surprising how accurately a person could run a line with a compass and pacing, for both alignment and distance. But any of them–myself and Perry and Woody too . . . in the morning for some reason your eye would be off, especially on diameter, and you'd have to measure for a little while. Then you'd be all right. Then on in the afternoon, when you started getting tired, your eye would start getting off again, and every once in a while you'd want to measure to be sure.

We got started on the timber survey around the first of May at lower elevations where we weren't bothered with snow, and we cruised all the timber clear on up to the head of the Green River and the head of Roaring Fork. There was very few times that I had to deliberately go out to check a type line or something that didn't jibe too close, but every once in a while I'd go out and check scale. They would know that I was check-scaling, but they wouldn't know how far back. I'd wait until they'd started from their station, maybe an hour or so. Then I'd go out and start from the same station, with them far enough ahead that I couldn't see them, so there wouldn't be any chance that I'd be influenced to take the same plots they had taken. Then at the end of the day we would compare our scales and see how close we came to each other. Some of them would check awful close on the estimated volume of the timber.

Lots of times a scaler would start out from a station, and maybe he'd finish up a line by three o'clock. If he did, he'd come in. I wouldn't want him to start another line, because if he had to stop in mid-line at the end of the day, it could be hard to find his place the next day. Maybe the next day he wouldn't finish up until five-thirty or six o'clock, so it never bothered me if a guy came in early. I never questioned him or anything. And so it would go on.

It got on into the summer and we had a lot of wind, especially in the afternoon. One day the boy from Ames, Iowa, came in early; I thought nothing of it. The next afternoon he also came in early, and the following day he came in early again. I thought, "Something is going wrong. Something is wrong here." So the next morning I told him, "I'm going to check-scale you today." Most of them didn't care one way or the other, but he seemed like he was just tickled to death that I was going to check-scale him. That kind of made me wonder. I let him get started, and then I started . . . and it was a windy day. Finally I caught up to him—he didn't know I was behind him, but I could see him. He was setting up his compass, and all of a sudden he jumped and started looking around, and I could see he was scared. I thought, "What in the world?"

He went back to his compass, and I moved up and . . . . "Boy," he said, "I'm sure glad you caught up with me. Now we can survey the rest of the day together."

So I thought, "Well, OK. I can go with him and let him do the surveying, while I just walk along and check him as he goes."

Many times out in the timber you'll have a snag blow over, and it can lodge on a limb of a tree or maybe in the crotch of a tree. Sometimes even a green tree will blow over on top of another tree. When the wind blows, the standing tree will sway and saw back and forth on the fallen snag that it cradles. A pitchy tree will make quite a screeching sound that will carry a great distance. We were going along when all of a sudden one of these cut loose off a ways. The boy whirled around and said, "Did you hear that?"

At first I didn't know what he meant. I said, "What do you mean?"

He said, "Do you hear that animal scream?"

I thought I was going to laugh; and then I thought I better not. So I told him, "That's no animal."

"Yes," he said, "the last two days that animal's been following me. And every so often it'll scream. It just scares the daylights out of me."

I could tell about where the sound came from, and it wasn't too far away, so I took him over and we found it. "Now," I said, "we're going to stay here. If we have to stay here for fifteen or twenty minutes before it sounds again, we're going to stay here." I told him what was making the noise, and it wasn't too awful long until a gust of wind hit and the tree screeched.

"Oh," he said, "that's what it is!" So we went back to cruising. This was all pretty heavy timber, and every once in a while one of these would cut loose someplace, but it didn't bother him now, and we went on through and finished up.

Things went well for maybe a week or better, and then he started coming in early again. The first day I didn't say anything, but I could see when he came in that something wasn't quite right. The third day in a row that he came in early, he came over to my tent. "Arch," he says, "I can't take it. I just can't take it. You showed me what was making that noise, and I know it's that. But," he said, "when I get out in the timber all by myself, it just scares the daylights out of me. I'm going to have to quit."

I hated to lose him, because as far as neatness in his work, he was the best of the four of them, but I knew there was no use trying to get him to stay. I wrote out his time and told him to take it in to the supervisor's office and turn it in. I didn't want to make the trip clear down to the station to call the supervisor, so I wrote a note and asked him if he'd try and get me another cruiser. They got me another boy out of Washington, and he finished up the season.

The Iowa boy went back home and about two weeks later I got a letter from him asking me if he could come back. "Now," he says, "I think I've got it out of my system. I realize how dumb I was and how foolish I was to let something like that bother me. But I'm sure that I got it out of my system, and I'd like to come back."

But I already had another man, so I wrote him a nice letter and told him I was sorry, but we had already filled the position. Whether that ended the boy's career in the Forest Service or whether he found work back east someplace or not, I don't know.

(Later, when I was on the Challis Forest we never did a timber survey, because there wasn't that much marketable timber. When we had a sale there and somebody was interested in timber, he went out and picked an area that he figured was feasible to log, and then we just went out and marked the trees. There was no previous survey of any kind; you just marked the trees for them.

No survey was made when I was in Nevada on the Toiyabe, either. I don't think that the Toiyabe in any place had sufficient timber that would justify this type of survey, and what they did was probably similar to what I did on the Challis. If somebody wanted some timber–wanted to start a sawmill or needed timber for a sawmill–they went out and located the area that was suitable for logging, and then the Forest Service marked the trees. We blazed the tree and the stump, and put the number on the tree and a number on the stump, and then entered it in a book so we could compute the board feet in the tree. Now they use paint to blaze trees. We actually blazed them and stamped them with a tool that was like a small, two-bitted axe with one bit cut off and a US welded on it that would indent into a tree a good eighth of an inch or better. You would just smack the tree with that end of the axe, and it would leave a US stamped on it. They also had a hammer that had US on either end, and with that one you could just stamp; but if you were marking timber where you wanted to blaze trees and then stamp them, you used the axe with a blade on one side and a stamp on the other.)

More bugs and fire When we finished up our timber survey on the Green River in early September they told us to go over to Greys River–they had had *another* outbreak of bark beetle, and we were to make a survey to see if it would be practical for another big bug project like the one in 1933. I took Perry Skarra and Lowell Woods with me. We completed the survey far enough that we estimated, as I remember, that it would cost around one hundred thousand dollars to treat the area. They had applied for the money from the regional office, and it looked like they were going to get it, and then for some reason it was canceled.

One afternoon when we were just about through with the bug survey (there may have been two days work left) we smelled smoke. One of the guys went out on a point where he could look down Greys River, and he saw a big smoke coming up. We were staying in a Forest Service stopover cabin close to where we were looking for bugs, and we had left our smokechaser packs at the cabin. We went back, picked up our tools and headed down the canyon. The fire was a big one that had been burning since early that morning, and it was already manned by CCC boys.

We worked on the fire a couple of days without making much headway. The second day started hot and windy, but by the afternoon it had clouded up and turned cold. It rained a little that evening, and the next morning there was about six inches of snow on the ground. What a mess, trying to move the camp out! They hauled out the CCC boys in one-and-a-half-ton trucks with stake sides. One truck with twenty-five boys in it rolled. Some of the boys were injured, but none of them got killed.

While I was on this fire the assistant Forest supervisor, Clark Miles, had come out. He'd received a telephone call from the supervisor's office notifying me that I was to report to the Black's Fork Ranger District on the Wasatch National Forest. I was to be the scaler on the Standard Timber Company tie sale on Smith's Fork, which flowed off the north side of the Uinta Mountains into the Green River. The Green River was in Wyoming, but most of Smith's Fork laid within the state of Utah in the Wasatch National Forest. The fire was out, but there was still a lot of work to do on it: getting pumps in and getting fire hoses in and thawing them out and rolling them up and gathering up tools and all this other stuff. But they wanted me to report as soon as I could, so I left and went over to Pinedale and picked up my wife and headed for the Uinta Mountains.

5

SCALING IN A TIE CAMP ON THE WASATCH FOREST, 1934-1935

JANE WAS PREGNANT, and we expected our child sometime in the latter part of October or the first part of November. Up to the tie camp I knew it was going to be isolated, so I found a place for her in Kemmerer, Wyoming. A women there had something like a private maternity ward in which she took care of women from outlying ranches and other places during their pregnancies and after. This was ideal for us.

Tie timber   After dropping Jane off in Kemmerer I reported to the ranger, Jay Hahn, and he took me up to the tie camp. This was in the latter part of September, and the Standard Timber Company had already set up their operation. They had hewed ties there the winter before, so this was at least their second year of work on Smith's Fork, and they had built a big commissary where you could buy anything you wanted. Quite a few of the tie hacks had built cabins around the commissary.

Jay took me out to the guard station—they called it the Hewinta Guard Station, probably a take-off on the Uinta mountain range. While I was there the guard station was used as the scaler's cabin, and after Jane gave birth to our child, that was where we and our infant son lived all the time we were there. It was a two-room log cabin. The kitchen was a fair-sized room, but the bedroom was just about wide enough for the bed, and that was it. We got our water from a little spring just off from the corner of the cabin. There was what they call a pitcher pump: it was a siphon pump, and you had to prime it with water to get it to pump, and then in the night–so it wouldn't freeze–you raised the handle all the way and it drained. Then the next morning you had to re-prime it to get water again.

The Uinta is one of very few (if not the only) major mountain ranges in the United States that runs east and west. The front, or north, slope of the mountain range in the country where we worked was very steep. You had to really climb going up there, but when you got to the elevation of our Hewinta cabin, at about ninety-five hundred feet, the country levelled off onto benches of a pretty good size. These benches went back at a gradual slope until you hit the base of the real mountainous part of the mountain; then it started climbing fairly fast. Merchantable timber went up this slope for a fair distance, and then it just petered out in the area the tie hacks called Lapland.

For hewing ties you had to have a special type of timber. You had to have fairly dense timber, so that the trees grew slim and tall with not too many branches, especially on the lower part of the tree. In a dense stand of lodgepole pine the trees don't grow to a very large size in diameter, but they grow tall. A good tie tree would probably run from twelve to eighteen inches in diameter at the stump, and you would get at least six ties out of the tree. (A tie was eight feet long, seven inches thick, and faced on two parallel sides.)

This Standard Timber sale area was strictly lodgepole pine. Lodgepole is a unique type of tree, in that the cones don't open while hanging on the tree. Lodgepole can drop their cones and they will stay closed for a long period of time. Of course, rodents–squirrels and chipmunks–open up a lot of cones and scatter a lot of seeds, but sometimes they stay closed until a fire goes through before they'll open. Sometimes they'll stay closed until the cone itself starts deteriorating, but the seeds stay viable for quite a long period of time, and after some disaster–like, say, a bug epidemic or disease or fire; fire being probably the primary one–these cones will open. After a fire, lodgepole will come up just as thick as hair on a dog's back.

I've seen a few places where lodgepole comes up so thick that trees only four inches in diameter would probably be over a hundred years old. They just stagnated, because they were growing so close together and the competition was so great between the trees that they'd never get any growth. Of course, it's natural that the hardier ones will survive and the other ones will die, and that way most stands will thin themselves. Lodgepole made excellent ties, because they were free of knots. The forest canopy screens sunlight from the lower reaches of a densely-packed stand of trees, causing them to drop their lower branches before they get any size. You wouldn't hardly have a branch as big as your finger for maybe four or five tie-lengths up. All your choice saw timber–those trees that are cut into clear lumber–conifer or hardwood, all become that way because those trees self-pruned. It's a necessary stage in the growth of timber if you're going to get good saw timber or good clear wood free of knots.

Lodgepole self-prunes very easily because they can grow so close together in such dense stands. In many groves I've seen lots of fairly large stands where four or five feet would be the maximum spacing, and I've seen a few cases where two trees a good foot or better in diameter stood right against each other. Trees that don't grow in groups like that, they call wolf trees. They're limbed right to the ground. They'll be big on the stump and short in height, and they'll have limbs on them three or four inches in diameter. These trees aren't waste trees, though: they produce a lot of cones. Practically every limb down just about to the ground has cones on it, and they're excellent seed trees, so we never damaged them or did anything to them at all. We always left them standing. Wolf trees always grow next to a meadow or in a clearing where there is no competition, and they get all the sunlight and water they want.

Hewing and hauling   In a timber sale, the Forest Service and the company had a signed contract that Standard Timber could go in and cut the timber, and they had an exclusive right to cut *all* merchantable timber that was on that big sale area. The first thing the company would do if they were going to open up an area for cutting was go in and build a sleigh road. They would try to build it through the center of the area, and they'd lay off strips of timber roughly at right angles to the sleigh road. These strips would run anywhere from two hundred feet to perhaps four hundred feet wide, depending on just how much timber there was on the strip. As soon as a tie hack was given a strip he would cut out a road through its center, and at the end of the road he would cut out what they called a round-turn, where a sleigh could come up and make a turn around and come back down the road before stopping to pick up hewn ties. (Some of these strips were on a slope, and it was a lot easier on the tie haulers' horses if they could load them coming down the slope rather than going up.)

There were a few tie hacks who worked alone, but most liked to work a strip with a partner, because two could work more than twice as fast as one man working alone. The company would assign the tie hack a strip, and he and his partner would cut out the sled road through the middle of it, and then they would start hewing ties. In hewing a tie the tie hack would first fell a tree–and I believe on that sale we had a stump requirement of twelve inches: no stump could be higher than twelve inches. He'd fell his tree, and then he'd limb it as far up as he thought he could get a merchantable tie, seven inches thick and eight feet long.

In our scaling for the Forest Service, a tie was a tie; it didn't make any difference what size it was, but the company scaled three sizes, I think: number ones, number twos and number threes, depending on the width of the tie. Tie hacks got a few cents more for the larger ones, and so on down to the smaller ones. I think the company required tie hacks to cut a log to six-inch top, because what was too small to be a tie, they sold as a mine prop. The Forest Service required that the tie hacks had to use all the tree up to an eight-inch top, and, as I remember, any tree that would yield two ties had to be taken. However, the more ties a hack could get from a tree, the better he liked it.

To fell a tree, a tie hack would saw an undercut, and then chop the undercut out. Whichever way he wanted his tree to fall, that's the side he'd make his undercut on. Then he'd go to the other side with a saw and cut in until the tree toppled over. All of the tie hacks that hewed up there were experienced men. To hew a tie they would stand on top of the log, and they'd always start at the butt end. I don't care if it was an eighteen-inch butt on the tree, or a twelve-inch, or a ten-inch, or what it was . . . they could tell how far in they had to start their cut in order for them to come up with a seven-inch-width tie. (Tie hacks would hew only two sides of a tree, leaving the other two naturally rounded. When a finished tie was laid, the hewn sides would face up and down.) About every six inches or so the hack would score the tree–he would cut in at an angle with his axe to just about the depth he was going to hew. (A good tie hack would never cut quite to the desired depth, but he'd come awful close to it.) He'd do that the full length of the tree, and then he'd do the same thing on the other side. He'd score the trunk as far up as he figured he could get a tie out of that tree, then he'd go and get his broad axe, stand on top of the tree again, and start cutting just a little bit inside of the depth he had scored.

Now, these broad axes came in different widths, but they had a blade that was generally about twelve inches wide. It was sharpened on only one side–in other words, the inside of the blade was flat. (You take an ordinary axe, they're sharpened on both sides, but a broad axe is only sharpened on the outside, and the inside is perfectly flat.) These broad axe blades were made out of very good steel. The bit was highly tempered, and they could sharpen those to such an edge that the tie hacks used to claim they could shave with them. I don't doubt it, because they were sharp! They always had a whetstone, and just about every time they'd stop you would see them dressing up their broad axe.

When a good tie hack finished the side of the tree, you saw very few axe marks. You didn't see the score marks, because he didn't score too deep, and you didn't see the broad axe marks, because the broad axe would always bite in just a little bit before where the blade had stopped the last time he swung. If you looked down the tie, there'd just be a faint ripple on the face of it. He'd hew the one side, and then he'd go back and do the same thing to the other side. Some of these men were so skilled that I've seen them hew the whole side of a tree, and that hewed wood would just lay out, all of it connected in one long strip.

After he'd hewed the tree, the hack would pick up his spud. A spud was a bark-peeling tool made from an old saw that had worn out. They'd cut a piece out of the saw about four inches wide, maybe fourteen inches long, where there weren't any teeth. They'd sharpen one end just as sharp as a knife, drill three holes in the other end, and attach it to a seven foot-long pole about two and a half inches in diameter. To do this, they would saw a slot in the end of the pole, and the prepared blade was set in this slot and riveted so it was solid, with about six inches of blade protruding beyond the end of the pole. After this was done, the end of the pole on either side of the blade was sloped down to the blade. That's what they used to peel the trees with.

With a spud, one man could peel long strips of bark. These trees were largely self-pruned; on the lower part there were no limbs, and they could take that bark off just as fast as they could push the spud. In the wintertime when the bark was frozen it was difficult to peel, and most of the tie cutting was during the wintertime, so they'd get all the bark off, but they'd often leave a little bit of the cambium. You could see that on the tie, but when the tie dried out most of that cambium would just fall off, anyway.

After peeling the upside of a hewn tree, the tie hack would cut the log into tie-length sections. The butt cut had an underslope on it, and there would be wood sticking out of the end of it, so he'd square that first. Then he'd start from that point and measure out eight feet—they had a stick cut to length just for that—and then he'd take his one-man saw and saw down through the tree and cut the individual ties. When it was sawed off, you had a tie that was eight foot long and seven inches thick, with two round sides and two flat sides. The two flat sides were seven inches apart. One round side had been peeled by the man who had started the job; the other round side hadn't.

Now, the tie hacks worked as partners, and that's about as far as the first tie hack would go on a tree. Then he'd start felling another tree, and his partner would come over and turn each tie to the round face that hadn't been peeled, and he would peel it. Then his job was to drag the ties over to the sleigh road that ran up the center of their strip. They called this "parking the ties," but what it was . . . he'd pile them. And for this they used a tool they called a pickaroon. To make a pickaroon they would take a single-bitted axe, and they'd cut out all the blade except about an inch of the outside part or edge. They'd heat what was left of this and put it on the anvil and hammer it out to a real sharp point. They could take that handled spike and sock it into a tie, even though the tie was frozen, and they could drag it. Those ties were heavy, but with a pickaroon they could easily drag one over to where they were going to park it.

When they parked the ties, the company required them to have the faces fronting the sleigh road pretty even; they didn't want the ties staggered back and forth. This was for two reasons: it made it easier for them and the Forest Service to stamp the ties; and it also made it easier to walk around on the other side to see if somebody had cheated a little and had a few short ties. With one look you could tell whether you had an honest pile or you didn't. (Well, with the Swedes and the Finns and Norwegians who made up the majority of the tie hacks you never had to worry about honesty too much. I don't believe I ever found a shorted pile with those three nationalities. Later on I'll go into another group that came in there that hacked ties . . . we found a number of things where they'd tried to cheat the company.)

A tie hack peels a tie with his spud.

Inspecting ties: "I had some trouble with the Okies.... Maybe they had a tie that had rot in it, so they'd cut two blocks off the good end. They'd put one end at the front and one at the back of the pile, and there wouldn't be anything in between. But they didn't very often fool you, because it had that different ring to it when you hit it."

Splash dams such as this were built when a stream had an insufficient flow of water to float ties. When the dam filled with water, the gates were opened and the ties were floated—or splashed—down on the high water.

The tie hacks would take turns, because the man who was doing the felling and the hewing had a tough job. Maybe one fellow would fell and hew two or three trees, and then the other fellow would do the same. There was one exception I remember to the custom of alternating jobs between partners: Big Gus was a darned good cook. He was a specialist on pastry–he learned that in Sweden–but he was an exceptionally good general cook too, and the company would hire him to cook in one of their camps. Big Gus would cook until he got to weighing about two hundred and fifty pounds or so and then he'd go hack ties for awhile. He had a helper they called Little Gus, and I don't believe Little Gus ever weighed over one hundred and thirty-five pounds. Big Gus would cook and little Gus would be the flunky, and they'd cook until Big Gus got pretty overweight, then they'd go back to hewing ties. Big Gus would do all the felling and hewing, and little Gus would do the peeling and pulling the ties over to where they were going to park them–they'd worked together for years and years that way.

Standard Timber hired its own tie haulers. The ties were hauled on a heavy sled upon which they built a platform . . . or a bed, you might as well call it, a flatbed that was the width of a tie-length–eight foot wide–and had a front on it probably about three-and-a-half foot high. Ties were piled on the flatbed and then hauled to the banks of Smith's Fork, the sled pulled generally by two horses. (Once in a while they'd put four on, but generally the team was two horses. These were not real big workhorses, but they were pretty-good sized and well-trained.) They'd stack the ties along the banks of Smith's Fork, and the piles would be strung out for close to half a mile. There the ties would stay until the spring runoff. Then Smith's Fork was driven at high water to the Green River, where the ties were boomed at the mouth of the creek. From there they were put on a railroad car and hauled out to the treating plant.

Now, Smith's Fork as a rule didn't flow quite enough water to carry the ties down, so they had dammed the creek to build a reservoir. The reservoir was on Forest Service land, but it was built by the Standard Timber Company. It wasn't too large, as reservoirs go, but they'd fill it and then open the gates, and they could control the flow of water that came out. This would give them enough head to carry a group of ties clear down to the Green River; then they'd let the reservoir fill again, open the gate and start the water down, and then dump more ties in. Each time they made a drive there was loggers or tie hacks that followed the drive down, so if they jammed or hung up anyplace, or ties got washed out in the backwater, they could bring them back into the stream. They'd follow the drive down to the Green River, and then they'd come back, while another bunch took another drive down. And that's the way they transported ties from the tie camp to the Green River.

There were several reasons for hewing ties in the winter: One was that the tie hack had to deck his ties next to the haul road, and where he'd cut them sometimes might be as far as two hundred feet or better from the haul road. In the summertime, with all the trash on the ground, it'd be very difficult to move ties that distance. In the wintertime you slide them on top of the snow.

Second, for the tie hauler it was far easier to move those ties with a sleigh than it would have been with a wagon. In fact, it would have been impossible to even think of moving ties with a wagon through some of those skid roads that they'd cut out. Even on the main road where they hauled the ties down to Smith's Fork, it would have been next to impossible to move with a wagon.

Third, the tie hacks told me that a tree hewed a lot better when it was frozen than when it was green. They hated a warm spell that would warm enough to thaw the trees out. They wouldn't quit cutting, but the trees were a lot tougher to hew. Of course, those warm spells didn't last very long. [laughter]

Perhaps most important, the ties were all transported to the Green River on river drives, and in the summertime they wouldn't have the necessary volume of water to carry the ties down. They had to have that spring runoff to have enough volume of water, or sometimes, when the volume wasn't enough, they'd create a splash-down with water held in a reservoir to carry the ties down on a head of water. It wasn't until they went to *sawing* ties that it became a summer operation. They went to sawing shortly after they finished the Horse Creek sale in the spring of 1936. The Horse Creek sale was the last hewing operation in Region Four, and I wouldn't be a bit surprised if it was the last in the Forest Service.

"They took everything"   On this sale area, there was no marking of the trees–they tried to completely cut an area out as they went, and they took everything that was thick enough and tall enough to make two ties or more. Once in a while you'd kind of check a strip over after they got through cutting, and very shortly after I came onto the tie sale I was checking a fellow's strip out. He was working up ahead, and I saw one good-looking tree still standing in the area that had already been cut. It was just the right size and a nice, tall tree, and I thought, "Why in the world didn't he take that?" I walked around it, and it looked like a perfectly good tree to me, so I went on up and said, "You missed a tree back there." He told me where it was. I said, "Yes, that's the one." He didn't say anything; he picked up his saw and his axe and his peeler and said, "Come on down. I want to show you something." So I went down with him. It didn't take those fellows very long to knock a tree down, and he knocked it down and took his peeler and started peeling it. He peeled quite a good section, but once he started peeling it I could see what he was going to tell me. It had a twist, and I imagine by a height of ten feet or so that tree had made a complete twist. What causes it, I don't have any idea, and it isn't very common, but the log wouldn't be worth a dime for a tie. Those fellows could see the flaws in trees. There were a few of them I could tell by looking at the bark if they were twisted, but there were a lot of them I just couldn't tell.

I never disputed the tie hacks any more. If they left a tree, I figured there was something wrong with it. [laughter] They were cutting all of the lodgepole pine that were healthy and of the right diameter and height. It was clear-cutting as far as commercial timber went, but as dense as that timber grew, there were still quite a few small trees–too small to be merchantable–that were left. Some of those were real tall; but after the other timber was taken out, whether they could stand the wind and not be blown down, I don't know. There would be places where they'd take just about everything, and there'd be other places where you'd have a fair amount of small trees left. Of course, a sale area was all open for cutting, and there would be no strips of it set aside that couldn't be touched, either as seed trees or just to preserve the forest environment. They'd never get by with that now, but at that time they took everything that was merchantable. This wasn't as bad as it sounds, though. Lodgepole pine reproduces real fast and I imagine there is a heavy stand of reproduction over most of the area now. That was over fifty years ago, so they would be fair-sized trees.

One thing about lodgepole: you never have to worry about reseeding it because, as a rule, it will come up a lot thicker than you really want—it's one of the few trees that you don't ever have to worry about replanting. And there's very little invasion of a lodgepole stand by other tree species, even after a burn or after it's been logged off—other tree species just can't stand competition with the lodgepole. In fact, many times if you have a mixed stand of, say, lodgepole and Douglas fir (which you have a lot of up in Idaho and Montana), and you have a fire go through, a stand will come up pure lodgepole. The fir is finished. You don't have to worry about lodgepole reproducing itself!

Another difference between then and now was that the slash and the branches and the tree tops and everything from the tie-hewing activity was left right where it fell. There was no cleanup or anything afterwards. On *some* timber sales, especially logging sales, the Forest Service would go in and burn the slash after an operation. In the fall they'd go in and pile the brush and burn it, and a certain amount of the fees that were paid in for the timber was set aside for cleanup and burning of the slash. That didn't happen up there on the Standard Timber sale. This was a high elevation, and probably one of the reasons they never worried about the slash was that the country should have gotten a lot of rain during the summer, so maybe there wasn't much of a fire hazard. Maybe the country kept damp enough. Even though at that elevation there was quite a bit of lightning, I don't remember hearing of any big fires that originated in any slash that was left from the logging and hewing.

Among the tie hacks   Practically all of the tie hacks worked with partners, and many had worked together for years. When they'd go into a cutting area the first thing they'd do was build a cabin out someplace close to their work, and, boy, it didn't take those fellows long to build one. And they never bought a thing for that cabin—all of their cupboards and everything else was . . . if they wanted a half-inch board, they hewed out a half-inch board. If they wanted a two-by-four, they hewed one out. Everything in there—their bunks, their chairs and everything—they made themselves. I was talking to somebody not too long ago who was up in there, and he said some of those old tie hack cabins were still standing. They were well-built.

The woods boss decided who got what timber to cut. He designated certain strips, and some of them were real tough to work on and some of them weren't. If he gave a hack and his partner a tough strip one time, he'd try to give them an easy strip the next time–it was possible to cut more than one strip during the season. Some of those strips were real long, and they could probably spend the best part of the winter on one; and then some other strips were short. A hack and his partner could build a cabin adjacent to the first strip they were going to cut and find out later on that they had to walk a considerable distance in order to cut the next one they were assigned. If they had to go too far, they built another cabin. You see, those fellows could knock up a cabin in no time at all. [laughter]

Building another cabin was no problem for them, but, of course, some things could not be duplicated. For chimneys they'd use ordinary stove pipe. If they had to abandon one cabin and build another one they'd just take their chimney and stove with them. (They had a small cooking stove, but no heating stove of any kind.) If they had hinges on their cabinets, they'd take the hinges with them. The doors were handmade, but they'd take the knobs and the hinges off the doors and windows.

There were tie hacks who, for some reason, didn't want to build a cabin . . . or maybe they thought it was cheaper to live in the company camp. The company had two camps where they had built bunkhouses and a big cook shack, and they hired cooks and flunkies, and the men ate their meals there. The cook and his flunky also prepared lunches for them, and they were taken out to them at noon. I was welcome to eat lunch in their camps any time I wanted, and I'd generally eat dinner at the cook shack (mess hall). If I'd stop there and have a cup of coffee and a sweet roll, they'd never charge me, but they'd bill me at the end of the month for my dinners. They fed excellent, extremely good.

I have never been in a timber camp yet–even up in Montana, where there was all saw timber–where they didn't feed exceptionally good. A timber camp is a lot like a fire camp: if you feed good in a fire camp or a timber camp, you won't have any problems. You can work them and it can snow and the weather can be bad and cold, but if you feed them good you have no problems. If you don't feed them, you're in trouble. More men have walked off fires and more men have quit timber camps due to the food–or lack of food–than probably for any other reason.

It didn't cost much to operate the camp, because there was only a handful of people who were actually employed by the company: the woods boss, the clerk in the commissary, the two scalers, cooks and cooks' helpers . . . and then they had flunkies and a few men working around there besides that. Tie hacks would drift into camp come fall and the company would give them work. They weren't hired; they just went to hewing ties and the company bought their ties. (Loggers like that were called gyppos.) Tie hacks were strictly individualists, and they would often come into a camp dead broke. All they'd have would be the clothes on their back, and they would go to the commissary and buy clothing; they'd buy a broad axe; they'd buy a saw; and they'd buy a double-bitted axe. These men that had their own cabins, where they bached themselves, could also buy all the grub they wanted, and everything was on the cuff–every bit of it on the cuff. And if they wanted to send out for whiskey or medicine or anything, they could go to the commissary and draw whatever cash they wanted. The storekeeper would enter it in his book, and then in the spring when they were all through the whole thing was settled up. They'd walk off with what money they had left, and that'd be it for that year.

Now, tie hacks were kind of a funny bunch of men–they very rarely came back to the same company two years straight. There was a few of them that come back to Standard each year, but there was two other tie operations going at that time–one in the Big Horn Mountains in eastern Wyoming, and another out of Steamboat Springs, Colorado. The tie hacks would rotate. They'd go to one operation one year, and they'd go to another one the next year; then they'd come back to Standard and they'd make that circuit. When a tie operation was over, most of the hacks would go on a big binge . . . except the Finns–the Finns never did; they're a different bunch of people.

After each season was over a lot of the Finns would go to Rock Springs or Kemmerer and work in the coal mines until it was time to go back to hacking ties again. Many of the other tie hacks, however, more or less took their summers off, and by the time they got back to hacking ties they were broke, and the credit process started all over. After a season, what tools the tie hacks had left, the company bought back. The next fall a tie hack comes in and looks over these used broad axes and says, "Oh, this one would last me half the winter." So he'd buy it at a considerable reduction. Of course, the prices on most items from the commissary were higher than if you went to Evanston and bought them in a store there. And they should be higher, because of the cost of transportation and spoilage and freezing of certain things. Jane and I bought a little from the commissary, but not very much, because we bought most things that we needed before we came in. There were a few perishables, like potatoes and vegetables and eggs and stuff like that, that we'd buy over at the commissary. I felt that it wasn't right, really, not to patronize the commissary some, because I was working right with those fellows.

I guess there must have been about one hundred and fifty men in the camp, and the average tie hack would figure on hewing about twenty ties a day. In fact, if he hit his twenty ties at about four o'clock in the afternoon, he'd call it a day. But one day after the first of the year a bunch of tie hacks got to talking—I don't know whether they were drunk or not, but they got talking—on who was the best tie hack, who could hew the most ties. One of the fellows scaled part time for the company and he hewed part time. (Whenever they needed an extra scaler he'd scale for the company.) He wasn't a very big man, but he was well built. He bet that he could hew a thousand ties in a month . . . I don't know what the wager was. Anyway, he started in, and he hewed his thousand ties in a month. He hewed Saturdays and Sundays and during the week, and it if wasn't storming too bad he was out hewing ties in the storm. He had to hew better than thirty ties a day to make it, but he did it. I don't think there was another tie hack in the camp that could even come close to that.

Most of the tie hacks were either Swedes, Norwegians or Finns. I worked around Finns in farming country and other places, and I felt they were a more hardy and hard-working bunch of people than your other nationalities . . . at least that's been my experience. Finn tie hacks would be out at first light in the morning, and they would work as long as there was daylight; and sometimes on a full moon I wouldn't be surprised if they'd work maybe a little bit longer than that. Sunday was the only day they didn't work. The Finns were more productive than the Swedes and Norwegians, but there was no difference that I could tell in the *quality* of their work as tie hacks. Swedes and Norwegians were expert at it just the same as the Finns—they were all about equally expert at it.

Now, the Finns were kind of clannish. They were a good bunch of men, friendly and agreeable, but they always had their own camp and their own cook. (They also had their own tie hauler, if I remember rightly.) Finns lived a little bit different from the Swedes and Norwegians, and I presume they probably ate a little different. One of the differences was their Finnish bath. (I guess you'd call it a sauna now.) If the Finns had worked from any camp other than their own, they wouldn't have had their Finnish bath, and that played quite an important part in their living–I wouldn't say a Finn took a Finnish bath every night when he came in, but they took them very frequently.

The Finns took care of their own people. If a Finn got too old to hack ties, he stayed in the camp and he was a flunky. He peeled potatoes, he brought in wood, did whatever he was able to do, and they just took care of him. (I don't know whether they paid him any wages, but he got his meals and a place to stay.) Preparing the daily Finnish bath was always the responsibility of one of these flunkies.

The Finnish bath was actually a separate structure that the Finns put up in their camp. They built a log cabin with a low roof; it'd be head-high but not an awful lot more. It was chinked and practically air-proof and moisture-proof, and in the end of it they had a big fireplace. They rolled fair-sized rocks into this fireplace; maybe some of them would be eighteen to twenty inches in diameter. Early in the afternoon one of the old flunkies would build a great big fire and get those rocks really hot. Then just before the tie hacking Finns got in, he'd let the fire die down. They could close the door so it was . . . I wouldn't say it was airtight, because they had to have air to breathe, but boy, it would get warm in there!

There was no women in camp, and the men had their cabins around reasonably close to this Finnish bath. It'd be quite a ways below freezing, but they'd always undress in their cabin, and here they'd come a'running just as hard as they could run, through the snow to this Finnish bath. They had two decks that you could lay on, and they had fir boughs. You see, a fir has a very blunt needle and a spruce has an extremely sharp needle. They'd have these fir boughs, and they'd start sweating, and they'd kind of beat themselves with these fir boughs. They'd stay in there for quite a little while, and I guess they'd really sweat. When they came out, there'd just be a cloud of steam, and those guys would actually roll somersaults in the snow going back to their cabins.

They always wanted me to take a Finnish bath, but I told them, "Man, if I do that it will kill me!" And I think it probably would have, too. [laughter] But, oh, they were a hardy bunch of men. With the

Swedes and Norwegians, if you got a snowstorm—it wouldn't have to be a really bad snowstorm—you could take it easy for a day or so, because they wouldn't go out. But the Finns, it didn't make any difference how hard it snowed, they were out every single day. You *always* had to scale Finn ties, because they never missed a day. And I wouldn't say they didn't drink, but there was very little drinking in their camp. Maybe on certain holidays they drank some, but there was very little drinking.

The Finns were not small men, but most were smaller than the Swedes and the Norwegians. Among them was a big, husky man up in his sixties, I guess, or over. They'd given him a special contract, and he'd go in where some of those trees must have been close to two feet in diameter. He'd hew out twelve-foot ties, fourteen-foot ties, eight-foot ties . . . I don't think he cut any longer than fourteen-foot, but these were what they called switch ties. They paid a good price for those—a darned good price. Of course, to hew that much wood off of one of them, he didn't make too many, but he probably made as much money in a day as the average tie hack did.

There was a Finnish family, a father and mother and four boys, that had taken up an area off to the northwest corner of the cutting area. They were over there all by themselves hewing ties. The youngest boy, I think, was sixteen, and they had put him in high school in Rock Springs. One day on into the winter a little ways I went over to take up their ties and he was back, and he was out there hewing ties. So I asked his dad, "How come?"

"Oh," he said, "he got homesick. I'm going to let him hew ties for a few days and then I'm going to take him back." So he did. And they had to snowshoe or ski clear out to Robertson, where you could get a car to go to the highway. But it wasn't too long until here he was back again. I believe they did that three times [laughter] and finally the old man gave up. He told me, "Well, this summer, some way we can get him some more education. I would sure like to see him graduate from high school, because he's the only one of all of us who has ever been able to have any chance at an education."

In this family the father and the four boys would have breakfast and come out and start hewing ties. The older boys, of course, were old enough that they could hew right along with their dad, but the younger boy's job was mostly peeling and parking ties. The mother would come out about eleven o'clock with a big pack on her back; she'd build a fire, and she always served her men a hot lunch. You wouldn't call it a lunch–you'd call it a dinner. When they got through eating, they'd rest a while and then go back to work, and she'd pick up a pickaroon and start parking ties. Boy, she was tough [laughter] . . . she wasn't very tall, but she was well built, and she was just as stout as any of them. I used to marvel at her, watching her flip those ties around and stack them up. A stack would go up eight tiers high a lot of times–close to five feet. She'd park ties until about four o'clock or so; then she'd put her pack on her back and head for the cabin. The father and sons generally worked fairly late, but when they came in she had a good supper waiting for them.

Most tie hacks had come over to this country at a fairly young age, and very few of them ever married–at least, they never mentioned a family. (Now, the Finns were a little bit different, but the Swedes and Norwegians never mentioned family, so I assumed they were bachelors.) In talking to them, I learned that it had become rare for younger men to take up tie-hacking as a way of life–the work was just too hard for young fellows; they wanted to get a job that was a little bit easier. If one worked as a partner of one of the older tie hacks, in a period of time he could have acquired the skill (it wasn't that difficult if you wanted to really work at it), but I think the main problem was that the work was too arduous. All the tie cutting was in the wintertime; you were isolated from town and from society; and the younger men just didn't like that kind of job.

In the camp Standard Timber had what they called the woods boss–a Swede who was in charge of the whole camp. He was responsible directly to the superintendent, who had his office and what staff he had in Evanston. Under the woods boss was a commissary clerk, and sometimes he had a helper. Then there were two company scalers–one worked full time, and the other one was a tie hack so that whenever they got behind a little they could pull him in and he could scale. Once in a great while, the woods boss would also go out and do some scaling.

The commissary was quite a large building with a basement under it where they could keep stuff from freezing. Back of the counter they had storage bins for just about anything that you wanted to buy, but in front of the counter was a fairly large open floor. Open seven days a week, the commissary was a meeting place, and sometimes on Sundays when tie hacks didn't have anything to do a bunch of them would come in and sit around—they had a good heater in there and the place was warm. Of course, tie haulers would gather there, too. It was a room where if tie hacks felt like celebrating, they could all come in and have a dance. There were no women in the camp, so they would take turns on who would be women and who would be men for dancing. [laughter] Sometimes they'd tie a ribbon or a handkerchief around the fellow's arm, and he was a woman. Oh, they used to have some good times, and they always had good music! There were a lot of good musicians, mostly violin and accordion, among the tie hacks. They played all Scandinavian music and, of course, did Scandinavian dances. They had one dance they danced a lot—they called it Hambo, as close as I can pronounce it. They danced a lot, and most times they had a few whiskey bottles passing around, too, which added to their good time.

I can't remember any trouble of any kind among those men—not a single incident. In fact there weren't even prostitutes around, no gambling, nothing. It wasn't allowed. The operation was on Forest Service land, and I don't think the Forest Service would permit it, and the company sure didn't want that kind of people around. I can't even remember a fight. Of course, these fellows all knew each other. They'd traveled around to various camps for years, and the ones that would cause a problem probably had been eliminated quite a while before.

The only entertainment that involved any number of men was the dances over at the commissary. Of course, a lot of those fellows were musicians and sometimes four or five men would gather in some cabin and sing songs and have a feed or something. There was one pair of partners in particular . . . the darned guys, *every time* I took up their ties I'd have to go to their cabin and have coffee and sweet rolls. Or if was noon, I had to stay and eat. Boy, you'd better do it, too! If you offended them, you heard about it, and the whole tie camp heard about it. After we got through having coffee or we ate they'd have to play me some Swedish music. One played the violin and the other played the accordion, and they were good. Of course, the songs were Swedish, but it was enjoyable to listen to them. They'd play me two or three tunes, and then they'd go out to the timber and hack ties and I would go about my work.

Most of the jobs in the tie camp would be classed as dangerous professions. However, while I was there none of the tie hacks got seriously hurt. Once in a while somebody'd hurt an arm or a leg and be laid up for a while, but there were no serious accidents. But the year before, there was a tie hauler that got killed. He'd come down this steep grade, and there was a sharp turn, and he tried to make the turn but he was going too fast–the horses were running. They used rough locks on steep slopes, but this was on a south slope and they didn't have the desired effect. This was on a south slope, and it had been pretty warm that day, and the sleigh road had softened up. It was on late in the afternoon, but not late enough for frost or freezing to start. The tie hauler came off down this steep slope, and the rough locks weren't holding and his horses were running. He made the turn at the bottom, and the load of ties tipped over and he was crushed underneath them. He was buried right there where the accident had happened. The tie hacks put up a cross and carved his name on it and the date of his death. I hope somebody has put up a permanent marker there for him.... At the time there was just a wood cross; we went by it a number of times when we were scaling ties.

Scaling ties   The name *scaler* was probably used a long, long time before I was born. In my time they used the term for someone who estimated the number of board feet in a log. It was used in logging camps, and it might have originated from a tool that was used in scaling logs that they called a Biltmore stick, apparently invented by a man named Biltmore. This device had a scale of inches, and you could put it across the end of a log and determine the diameter of it; and then lower down there was a scale on which you could determine the board feet in the log. I kind of think that Biltmore stick, or scale, might have been where the name scaler came from. That terminology was used whether you were taking up ties or taking up telephone poles or taking up mine prop or even cordwood.

Both the company and the Forest Service scaled the ties. We found the parked ties, and we scaled them and checked them for rot and for short ties, which was very rare. Company scalers had double-headed hammers with either *STC* (Standard Timber Company) or only *ST* (Standard Timber) in raised letters on either end. When they would hit the end of a tie with one of these hammers, it would leave the impression of the letters indented in it. When they came to a deck of

Scaling a log with a Biltmore stick. With this device the scaler could determine the number of board feet in a log. When the log had been scaled, he would stamp it *US* with his hammer.

ties, first they would check it over to see if there was any rot showing in any of them, and then they would proceed to stamp the ties and count them at the same time. These ties were entered in the scale book under the tie hack's name. Then the scaler would go on to the next pile of ties.

In the Forest Service when we scaled ties, we had a very similar hammer—of course, ours had US on both sides. We also had a marking hatchet, which had a blade on one side and the US stamp on the other. Our stamp was a little bigger than theirs, and it probably indented into the tie a little bit further than theirs did. We also had a scale book, and whenever we took up a deck of ties we'd enter it in the book. At the head of the tally sheet was, of course, Standard Timber Company and the date, and I believe we had to enter the area in which we were scaling. At the head of the first sheet in the tally book was entered the accumulated number of ties since the tie operation started. At the end of each sheet, we totaled the number of ties on that sheet and we had a cumulative total at the bottom. For every day that we scaled ties we kept a tally sheet, and at the end of the month we had to make a summary of all the ties that were taken out that month and the cumulative total. When the tally book was filled, it was sent in to the ranger's office in Evanston.

Standard Timber was given the right to cut ties on this particular area, but they had to pay for the ties before they could put an axe to timber. From the accumulated total in the scaler's tally book the Forest Service could tell how close the company was getting to using up their last payment. When it got reasonably close, then the ranger would send the company what they called a letter of transmittal, requesting another payment. They were not paying per tree, however; they were paying per tie, and the money was paid directly into the Forest Service office in Ogden, Utah.

When I scaled ties and the company scaled ties, we scalers liked to go together, or if we couldn't do that, we liked to scale real close to the time that the other scaler scaled. The reason for that was that the ties couldn't be hauled until both of us had scaled and stamped them. That kind of aggravated the haulers sometimes because they'd maybe be up a strip that had piles of ties with only one scale mark on them, and they'd have to leave them there. They hoped there would be no snow before they could haul the piles out, because when they got a snowstorm in that country . . . boy, it could lay down the snow! The pile would be half covered with snow, and they'd have to dig it out. When they put ties with a lot of snow and ice on them on their load, they'd slide back and forth. So we always tried to scale pretty close together.

Okies at Archie Creek

Most of the tie hacks came into camp fairly early in the fall of 1934 . . . some of them started cutting before the snow even came. Later on, in the early part of December, there was another group came in; they had their wives and children with them, and they wanted to hew ties. These men were from the South someplace, and the tie hacks called them Okies. Now, I don't know whether they were the same as the Okies during the Depression or not, but the Scandinavian tie hacks called them Okies. Down at the very lower edge of the sale area was a creek called Archie Creek, and it was in an area where they could still get in pretty easy, because it was at a considerably lower elevation and the snow wasn't too deep yet. The company gave them that area to hew ties in, quite a ways separated from the rest of the tie camp. So they moved in and started hewing ties.

When the Scandinavian tie hacks hewed a tie, they stood on top of the felled tree and hewed down. These Okie fellows hung the head of their broad axe at an angle to the handle, so that they could stand on the ground and the head of the axe would lay flat to the side of the tree they were hewing. They did all their hewing from the ground, they weren't very ambitious, and they didn't do near the job of making a tie that the Scandinavians did. The Scandinavians used to laugh at them and told them, "You can stand on the ground for a while, but when we get five or six feet of snow, you'll be standing on top."

"No," they said, "we know how to hew ties. We have always hewed from the ground."

Well, the Scandinavians were right. It wasn't too long until the Okies rehung their axes and they were hewing from the top. They kind of had to learn the profession over again, because it was a different way of hewing. Ties hewed by these Okies varied in thickness and had a lot of axe marks on their faces. They weren't really the best ties.

I had some trouble with the Okies, and the company did too. One problem involved butt rot. They'd cut a tree and find it had butt rot, so they'd just move up and cut off another section above the rot. The rot would project a brown stain up the center of the tree for a ways, and our instructions and the instructions the company gave to their tie hacks was

Scaling Ties on the Wasatch

that they would not tally any ties that had brown stain in them because it indicated butt rot. (In a way, it wasn't quite right, I feel, because the company would still use those ties. If that tie was in the pile, they'd take it. It wasn't rotted, just had this stain in it, and if it was treated with creosote it would probably last just as long as one that didn't have the stain in it. Anyway, the company took them but didn't tally them; they got that tie for nothing.)

To get rot-stained ties by us scalers, the Okies would cut a potato in two and rub it over the brown stain. It covered that, and it was just the color of the end of the log and awfully hard to see, except that if you got sideways and looked across the end of it you could sometimes see what looked like white powder over the end of the tie. (I know I must have scaled ties that had brown stain in them, but they had rubbed the stain with a potato and I just couldn't see it.) If I ran onto one of those while scaling ahead of the company scaler, I had a marking crayon and I put an X on it, and they'd do the same thing for me so that we could pick them out.

Another thing that the Okies would do . . . . They would have a deck of ties ready to take up, and these ties were frozen. You had to hit them pretty good to put the whole indentation of the stamp upon them. When you'd hit them, they had kind of a ring to them, but you would go along stamping and sometimes you'd hit one and it wouldn't have a ring–it would be dull. So then you'd back up and you'd really swing on it and hit it again; and if the tie wouldn't move, you'd look around and find a pole if you could, and you'd hit the end of the tie. Sometimes you could drive a tie in two or three inches, so then you'd go around to the back of the pile, and you should have a tie sticking out three inches. But you wouldn't. What they'd done was maybe they had a tie that had rot in it, so they'd cut two blocks off the good end. They'd put one at the front and one at the back of the pile, and there wouldn't be anything in between. But they didn't very often fool you, because it had that different ring to it when you'd hit it, and we'd always mark those for each other so they wouldn't be taken up. We never ran into anything like this with the Scandinavians. The other Scandinavian tie hacks would boo them out of camp if one of those guys pulled a thing like that. They had so much pride in their work that they'd never *think* of pulling a stunt like that.

The worst stunt that those Okie guys pulled . . . . I'd generally check into the commissary every morning, and if the company scaler wasn't there himself, he would have left word with the clerk that they'd scaled such-and-such an area, so I'd try to make a point to go down there and scale it too. Well, one day I was told they'd scaled down near Archie Creek, and there was a big pile of ties; there must have been a hundred or better. (I don't know how those Okie fellows worked it with the company, but all their ties were scaled together. There was no scaling for individuals as with the Scandinavians.) So I went down, checked the pile over front and back, and started scaling. There was something that just wasn't right. I guess I'd scaled fifteen or twenty ties, and then it dawned on me that something was wrong with the ends of the ties. As I mentioned before, they'd fell a tree, they'd hew it, they'd peel the upper side of it, and then they'd saw from the round top surface through the tie to the round bottom surface. In other words, their saw marks were always at right angles to the hewn faces. You didn't *look* for this, but you just got used to seeing it that way. (Once in a great while the last tie from a tree might be sawed the other way, but not very often.) Well, these ties down near Archie Creek were all different. They were sawed from the flat surface to the flat surface. In other words, the tree had to have been turned on its side to have been sawed that way, and no tie hack would do that, because it was so much more convenient to just saw it from the round surface to the round surface.

So I thought, "What in the world?" And then it dawned on me: "I'll bet you those guys cut the Standard Timber Company stamp off." So I started kicking around in the snow, and finally I found a little piece of wood about half an inch thick or better that had been sawed on both sides. So I knew somebody was in trouble, but I finished tallying up the ties. That evening when I came in, the company scaler happened to be at the commissary, and I told him, "I think you've got a problem down on Archie Creek."

He said, "What is it?"

I said, "Well, it looks very much like they sawed your stamp off the end of the ties."

"OK," he said, "will you go down there with me tomorrow and I'll re-scale?" So we went down, and the minute he saw them he knew exactly what I figured, that they sawed the ends of the ties off. We kicked around some more and we found a few little blocks—they'd gathered up most of them, but they'd left a few. There was quite a fire circle where they'd had a fire to make coffee and keep warm, and that was the logical place for them to burn this stuff. So we went over there and we found a lot of pieces–in fact, some of them were a good third of the end of a tie. But we couldn't find one with a stamp on it, and to prove anything we had to find one with a stamp on it. So we dug around those ashes and dug around, and finally that scaler said, "I found it! I found it!" He showed me this piece, and it had the perfect stamp. It was charred, it was black, but it had the perfect stamp of the Standard Timber Company on it. Oh, he was mad!

If the Okies had gotten away with this, the Standard Timber Company would have paid them twice for the same ties. We went over to their camp, and, boy, that company scaler lit into them. He gave them a week to move out. He kicked them out of camp, lock, stock and barrel. In the time he gave them, the Okies were gone. They had their own sleighs, and they just loaded up their sleighs and hauled out to Robertson. I presume they must have had transportation of some kind at Robertson, but that was the last we saw of them. There had been about thirty hacks, plus their women and children.

The Okie tie hacks would have been doing good if they averaged fifteen ties a day, even that much. I didn't have to go down there very often to scale, because they just weren't that productive. And, of course, in any kind of a storm they were in camp. The company was glad to get rid of them, and in a way I was glad to get rid of them, too, because it was a long trek down there to Archie Creek. I'd take both snowshoes and skis, and where it was best traveling on snowshoes I'd travel on snowshoes, and if it was ski country I'd travel on skis. I had drilled a little hole through the tips of my skis, and I'd tie one end of a string through the skis and the other end into my belt, and I'd trail the skis. Then when I used my skis, I'd put the snowshoes on my back. It worked good.

There was a long climb coming back, and it was always at night. Sometimes it would be nine o'clock before I'd get back to Hewinta from Archie Creek, and it was all snowshoe travel coming back. My snowshoes were like a pair I had later on called the Pickerel. They were narrow and long and had a long tail on them. For traveling they were good, but taking up ties, you had so much turning and twisting that it was difficult sometimes. I also had a pair of snowshoes they called Bear Paws, and they weren't turned up in the front at all; they were just flat, no tail on them. They were like oval-shaped pancakes, and you could swing around on them with no trouble at all. Sometimes I'd pack those on my back, and when I got up to where I was working I'd put them on and scale the ties. Then I'd put my other ones on and go on from there.

It was a long climb coming up from Archie Creek. Of course, I always left for the creek early in the morning, but sometimes it would be just about dark by the time I got through scaling ties. I've been so tired at times that I'd set myself a quota of strides to take before I'd rest. I made up my mind that I'd never get below twenty-five, but, boy, there were nights that it was awful hard to make those twenty-five. [laughter] Jane always had supper waiting for me when I got back to our cabin, but I am glad I didn't have to make too many of those trips.

Winter at Hewinta The tie-hack camp was quite an operation; it was almost like a small town in a way. You had practically everything you needed up there. At the commissary you could get just about anything you wanted to buy, and there was a blacksmith's shop and stables. The grain and hay was all hauled in during the summer before the snows came. During the six months I was up there, the only time I went out (other than to see my wife when Jimmie was born) was to go to a rangers' meeting.

I went up to the camp the latter part of September. The baby wasn't due until November, but one evening at the end of October I came in and I got to thinking, "Well, tomorrow is Saturday. I think I'll go out and visit Jane." Saturday morning I got up real early and drove down through Robertson and Fort Bridger, and out to Kemmerer, about eighty miles away. I went up to the house where Jane was staying, and knocked, and a woman came to the door. I went in and asked her how Jane was . . . I didn't see her. "Oh," she said, "she's fine."

And I said, "Where is she?"

"Oh," she said, "she's back in the bedroom in bed."

And I thought, "In bed? She must be sick." Jane had heard my voice, and when I walked in, there she was lying in bed with a big grin on her face. And I thought, "She don't look like she's sick."

Jane pointed over towards the window, and here was a cradle. She said, "Look in there." I looked; there was my son. He'd been born a week or so before. The woman told me about when I could come and get Jane, and I came back in and got her and took her and the baby up to camp. She stayed there in the camp, and she didn't see another woman until she came out the next spring. My son had been born on October 26; he'd come a little bit early, I guess.

The Hewinta cabin wasn't a big cabin. The front room was a kitchen and living room, and there was a very small bedroom in the back that was just big enough for a bunk bed. The bed was built into the cabin, and we had a little heating stove in the bedroom that didn't hold much wood. Then, of course, we had a regular cooking range out in the kitchen. At night when Jane would put Jimmie to bed, she'd wrap him up good and then put a blanket over the top of the crib, and she had blankets all around the edge of it to prevent draft. It was always warmer in the kitchen, so she'd leave him in the kitchen, but in the mornings there would often be a little bit of ice in the water bucket. When I'd get up I'd take Jimmie out of his crib, and, of course, during the night he'd gotten wet, and there'd just be a big cloud of steam come out. I'd pack him back in a dry diaper and put him in bed with Jane. She'd stay there until I got the cabin warmed up; then she'd get up.

Jane was an outdoor girl, and she and her folks had camped out a lot. For my Forest Service career, I couldn't have married a better girl than she was. Once in the winter they called a rangers' meeting in the supervisor's office. They wanted me to come to it, so I had to leave Jane and Jimmie all alone at the Hewinta cabin while I was gone. We had this pitcher pump that would pump water. While I was gone the leathers in the pump went haywire, or the valve went haywire, and she couldn't get any water. Of course she could melt snow, but she had a lot of diapers and baby clothes to wash. There was a spring just a little ways off from the corner of the cabin, so she took a shovel and dug down through about five feet of snow to the spring, and that's where she got her water.

When I went up to Hewinta in September the clothesline posts were about eight or nine feet high, and about every eighteen inches there was a hole drilled in each post. I wondered why in the world they had them so high. Well, before the winter was over we found out that after every storm you'd have to raise the cross arms on the clotheslines a foot or two. We just kept moving them up, and when Jane wanted to hang the clothes she'd put on a pair of snowshoes, go out and hang the clothes, and come back in. [laughter]

The clothes would freeze dry. It was often way below zero, and it was very rare that it got warm enough to dry clothes. But we would hang them on the line, and the freezing would take a lot of the moisture out of them. It wouldn't freeze out all the water–they'd still be a little bit damp. Jane also had a clothesline strung across inside the cabin, and she'd bring them in frozen and hang them for a little while on the clothesline, and they'd be perfectly dry.

A wild ride   By the early part of April, 1935, the sales area was getting pretty well cut out. All of the merchantable timber up at Lapland had already been cut, and they had moved out of that area, and most other areas were getting pretty-well cut out. There was still a little timber left down in Archie Creek that the Okies hadn't gotten, but we didn't worry too much about that. By the middle of April some of the tie hacks had already pulled out or were ready to pull out. They still had the river drive to make, and they were holding quite a few of the tie hacks back to make the river drive, and then the operation would be completed. About the middle of April I was notified that I was going to be transferred to Evanston, Wyoming, to take over the Black's Fork Ranger District, but I wasn't given a date. The ranger at Evanston, a fellow by the name of Jay Hahn, was being transferred to the Loon Creek district on the Challis, but they didn't know exactly when the transfer was going to be completed.

I was worried about getting Jane and the baby out, because there was still quite a lot of snow. One day I was over to the commissary. Things had slowed down pretty well, and there was quite a few tie hacks there. I made the remark that I was afraid I was going to have to leave my car at Hewinta because there was no way I could get it down to Robertson. (From Robertson, we could drive out; the road from there to the highway was always kept clear of snow.)

A tie hack said, "Oh, don't worry about taking your car out. All you do is tell us when you want to go, and we'll take your car out."

I asked him, "How are you going to take it out?"

He said, "We'll load it on a tie sled. And your family . . . we'll take them out too. We'll take another tie sled and nail a chair down on its bed and have plenty of blankets, and your wife and the baby will be nice and warm." There was one man who was hewing ties in an area out away from the commissary who had brought his wife up, and that was the only other woman in camp besides Jane. She wanted to go at the same time, so when it came time they took both of them out.

It came to a weekend, and I thought, "Well, I might as well get them and my car out and find Jane and Jimmie a place to stay in Evanston. If I'm going to be a ranger there, I want to get that taken care of before I go out." So I went over this afternoon and told the hacks I would like to take my car and family out the next morning.

One hack said, "Fine. We'll be over there." So I shoveled out in front of the garage door so we could shove the car out a little ways, and sure enough, the next morning here came two tie sleds. They'd notified this other tie hack to have his wife down there, so they had two chairs nailed down on the tie sled. Jane carried Jimmie, and she and the other woman went out and sat down on the chairs. The hacks wrapped blankets around them, and they took off.

The guys on the tie sled that was to carry my car had been drinking quite a little. They'd gotten a bottle some place, and they didn't seem in any hurry to move out. Finally I said, "Well, when are we going to load the car?"

They said, "Don't worry." They went over, and . . . I don't know how many there was–six on each side or what–but they just lifted the car up and set it on top of the tie sled. They blocked the wheels and put a chain binder on to tie the car down, and I locked the cabin up, and we were all ready to go.

I was worried about these darned guys being drunk. The driver wasn't–he'd been drinking a little, but he wasn't all that drunk. Some of the rest of them, though, they were pretty well in their cups. We started down, and they handed me the bottle a couple of times, and I passed it up. But the first steep place we hit . . . . I guess the horses couldn't have held the sleigh back, anyway; it was too much for them. But the driver put the whip to them and let them run. Man, we went down that steep slope just as hard as those horses could run, and if one of them had stumbled they'd have wrecked the car and probably killed a bunch of us. I reached for the bottle, and I took a good big swig out of it. There was a number of these real steep places, and I wouldn't have given a nickel for that car going down some of those, but we made it through.

Jane did not have the same kind of experience. They didn't have as much weight on her sleigh, and the horses could hold it back. We got down there and unloaded the car, and Jane and Jimmie and I and this other woman took off for Evanston. We hunted around and finally got an apartment, and I got her all set up. I was gone a couple of days, but Jay Hahn's transfer still hadn't come through, so I went back up to the tie camp.

I had handled all the scaling by myself, until they moved up to Lapland. For a short time I scaled up there, too, but it became way too big a job; I just couldn't handle it. So the Forest Service sent in a young fellow by the name of John Parker. He took over the Lapland scaling, and I'd go up once in a while to spend a day with him. His tally books were sent down to me every month, and his tally was included in my tally when I made the monthly reports.

About the time that I took Jane and the car out they had finished up the Lapland cutting and closed the camp. John Parker moved down the slope, and he and I bached there at Hewinta. I was there for not over two weeks when they told me to move on out and leave the rest of the tie camp operation to John. That was the last I saw of the tie camp. The following year Standard Timber moved their tie hewing operation to Horse Creek, and a lot of the hacks went there and hewed ties the winter of 1935-36.

6

Kamas Ranger and Horse Creek Tie Scaler, 1935-1936

FINALLY I LEFT the tie camp and took over the Black's Fork Ranger District. The ranger stayed long enough to get me familiar with some of the problems of the district, and we had to check the property so I could receive it. As I remember, we didn't actually go out on the district to do this. All the getting acquainted with the district was done in the office, through the files and through just talking back and forth. Then the ranger took off for the Loon Creek district on the Challis.

A brief appointment   After three weeks I still hadn't gone out on the district. I was getting familiar with the files and with a lot of the people around town, and some of the permittees had come in to talk with me. One morning the supervisor called me and asked me to stay in the office, saying that he wanted to drop by and talk to me. I waited and he came over, and after visiting back and forth, he finally asked me, "How would you like a transfer?"

And I told him, "Chet, I just got on the district! Why transfer me now?"

"Well," he said, "I've just come over from the Kamas district," (the adjoining district) "and we've got a problem over there. Morgan Parke, the ranger, has been on that district his entire career. He's only got a couple more years to retirement, and we had hoped we could leave him there until he retired. But we can't do it. We're going to have to transfer him. We offered him a district down in southern Utah, and he refused to take it, so we asked him what district he'd take. He said that inasmuch as Evanston is just a short distance from Kamas, he would take the Black's Fork district. So," he said, "it looks like we're going to transfer you to Kamas. There's a lot of problems on that Kamas district," he told me, "and I'd like you to solve as many of them as you can. But this is just a temporary assignment."

Back in those days the Forest Service had starting districts, and the Black's Fork district was a starting district. When a man got his first appointment as a ranger, he was put on a starting district. These were districts that were not too complicated and the workload wasn't all that high. You'd gain experience on a starting district; then they'd move you up to better and more important districts until you'd get to your top districts. Then when you had served a few years on that, you were ready to go be an assistant supervisor or take a job in the regional office or bigger, more important jobs. This Kamas district was a finishing district for a ranger, so the supervisor told me, "It's just a temporary assignment until we can transfer another ranger to this tough district, and then, of course, you'll be moved."

I said OK. So I was transferred to the Kamas district, and Morgan Parke was transferred to the Black's Fork district. Parke had his home in Kamas and he left his family there, and about every weekend while I was stationed there he was back . . . and lots of times during the week he was back.

Kamas When I moved over to Kamas I left my family in Evanston, because I didn't know anything about housing in Kamas. When I got there I discovered that houses were very scarce. I hunted and hunted and couldn't find one, and I stayed in a room in the hotel for maybe two weeks. One day I was down to the post office and the postmaster, a man in his late forties, asked me if I was looking for a house.

I told him, "Yes, but I haven't been able to find one."

"Well," he said, "my wife passed away not too long ago, and we have a nice house. I just don't care to live in it with her not being there. Maybe you'd be interested in renting it." He took me over, and it was a real nice house; and everything was there in just the way she'd left it.

We didn't have too much; all we had, we could move in the back of the car. So the landlord took his linen and towels and left about everything else. Before we got into it, I told him, "I sure like your house, but I don't think I can pay the rent on it."

"Well," he said, "the rent isn't going to be very high. All I want is to have someone take care of the house."

Kamas, Utah, was quite a small town. It was settled shortly after Salt Lake Valley was settled, but at the time we lived there I doubt that there was over two hundred people who lived in the town. The people had lived there for a long period of time, their families were intermarried, and they were kind of clannish. It was difficult to get along with some of them—not so much to get along with them, but to become friends with them. To add more to my problem, Morgan Parke had been there his entire Forest Service career, and on top of that he had had a problem with alcohol. For a number of years the district hadn't been run the way a district of that importance should have been run. That, alone, is what caused many of the problems on the district.

If Kamas wasn't *the* most important district on the Wasatch, it was one of the most important because it had a lot of timber activity on it, and it was a district that was very high in recreation—it was right close to the big population centers of Salt Lake City and Provo and all those little towns along that Wasatch front. Of course, your recreation load was seasonal, and although the timber operation sometimes could run all winter, if you had a really heavy snow winter, then that ceased. The year that I was there I had two fires, including a bug-job fire, but normally fire didn't play too big a part on the district. During the summer we had rain, and a lot of the country was fairly high and pretty heavy snow country with moisture held over on into the fire season. You didn't worry too much about fire.

The Weber River bounded the district on the north side, and the Provo River on the south. A lot of tributaries from the main rivers went up quite a ways into the district, and were followed by roads. There were also a lot of lakes, especially on the Provo side. Utah Power had reservoired a lot of these lakes to maintain water flow for their power plants on down the river, and they were still in the process of reservoiring some of the smaller lakes when I was there. They would get a special-use permit from the Forest Service to put in a dam and raise the level of the lake. Of course, to all these lakes they'd built usable roads–not fancy roads, but you could get in and out with a car all right. Around quite a few of the lakes the Forest Service had issued summer home permits. The permit users weren't too happy when Utah Power would lower the water level of a lake, but the utility cooperated pretty well with the home owners, and if they needed the water, they'd be the last lakes they would lower.

One of the problems on the Kamas district (and it seemed like it was a problem on all the districts that I had) was that you were always short of manpower. You could go out and hire men to do general labor, like working on a road or working on a fire, but there was always a shortage of men that could help you with *your* job. I really felt it there on the Kamas district. At first there was nobody to help me with the timber and grazing jobs, and there was nobody to help me with the recreation job other than two recreation guards. I had to try to handle all that pretty much by myself. Later on into the summer they sent in a young fellow who did a lot of the scaling, especially in the Great Lakes Timber sale, which really helped me. But even with that, it seemed like you were behind the eight ball all the time [laughter] in trying to get your job done.

"We've never paid for mine prop" Mine prop cutting was big. Some years before they'd had quite a bug epidemic on the district. It had taken big trees, little trees–it didn't make any difference; it was a real heavy epidemic. Woodcutters would cut both green trees and dead trees, but they were cutting most of the dead trees for firewood or mine prop. (They'd take their prop over and sell them to the mines or to the mine-prop dealers in Park City.) So along the haul roads you had mine prop cut from dead timber stacked up, and you'd also have firewood being piled up.

One of the problems I encountered was that of mine prop that had been cut without a permit. Now, for firewood you didn't have to have a permit; you could cut firewood for your own use wherever you wanted to cut it. But I would be driving up a haul road, and here would be a stack of timber the same size and length as for mine prop. I'd ask the cutter, "Are those mine prop?"

He'd say, "No, that's firewood." Well, there was no way that you could dispute him. If he said that was firewood, that was firewood. But in your heart you knew that they were mine prop. Sometimes the pile of timber would disappear, and you'd drive around by his house . . . no firewood piled up, and you'd cruise around, and nobody knew what had happened to it. But they'd gone over to Park City and were sold as mine prop. Apparently this had been going on for a long time, and it was just an out-and-out trespass.

Well, one day I was going up the canyon and I came upon Bob Pack, who years before had been a Forest supervisor–I imagine shortly after the Forest was created. He was cutting, and he had about a truckload of this supposed firewood piled up. I stopped and talked to him, and he told me, "I'm cutting them for firewood."

(Mine prop are eight feet long, but they would look you in the face and tell you it was firewood. [laughter] Well, how are you going to tell them it isn't? See, that was the kind of people I was dealing with.)

I said, "OK."

A few days afterwards I was up the canyon, and the pile was gone. So I checked around his house like I generally did, and no firewood. There was a young fellow cutting mine prop that I got along with pretty good, so I asked him, "What happened to Pack's timber he had piled up there?"

"Oh," he said, "he hauled it to Park City."

I says, "Well, what mines do you fellows deal with over there?"

I can't remember the name of the mine, but he said, "Most of us, if we have any to sell, we'll sell it to that mine."

The next day I jumped in my car and went over to the office of the mine. I asked them if they'd bought any mine prop from Bob Pack over at Kamas recently, and they said, "Yes, we bought a load just a few days ago." I thanked them, and I went back and wrote out a timber trespass against Pack. The next thing I knew, the Forest supervisor, Chet Olsen, called and told me Pack had been in there with a complaint and wanted a hearing. So he set the time and date and told me to notify Pack to be there, which I did.

The supervisor wanted to know what the problem was–my side of it–and I told him exactly what the deal was. And he asked Pack, and Pack admitted that he'd cut mine prop, and he'd taken them over to Park City and sold them. "But," he says, "that's the way we've *always* done it." He said, "We've never paid for mine prop before we sold them, because sometimes we cut them and can't sell them, and we have to cut them up for firewood." So the outcome of the trespass was that inasmuch as this had been the policy in the past, the Forest Service would just overlook it. Chet warned him about doing it again, and I warned him about doing it again, and later on I wrote him a letter stating exactly what the policy was.

If that had been all of it, it wouldn't have been so bad . . . but I don't know how many of exactly the same kind of trespasses I had after that! Different men, but every one, when I suggested trespass, said, "No, you can't. You didn't trespass Pack; why are you picking on me?" So that halted my trespasses for cutting mine prop. There wasn't a thing that I could do until I sat down and wrote a letter to each and every user on the Forest that was cutting mine prop, and stated in writing what our policy was, and stated what I was going to do if I caught them in trespass again. Well, that ended it as far as selling illegal mine prop. They'd been duly warned, and this story about how they did it before wouldn't hold any water.

"It's the fault of the damned haulers"

Two years before I came to the district there had been a big blow-down in an area called the Broadhead Meadows. This was a tremendous blow-down! Whole sections of lodgepole timber were more or less laid flat. Although it was an extremely strong wind that did it, it wasn't a tornado: the timber was all laid in one direction, and if the wind had a whirling action the trees would have laid in all different directions. In a blow-down, the roots of most of the trees are still anchored to the ground; that was true of this one, and even though the trees were laying more or less flat, most were still alive.

One of the biggest timber operations on the district was the Great Lakes Timber Company. They were cutting the smaller downed timber for mining props, but their big operation was with power poles and telephone poles. These were cut and decked along the haul roads, and the hauling was all done by truck. About halfway between Kamas and Park City, Great Lakes had a treating plant where these power poles and telephone poles were butt-treated. (Most poles now are treated full length with creosote, but these were just butt-treated up to probably around ten feet, and then they were sold to different power companies and telephone companies that needed that kind of pole.)

Great Lakes' cutting of the blow-down was a total cut. Anything that was merchantable, they'd take—clear cut, and although the blow-down wasn't 100 percent cut, it was getting close to it. Even where a tree was too small for a telephone pole, they'd take it and sell it as a mine prop.

And I'm afraid the trees that were left wouldn't stand too long, because with everything else around them taken they'd blow down, too, in a short time.

Not too long after dealing with the illegal cutting of mine prop I was scaling poles up to this Great Lakes Timber Company. There was a pretty-good sized pile, but they were still adding to it, so I decided not to scale it. A few days afterwards, I figured they ought to be through piling on that one pile, and I went up . . . and the pile was gone! I'd never been to their treating plant, but that Sunday Jane and Jimmie and I went for a ride and drove down there. There was nobody there. The gate was closed, but it wasn't locked, so I went in and started looking around, and I found a lot of poles that didn't have the US stamp on them. There were some that they'd just hauled out of the Forest, but there were also other poles that went back for as long as they'd been cutting, I guess, because they were bleached out to a dark gray color. At first I didn't know what to do, but then I thought, "Well, the only thing I can do is come down tomorrow and scale them."

The next morning I came down and went to the treatment plant manager and asked him, "Are there any poles coming into this plant other than the ones taken off of the blow-down?"

He says, "No, that's our only source of poles."

So then I told him, "Well, there's a lot of poles out there that have never been stamped. They've been hauled out illegally."

He kind of hemmed and hawed, and finally he admitted it; he says, "Well, it's the fault of the damned haulers. When they bring them down for treatment we can't pay any attention to whether the poles are stamped or not–we just process them."

So I took them all up. The arrangement with the Great Lakes Timber Company was about like the one with the Standard Timber Company, and I was going by our Forest Service policy: No timber could be cut before it was paid for. Great Lakes made periodic payments, and we always kept the payments ahead of what their cut amounted to. They had just made a payment a very short time before. With the poles that I'd taken up, their cutting didn't exceed the payment, but it came awfully close to it.

When I got back to Kamas I went to the Great Lakes Timber Company's operations manager and told him what I'd found out. Of course, he blamed the haulers too; but I know darned well that he knew what was going on. There's no way it could happen that he didn't know what was taking place. I thought, "Well, it's a trespass again." Then I thought, "This goes way back a long time before I was here, so I'm not going to say anything about it (I had learned my lesson from the firewood/mine prop controversy) *but* I'll have to write in and request another payment just as quick as I can." So I did. As soon as my request came in, the administrative assistant got on the telephone; he said, "What's going on out there? They just made a payment a week or so ago. They're not cutting timber that fast, are they?"

I told him, "No." I said, "I've just decided that I'm going to scale up everything. I'm going to scale up everything on the landings and everything in the timber," which I did; I had to follow through with what I had said.

Although I thought I could maintain that practice of scaling out in the timber, it didn't work too good. It was just too much work, too much running around, and I couldn't spread my hours long enough to get the job done. So I was going to have to go back to scaling only the poles that were decked. Finally I realized that if the district was going to be run the way it should be run, and all functions–whether grazing or recreation or timber–were to be handled the way they should be handled, I was going to need help of some kind. So I told the supervisor, "There's no way that I can handle this. I'm going to have to have help, especially on the timber end of it."

The Forest Service sent a young fellow out–he was a college graduate, trained in timber management. The CCC camp was close to the pole operation, and the new fellow stayed there where he could get his meals and easily run back and forth to the cutting area. After that, he took care of most of the timber-scaling work on the district. Having an assistant relieved me of the timber work. I could concentrate on the recreation and range part of the job, and the activities of the CCC on the district.

CCC
trouble

We had a CCC camp on the district, and I had some problems with it. A lot of it went back to the type of superintendent we had, and also–to a certain extent, I think–to the foremen in the camp. The boys weren't getting out on the job like they should, and at the least little excuse they were back in camp. In other words, you didn't get the production out of that camp that I've seen out of a lot of CCC camps.

It didn't last too long, but at that time *all* CCC camp superintendent appointments were political appointments. When you have political appointments, lots of times there's no thought of experience or anything–you're just a political appointee, and that's it. Well, the superintendent of the CCC camp on the Kamas district was a singer out of Salt Lake, and he had *absolutely* no training in timber management or the handling of men or anything like that. All in all there were in the camp 150 to 200 men, and most of their foremen didn't have all that much training as far as forest work went, so they were having problems on their various jobs.

They'd had beetle epidemics in that district for a long time, and late in the spring after the snow had left there was a new breakout not too far from the camp. The Forest Service decided to have the CCC boys handle the treatment. Well, it's kind of a dangerous job, and I didn't care too much to be working boys on it, but they got by all right–they didn't have any trouble; nobody got burned. But they burned too late in the spring or the early summer, and they had some fires get away from them. One became a pretty hot fire and a fairly good-sized fire, but small enough that we could handle it with the boys in the camp if we could keep them on it. They put their full crew out on the fire, and we were getting pretty close to controlling it. I went around and told all the foremen that we were going to stay out until we had the fire controlled, and then I wanted one crew to stay out to watch that it didn't blow up and take off during the night.

It got past the normal CCC quitting time for the boys to head back to camp–probably something like four o'clock–and I was going around, and here was this sector of the fire with nobody on it. I asked an adjacent foreman about it. "Oh," he said, "the foreman decided that was enough, and he took his men back to camp." This was strictly against my instructions! We about lost the fire, but we moved the other crews around and finally got the thing beat down enough that I figured it was safe to leave it with the crew that I'd set up to patrol the fire during the night.

I couldn't believe that a foreman would *deliberately* pull his men off a fire after he'd been instructed to keep them on. I was teed off! As soon as I could, I went down and told the foreman that he was fired. Well, I didn't have authority to fire him, because they're employed directly out of the supervisor's office. The foreman got in his car and headed in to the office and told the supervisor. When I got into Kamas later in the day, I called the Forest supervisor and told him exactly why I'd fired the foreman. They didn't send him back. If they had, he'd have gone down the road again, or I'd have . . . *somebody* would've gone down the road. They transferred him to another camp–which I didn't care too much for, either–but they sent a man in to take his place, so we had a full complement of foremen on the CCC camp.

We also had a little trouble with the CCC officers. There was split jurisdiction in the CCCs–the military took care of the boys while they were in camp, and the Forest Service took care of them when they were out in the field on Forest land. The CCC boys on my district were building a dirt road up to Mirror Lake, and some of the grades were fairly steep. One day I was coming down from Mirror Lake and saw a car stopped up ahead, and there was a CCC dump truck with two wheels off in the borrow pit below the road. Two Army officers were standing by the car, and both of them had been drinking. One of them, the captain, was pretty-well oiled. The left fender on his car was smashed in.

What had happened was these two Army officers were coming up the road, and there was plenty of room for them to have stayed on their own side, but they were hugging the curve. Well, the CCC boy was coming down the road with the truck, and when he saw the car come around the curve he threw the truck into the borrow pit. They'd bladed off a bunch of small rock in the borrow pit, and his brakes didn't help him an awful lot. If the Army captain hadn't been drunk I'm sure that he'd have pulled over to let the truck go by; but instead, he slammed on the brakes and stopped. There was no way for that boy driving the truck to get around that car–no way in the world. The officers were just lucky the CCC boy threw the truck into the borrow pit, or it would have been a head-on collision.

I looked it over, and I took some measurements and got what information I could out of the Army officers. I could talk to the lieutenant, but the captain, I couldn't get much out of him. And I got the CCC boy's story of exactly what happened. The officers finally turned the car around and went on down, and I followed the CCC boy down to the camp, because I knew that as soon as he got in camp, they'd jerk his license . . . which they did. I made my report to the superintendent, saying I didn't feel that the boy should lose his license. But I guess CCC policy was that any time a boy was in an accident, whether he was the cause of it or not, his license was automatically pulled.

I didn't stay very long at the CCC camp, and as I was driving on down the road to Kamas I saw these two Army officers taking pictures of their damaged car. I thought nothing of it, really. But they filed a claim against the company that insured the car, and the company took action against the Forest Service. The first thing I knew, the Forest supervisor called and told me that the insurance company was bringing charges against the Forest Service and he wanted me to appear at a hearing in Provo, Utah, on such-a-day and such-a-time.

I brought the boy over, too, and I was a little bit late getting there. In fact, the meeting had already started. They were presenting their case, and they were passing a picture around to show the damage and everything. The picture came around to me, and I looked at it, and I told them, "This isn't a picture of the accident. I know exactly where this picture was taken–it was taken on the road going into Kamas below the CCC camp. This picture was taken on a curve of a newly constructed road, and this isn't anywhere near where that accident occurred."

Of course, the insurance man grabbed onto that right away. Finally they pinned the Army officers down, and they finally admitted that that wasn't where the accident occurred. All they wanted to show, they said, was the damage to the car. They were trying to weasel out of responsibility for the accident, and pin it on the CCC boy.

WPA: "A rough, ringy bunch"

About the last of June a WPA camp had been set up on the district. This WPA camp was made up of men mainly from Park City–miners that were out of work. The Forest Service set up the camp to run a sawmill, and the main job was to saw out material for picnic tables. At that time all the Forest Service's camp tables, picnic tables, were made out of natural logs. They'd saw out what would constitute a whole kit of material, and this'd be bundled up as a unit with the bolts and everything necessary to put it together. If a Forest wanted, say, fifty tables, they'd just load fifty of these units on a truck, and off they'd go. The tops of the tables would be made out of logs that were probably between twelve and fourteen inches in diameter, faced on three sides, and the seats were made of single logs, faced on one side so you had a twelve-inch or better seat. The supports and legs were made out of smaller timber. These pieces were all sawed, and wherever it was necessary to cut out where two logs fitted together, all the cutting and everything was done at the sawmill camp. These kits were produced primarily for the Wasatch Forest, but if the Wasatch had a surplus of tables and there were other Forests that wanted them, they could get them from the Wasatch.

A portable sawmill was sent out, all torn down. They had a brand-new blade for it, maybe forty-eight inches in diameter, a far bigger blade than what we really needed for that kind of work. They had sent out a sawyer when they started setting up camp, and I asked him if he could set up the saw. "Yes," he said, "I've worked on a lot of them."

They got the sawmill set up, with the blade on the mandrel ready to go. The next day I went up to see how they were doing. They had begun operating the mill, but the guy hadn't really known how to set it up, and the blade wasn't running true with the carriage. He got that thing so hot that it turned purple and had waves in it. (When he shut it off, it finally cooled down to a blue hue.) I was sick, because the blade was completely ruined. I thought, "If I tell the supervisor what's happened, he'll fire me."

I was feeling pretty low, and I came on down into Kamas, and pulled into a service station to get some gas. I was telling the attendant my problems (I knew him pretty well), when a guy standing near us said, "I'm a sawyer. Right now I'm hauling pipe from Salt Lake out to Kamas, but I got just one more load and I'll be out of a job." And he says, "I have a family to support; I'd really like to get that job on your sawmill." I told the attendant that I figured the blade was completely ruined, but this fellow told me, "If you want to take me up to your mill tomorrow, I'll look at it and see if I can fix it. Where do you want me to meet you?"

I said, "You might as well come down to the service station; I'll be here." The next morning, bright and early, I was down there, and he was there waiting for me; he had a big ball peen hammer with him. We went on up to the mill, and the blade was still on the mandrel. It had been so hot it was actually blue, but he looked at it and turned it around a few times and says, "I don't think it's ruined. I think I can fix it."

He took it off, and hunted around and found quite a large stump that was flat, and he laid the saw on this stump. Then he cut a thick stick and drove it into the ground next to the stump, and kept driving it in until he could lay that big saw blade flat on the stump and the stick would support its outer edge. He started hammering, and he knew where to hammer. That guy knew saws! [laughter] He knew where to hammer, and he'd hammer here for a while; then he'd hammer over there for a while; then he'd hammer another place; and then he'd tell me to turn it, and I'd turn it a little bit, and he'd hammer some more, all the way around the whole saw. We turned it over, and he did the same thing on the back. He could tell by the bulges, I guess, where he should hammer and where he shouldn't. Finally that saw started straightening out, and the blue coloring started going out of it. We worked on it on into the middle of the afternoon, turning it and hammering it. Finally he got down, looked straight on at the edge of it, and said, "Well, it looks like we got it pretty close."

He put the blade on the mandrel, and then he took a board, put a pencil mark on it, and laid it across the carriage of the saw. As I remember, he nailed it down. Anyway, he put it so the mark was at a certain place on the saw, and then he turned the blade three or four times, watching how it crossed the mark so he could judge where it still needed trueing. When we turned the saw, if it wavered a little at that mark, he'd have me . . . . We had a block of wood maybe ten inches across which I'd hold against the back of the saw so he could hammer it there until that section of it came true to the line. He went around the whole saw that way. There wasn't too many places that he had to hammer, but he kept hammering them until you could spin the saw and it wouldn't deviate one iota from that line. In this hammering, he told me that it actually re-tempered the steel in the saw.

Afterward, that saw was just as good as a brand-new one. So I hired him. He was a terrific sawyer, but hard of hearing . . . which was good for him, because he couldn't hear the saw when he wasn't wearing his hearing device [laughter], but it was hard for the men to communicate with him. Sometimes they actually had to go up and touch him to get him to understand what they wanted to tell him. But he remained the sawyer as long as that camp ran.

We had around thirty men in the sawmill camp, and they worked ten days on and four days off. When they'd return from their days off, you were never sure that you were going to get the same men back. Say, if a couple of men went back to town and found jobs, the WPA office would send us two new men to fill in. When they first started out, it was kind of a tough operation there; they were a rough, ringy bunch of men. Similar to the problems we had had in the CCC camp, the biggest trouble was that they wouldn't get out in the morning. The cook'd have breakfast ready, and some would get up and some of them wouldn't, and they wouldn't obey the foreman when he ordered them to get up. Same way in the evening: when they got tired, they'd come in; they wouldn't work a full shift. They weren't paid very much, but you couldn't fire them. On their part, that was the beauty of it—you couldn't fire them. That was the *main* problem. There was also animosity between some of them and the foreman. They were just looking for a fight, is really what they were doing, and the foreman couldn't control them.

One day a good-sized young fellow named Sessions—not an awful lot taller than me, but built a lot heavier—came to the district office and said he was looking for a job. The only job I could think of was to replace the foreman of this WPA camp, and he looked like he could do it. I told him what the problems were, and I told him that he could have the job if he wanted it, but it was going to be a tough assignment for a short time. I said, "Do you think you can handle that bunch of men?"

"Oh, yes," he says, "I can handle them."

So I hired him and fired the other guy. The other guy was . . . I'll call him a civilian [laughter], appointed by the Forest supervisor. So I fired him and put Sessions on. (Of course, Chet Olsen, the supervisor, already knew the problems I was having with the other foreman.) I hired Sessions, and I took him up to the camp and introduced him around and left. It couldn't have been more than three to four days later when I went out there, and boy, everybody's out working like everything! I stopped and talked to one of the fellows. He says, "That foreman you brought up, he's tough!"

I said, "What do you mean?"

"Boy," he said, "he gets you out in the morning and says, 'You work!'" He told me there was a couple of them decided to take Sessions on, and he rolled up his sleeves and says, "OK!" I don't know whether they actually fought or not—Sessions would never have told me if they had—but he got his bluff in, and after that he had a dandy camp that ran just as smooth as could be.

An extra-period fire

A fire started the day before a Mormon holiday, toward the middle or late part of July, 1935. It started along a trail just above a ranch, so I am pretty sure it was man-caused. A rancher reported the fire to me, and I started rounding up my road crew, which I thought was probably adequate to handle this small fire. We went out and worked on it, and we got the fire controlled by evening—it was still burning inside, but we did have a good line around it. The next morning we got on the fire, and we thought it would just be easy going from there on out, but on about ten o'clock we got a heavy southwest wind, and boy, it blew! The first thing we knew, it started crowning. See, up until this time it was all a ground fire, but it started crowning back in the fire, and here it came! When it came to our trench, it didn't even know it was there. And then I had a *fire* on my hands.

This was on a Mormon holiday, and just about everybody in the valley around Kamas had gone to Park City to a big celebration, so I could get no local fire crews. As if that weren't bad enough, my CCC camp was in the process of changing groups of boys. They were down to maybe fifty, and the Army always required twenty-five men present at all times, so that in case of fire in camp they'd have somebody to fight it. (You see, these were all wood buildings; and why more of them didn't burn down, I don't know.) The other twenty-five or so boys were on leave, gone to Salt Lake or Provo or some place, and my WPA camp was on their four days off, and they were gone. All I had was my road crew, which had already put in most of one day on a fire shift. I knew I needed help; I needed help, and I needed it bad.

I didn't dare leave the fire too long, but I went down to a ranch just below and called the dispatcher in Salt Lake and told him what I had, and that I needed at least twenty-five more men. He says, "Well, I can't spare them." We had received a written policy from the regional office just a very short time before that we would man *every going fire*. Period. Well, they didn't have any other fires on the Wasatch front. They had at least three CCC camps, and they didn't have a going fire, but he said, "We don't dare let anybody go, in case we might get a fire." I tried to talk him into it, and, "Nope; no."

So I went back, and there was no chance in the world that we could even think of stopping the fire. It was still crowning when I got back. The policy then was that it was just about mandatory that you get a fire stopped in the first burning period. If you didn't get a fire in the first burning period, you were called before a fire board review. I saw that if I was going to get that fire in the first burning period now, I was going to need fifty men, so I turned around and went back down to the ranch to use the telephone again. I told the dispatcher that I needed fifty men now, but he refused me again. So, I says, "Well, will you call the Forest supervisor? I want his decision on this." He told me that he had called the Forest supervisor, and the Forest supervisor told him the same thing:

not to turn any CCC boys loose on the Wasatch front. So I was turned down a second time.

I got back to the fire not too long afterwards and found that Morgan Parke, the ranger that originally was there but was now at the Black's Fork district, had come up. I told him what the situation was, and that I needed help but didn't dare leave the fire, which would have been very poor policy. I would really have been in trouble if I had left that fire. So I asked him if he would go over to Park City and try to round me up as many men as he could get, up to fifty. He said he would, and he took off. Well, I felt a lot better then, because I knew that help was coming in a short time. We waited, and no help came, and it came dark and my crew was completely shot. Finally we sat down around a camp fire.

Early the next morning the assistant supervisor, Felix Koziol (they called him "Kozy"), came out, and the minute he saw my situation he knew darned well that I had to have more men. He asked me what I had done, and I told him that twice I had tried to get men from camps out of Salt Lake or Provo, and I had been refused both times; and that I had sent Morgan over to Park City to get men, and he hadn't come back. So Kozy jumped in his pickup and took off. He said, "I'm going to do some checking."

He got in to Kamas, and found that Morgan had had a few drinks and was asleep in bed and had not even tried to round up any men for me. During the late evening, though, the twenty-five CCC men who were on leave came back to their camp. So Kozy went up and got them and had the cooks make them up some good lunches. We had been living on rations and were getting short of grub, too, so he stopped in Kamas and got grub for us. These CCC boys showed up on the fire, and they were fresh and ready to go, and they kind of rejuvenated my road crew and me, too. By then the wind had started dying down, and we tied into the fire and by evening we had it. We got it stopped and started mopping up.

The next day I had the CCC boys back on it, and I before long it was completely safe and it was practically out. It had been cloudy all day, and on to afternoon we got a cloudburst. Man, it poured down on that fire, and it washed the ashes and burned wood into the creek, where eventually a lot of it washed down into the Weber River. Of course, that ended the fire; there were no problems after that.

That fall, the latter part of October or the early part of November, the region held its annual board of fire review in Ogden. Any ranger who had had an extra-period fire during the season was called in to explain the circumstances. I think it was just arbitrarily decided, but at that time a burning period was calculated from ten o'clock one morning to ten o'clock the next. If it started at eight o'clock or nine o'clock a.m. you still had this burning period to catch your fire, but if it started, say, at noon, then you'd have the rest of that day and night and a full burning period afterwards to catch it. [laughter] So you were lucky to have a fire start about noon or later.

Any ranger who had an extra-period fire during the season was called in to explain it. If it was lack of manpower, then you would explain why during that period of time you didn't have adequate manpower on that fire to catch it. Sometimes it was poor judgment; sometimes you didn't order enough men. A lot of rangers got into serious trouble because they just used poor judgment, and maybe they didn't look ahead and see what their problems were going to be four or six or nine hours later. From my fire experience in Montana I knew that when you order more men, figure what it's going to take to stop that fire now, and then order enough more to take care of what that fire will have become by the time the men get there, and probably a little bit more. You always wanted to order more. You never got in trouble by ordering too many men, unless you were *way* out of reason; but you sure got in trouble if you didn't order enough men and the fire went uncontrolled.

The regional forester we had then, he was tough. If you had an extra-period fire, somebody was to blame! It didn't make any difference about the weather or what–if you had an extra-period fire, somebody was to blame. I was among the last of the rangers who were interviewed that day, and when I came up I was between the devil and the deep blue sea on whether to bring out the fact that I had ordered men twice from the dispatcher in the supervisor's office and had been turned down. There were so many things I didn't know . . . . The kind of an individual that the fire dispatcher was and the way he turned me down the first time, I was not too sure that he had actually called the Forest supervisor; and there was a possibility that he hadn't, but I decided I wouldn't mention it.

Of course, the assistant Forest supervisor knew the whole story, the same as I did. And I thought, "If he don't mention it, then I'd better not," because I could get myself into awful serious trouble if I started accusing the Forest supervisor and the fire dispatcher for my mistake. So I decided I'd better take it as it was, and I told the board that I just couldn't get adequate manpower until my CCC crew got back, and I couldn't stop it. Period. Well, they kind of accepted that . . . and then there was the fact that it was a Mormon holiday. If it hadn't been a Mormon holiday, they would *really* have taken after me. But they knew what the situation was on those Mormon holidays. They really celebrated, and even if I had scoured the whole valley I doubt if I could have picked up four or five working men. Everybody was gone over to Park City. Anyway, I took quite a roasting on it, but I charged it up as experience, and it was a very good experience.

I learned one thing in particular on that fire, and that was to keep notes. See, on the fire I hadn't kept any notes. I kept them in my head, but I hadn't kept any written notes. After that review I made up my mind that I was going to keep very accurate notes on decisions that other people made, and on requests that I'd made and decisions that I'd made. Now, you take a going fire, there's times you've got to make a decision, and you've got to make it right now. And you hope your decision is correct. Sometimes it is and sometimes it isn't. But when you make that decision you want to put down as to why you made it, and then if you go in to a board of fire review and they say, "Why in the world did you make that decision?" you can say, "These were the factors."

I'm sure the decision that the fire review board made on how I handled the fire–which was that I had failed to request sufficient manpower–went into my personnel file. But it did not seem to have an effect on my career. In just a year's time I was given a fire ranger district on the Challis National Forest. In fact, I had *two* fire ranger districts, and I was considered at that time to be one of the better firemen in the region. I was just a victim of circumstances. I had enough fire experience in Region One to have enabled me to handle that fire with no problem if I'd had the manpower. But I didn't have the manpower, and you can't fight fires if you don't have men . . . especially back then, when it was all handwork.

Just a punk ranger

That one year at Kamas was by far the worst in my Forest Service experience. There were more problems and misfortunes in that period–it wasn't even a full year–than I had during the rest of my Forest Service career.

There were some tough problems on that district–especially trespass–that *had* to be solved. I'm sure if the regional office had known what was going on there before I arrived, somebody would have been in serious trouble. There must have been thousands and thousands of dollars' worth of timber–telephone poles and mine prop and firewood–that were sold off of that district without a penny being paid for it, and it had to be straightened up. Well, my efforts to deal with the situation created quite a bit of animosity, and my job wasn't made easier by the fact that I was temporary on the district, and this was my first district as well. I was what they called a punk ranger back then. I had so many strikes against me that I still have a feeling that due to my age and being temporary, I didn't get the support from the Forest supervisor and from personnel in the supervisor's office that I would have received if I had been like the man that followed me . . . or if I'd have been a ranger with fifteen, twenty years' experience. I didn't get that support. It wasn't right, because I had a job to do, but it was just the way things go. For example:

When they sent this young fellow out to help me with my timber job, not only could I give more attention to the CCC camp, I also had more time for recreation and range management supervision on the district. One day I found a bunch of sheep that weren't permitted on the Forest at all. They were in trespass, so I saw the herder and told him to get them off. He told me, "No. I can't move them until the boss tells me." So I got ahold of the owner and told him to get them off, and he said, "Well, the boundary isn't posted too well. We can't tell where the line is." But he got them off, and in a few days I went up there and re-posted the boundary so anybody could see it.

It wasn't too many days afterwards I was out there, and they were in trespass again. They were grazing, but they were moving around real slow, so I climbed up on the side of the ridge and counted them. I might have been off a few head one way or the other, but I got a pretty good count. So I trespassed the owner. As soon as he got my letter, he goes in and starts complaining to the Forest supervisor, who called and told me that there was *another* complaint on my district; he wanted to know if I'd be in my office at such-and-such a time. So I was there, and the sheep owner came in, and we had a talk. Finally he admitted that he was in trespass, but that in the past Morgan Parke would let him graze that area whenever he was in country. He said he figured Morgan's word was what he was going on, and that he wasn't paying any attention to me because I was just a young sprout. So the outcome of it was that they tore up the trespass and let it go.

My generation of rangers who were college educated met some resistance from the old-timers in general. At times you weren't too well accepted by the old rangers that had been cowboys or miners or whatever, some of whom had never even finished high school. Very few of your livestock permittees had any college education, either, and it was tough for these men to take instructions from a young ranger who they knew couldn't have had very many years' experience. Education and training received in college didn't mean a damn to them; it was on-the-ground, hard experience that they wanted and depended on.

It was the initial contact that was difficult. For the first year or so you'd have problems, but each year you'd gain more experience, and after you'd been on a district a few years, this education business didn't come into it anymore. Finally, they would begin to listen to new ideas that you had. Sometimes they wouldn't adhere to it, but you could talk to them and explain and show the advantages of new systems of grazing, and sometimes they would go part way with you. Of course, it never reached that stage for me at Kamas because I was there less than a year.

Even with all my problems on the Kamas district, I gained a wealth of experience, especially in dealing with trespass and fighting fire. It was good experience that lasted me through my entire Forest Service career. I became more careful in getting enough facts, and making sure that they *were* facts, on a trespass before I wrote out the trespass report; and in fighting fire I learned to keep accurate, complete notes, not only on what I did or said, but also on what somebody else did, what instructions they gave me, and what the conditions were when I made a decision. I never worked again on a fire that I didn't have a notebook in my pocket. Man, I kept notes! I kept accurate notes, and it paid off–it paid off many times.

Preparing to leave After the extra-period fire, things had kind of leveled off. We didn't have too many problems, and everything went on pretty easy. I had not been back home since I was married, and my folks had never seen Jane or their grandson, so I wanted to go back home for a visit. Probably about the last of November I talked to the supervisor, and he said, "Sure. You decide when you want to go, and put in your application for leave, and I'll grant it."

I think I applied for leave about the twentieth of December, and it was approved. Everything was running smooth and no problems, so Jane and I and Jimmie took off; we went back home and spent about a week. We got back to Kamas sometime after the first of January. Not too long after we'd left they'd had an awful snowstorm, and since this was early in the winter, and they knew they were going to get more snow, they decided to pull out the WPA camp and the CCC camp. When I got back there was a good twenty-four to thirty inches of snow on the road going up to the CCC camp. No way to get up there unless you went on skis or snowshoes.

Very shortly after I returned they notified me that a ranger by the name of Alonzo Briggs was coming to take over the Kamas district. Right around the early part of February, Alonzo showed up, and we checked all the property around the office. That was about the only place we could check property–because of the deep snow we couldn't get up to any of the guard stations or the CCC camp, and we couldn't get up to the WPA camp. There was a property transfer done anyway, for all of the big equipment (like trucks and Caterpillars and graders) that had been moved out from the WPA camp and the CCC camp. I was informed that the rest of the property was still there at the various camps. (I think our routine property returns had to be submitted around the last of November, and I had made a complete property check then. I had made a list of where everything was–at that time if you had a shortage, you had to pay for it–and I remember some hand axes or something were short, but everything was brought up to date.)

Briggs was a man of probably fifty years then. When we checked the property, I wanted to snowshoe in to the various camps, which we really should have done, but he said no. He said, "That's too long a trip. As soon as the snow goes, I'll check the property. I'm sure everything is there. You've got your list showing that everything is there." I was foolish to leave the district without making a complete check, but I did. (Later

I would suffer for this.) They told me that I was temporarily being transferred back to the Wyoming Forest. Jane and I moved over to Kemmerer, and for a short time that winter I worked in the office completing maps and timber estimates from the timber survey that we'd done there in 1934.

Tie jam at Horse Creek   In the meantime, the Standard Timber Company had moved its tie hack operation from Smith's Fork in the Uintas to Horse Creek on the Wyoming Forest. Horse Creek came in to the upper part of the Green River from the west. Standard Timber had cut ties in there a few years before, but they had never finished, and the Forest Service required them to go back and finish the sale, which they did that winter. It had been an awful tough winter, and there was a lot of snow. The road to the upper Green River was closed a lot of the time, and so was the road between Kemmerer and Rock Springs. There were two young fellows scaling on the sale–Mel Coonrod and the Lowell Woods who had been one of my cruisers when we cruised timber in the head of Green River in 1934–and in the latter part of March or early April, Mel's wife (who was expecting) wasn't feeling too good, so they decided that she'd better come out. I made one trip up to get her, but it was storming so that they couldn't bring her down from the tie camp. With the next trip I made up I brought her down, and she went back to where her folks lived.

Mel and Woody were winding things down pretty well at the camp. Everything was just about cut out, and it was finally decided to move both of them out and send me up to the tie camp to finish out the work. I was supposed to stay there until they started the tie drive. There was some cutting still to do, but most of my job was just snowshoeing around through the sale area, picking up odd stacks of ties that either due to being covered with snow or for some other reason the company hadn't scaled.

There had been very few women in the tie hack camp on Smith's Fork, but at Horse Creek there was a lot more tie hacks with families. These tie hacks had moved their families in and built more or less permanent cabins and spent the winter in there. As far as I know, there was no company policy regarding bringing in wives and families, and if a hack wanted to bring his wife or family in, the company would work it out some way so they could give him a patch of timber where he could build a permanent cabin and spend the winter. He might have to walk further than some of the other tie hacks, but they'd still give him enough timber to keep him busy all winter.

At one place, there were three cabins spaced a little distance apart around the edge of a big meadow. It was an area that was kind of isolated from the other tie hacks in a fair-sized body of timber that they wanted to cut out. In that particular area, the snow must have been close to nine feet deep, and in deepest winter it came up over the eaves of the cabins, and maybe a little higher. For a place for the children to play they had shoveled the snow off of the roofs of the cabins, and it was banked up around their edges. All of the cabins had a low slope to the roof and were dirt roofs, and that's where the kids played.

One time when I was going along I heard children's voices. I heard these voices, and all of a sudden one of these kids stuck his head up from behind a snow bank, and they were playing on the roof of this cabin. You'd look off across there and you'd see a black smokestack coming up through the snow, and there would be a cabin beneath it. They would just shovel out around the door and a window so they had light and could get in and out all right.

The camp commissary was on a bench about one hundred and fifty feet above Horse Creek on the south side, and the Horse Creek ranger station was on the north side. During the winter the company had dumped a couple of thousand ties off the bank down into Horse Creek, and they'd slid down and piled up clear across the creek. Some water was seeping through this pile, but not an awful lot. With all the snow melting, high water was coming down, and these ties had dammed up quite a body of water back of the tie pile. One evening I was coming back walking along the skid road, and I came to this big pile of ties; the top of them was right level with the road. I could hear somebody working down on the tie pile, and I thought it was kind of late for that, so I walked over and looked down. There was a young fellow by the name of Jim Black down there. When they were at Smith's Fork he was scaler for the company, but when they moved over to Horse Creek they appointed him woods boss–he had been promoted. He was inspecting the pile and trying to break some ties loose.

Now, during the winter they'd dump a bunch of ties down and then it'd snow. And more ties and more snow, and the snow would turn to ice, and that whole mass was practically frozen solid. He was trying to get the front of that pile broke loose so it would slide down into the creek. He swore and said, "We tried to get this tie drive started today, but we can't bring any ties down from above until we move out this pile." There was a pickaroon stuck in a tie there, so I yanked it off and worked my way down to where he was. We worked there for maybe half an hour; it was getting pretty late. You'd get one tie broke loose and then you'd have to work on another one. It was an awful mess.

Finally he said, "Let's forget about it. I hate to do it, but tomorrow I'm going to have to blast this pile, and I'm going to have to load her really good. I don't want to do it because I'll damage so many ties." So we started climbing up the pile. It was a fairly steep face, and you had to go on all fours most of the way. There was a tie sticking straight up about two feet out of the pile, and I reached out with my pickaroon and socked it into the end of it to pull myself up. When I did, it gave, and the pile kind of quivered. Jim was below me; he said, "Arch, what did you do?" I told him, and he said, "Stay there. Let's pull that tie out and see what happens." He got up there, and both of us would sock our pickaroons into the tie on either side and pull, and we could slowly pull it up. Finally it came out, and the ties started shifting. Pretty soon a bunch of them slid off into the creek.

Jim said, "Let's get out of here!" So we scrambled to get out of the tie pile while first one bunch would go and then another. We climbed up and stood on the road and looked. Oh, Jim was happy! "Boy," he said, "look at those ties go!"

They had planned to drive Horse Creek to the Green River, and then they were going to boom the ties at Green River and truck-haul them from there to the treatment plant. They had a crew down there working getting the boom strung out, but when they couldn't get the drive started, Jim had told them to call it off—that the ties weren't coming that day. Now he said, "I've got to get on the telephone and tell those fellows down the river." So he hurried and called, and they got out of camp and got the boom just about completed. They lost a few ties when they came down, but not too many.

As the pile broke loose, more water kept coming through. Finally the bottom of the tie pile weakened enough that it took the bottom out, and then water started undercutting the pile. Soon that river was just solid ties from one bank to the other! The main road going up Horse Creek went up the north side, and the tie hack camp was on the south side. Below the commissary was a temporary bridge across Horse Creek. The ties took the bridge out . . . the next morning they went down and the bridge was gone. [laughter]

The whole pile wasn't gone by morning, but there wasn't an awful lot left of it. They worked a few ties down off the bank, and then they could start dumping ties into the stream further up. There was plenty of water to carry them down, and it took them a very short time to complete the drive.

The Forest Service had told me that as soon as the drive got going I was to come on back to Kemmerer. When I got back there (this was the spring of 1936) they told me that I was going to head up a timber survey crew at the next cutting area for the Standard Timber Company, which was going to be LaBarge Creek. LaBarge Creek was four or five streams on down from Horse Creek, also coming in to the Green River from the west. There's some big basins in there, and the timber was just the right size, ideal for tie-cutting.

I don't think I was in the supervisor's office over a week before going out to the Big Piney ranger station. The ranger's name was Ed Cazier. We were going to have to do our own cooking on the timber survey, so Ed and I worked up a camp with tents and cooking utensils and stoves and all we needed. We didn't have any pack mules, but there was a CCC spike camp up in the head of LaBarge Creek that was treating bug timber, and they had a pack string, so whenever we had to move camp we'd go up and tell them, and they'd come down and move us.

I had a very good crew made up of Andy McConkie, Glenn Brado (we called him Bull Brado), Henry Ketchie and another young fellow from Oregon or Washington. McConkie, Brado and Ketchie went on to become forest rangers, and McConkie later was supervisor of one of the Utah Forests. That crew was just as good or maybe better than the crew that I had in 1934 surveying in the head of Green River. I left the crew on the Fourth of July. We hadn't completed the survey, but we were getting awful close to it. After I left, I don't think that they were in there more than three weeks and the job was finished.

Sawed ties   In the meantime the Standard Timber Company had decided to go to sawed ties rather than hewed ties. Sawed ties had been on the market for a long time before I was in the tie camps, but the tie camp on Horse Creek really ended hewing in the region. In fact, I wouldn't be surprised if it was one of the very last hewing jobs in the United States. When Standard Timber decided to quit hewing timber, a lot of the tie hacks went together in groups of three or four and bought little portable sawmills, and as soon as the survey was over they moved up and started sawing ties. Now, sawing ties was a summer operation. Of course, in that country they could probably work on up until maybe the first of the year before the snow got too bad. A little bit of snow would actually help them–it would be easier for them to skid the logs, and they'd get less rock and dirt on them, so it would also be easier on the saws. But as the winter progressed, the snow got deeper and deeper, and eventually it just drove them out. With deep snow it would be absolutely prohibitive to even think of running a portable sawmill.

The difference between when the company ran the hewing operation and the portable sawmills was that each sawmill operator took out a permit from the Forest Service for so much timber. He owned that timber and he sawed the ties, and the company bought the ties from the group that sawed them. Most of the time there were at least four men that ran a portable mill. So, it was a private operation that up in Montana and northern Idaho they called a gyppo operation. (Where that name comes from, I don't know, but it's a gyppo operation.)

A number of years after I was at the LaBarge Creek tie-sawing camp, I met one of the portable sawmill operators in Salt Lake City, and we had a long visit. He told me about their operation. He said that they were really happier sawing than they had been hewing, because they were their own boss, and if they wanted to take four or five days off, they could. Of course, in a sense they were their own bosses when they were hewing, but they didn't have the freedom they had in sawing ties. And he said they actually made more money selling sawed ties to the Standard Timber Company than they did when they were hewing ties.

They'd have a team of horses to drag trees in to the mill, and if they got to a point where they had to drag them too far, they'd just move the mill. (It didn't take much to run a portable mill.) They'd just fell a tree and limb it up and cut it in sixteen-foot lengths, and drag it in to the mill and run it through and saw the ties into eight-foot lengths. That would be it, and when sawing ties, they no longer waited for the spring thaw to get the ties out. Everything was truck-hauled out.

These were all gasoline-powered mills, many run by Model A Ford engines. The old Model A motor was used an awful lot to run sawmills and compressors. A lot of the miners would get an old Model A engine and convert two of the cylinders to compress air and two of the cylinders to run the motor. I don't know how they did it, but they'd run on two cylinders and compress air with the other two, and those were used a lot. That old Model A motor was one of the best motors that Ford ever made. You just couldn't wear one of them out.

Ties hewn from round timber were probably still the most common type up through the mid-1930s, at least in the western part of the United States. In fact the tie hacks always claimed that the hewed ties were by far the best tie, and I kind of agree with them. There was more wood in a hewed tie on an average, and the weight of the rail on the tie was supported by the concentric circles of the annual growth, so the pressure came down on a far firmer surface than on a sawed tie. Some of your sawed ties were all right, but when they got into a big tree where they were taking two or sometimes three ties out of a butt cut, you didn't have the strength of hewn ties. You had your grain in a quarter circle, and the pressure from the rails caused the sections of wood on the side of each annual ring to split off–they would develop cracks very readily in their ends. But squared ties were a lot easier to lay because they were all a uniform size and shape, and workers knew exactly the size of hole that they had to dig to slip a tie underneath a rail.

7

ON THE CHALLIS FOREST:
RAPID RIVER AND LOON CREEK, 1936-1942

I WAS IN CHARGE of the timber survey camp on LaBarge Creek until the Fourth of July. On weekends we had the policy of always leaving one man in camp; we'd never leave the camp unattended. The Fourth, it was Andy McConkie's turn to stay in camp. My family was in Pinedale, not too far away, and I'd gone in to visit and spend the holiday with them. We'd taken a ride and come down to a little town by the name of Daniel, north of Big Piney. I knew everybody up and down the highway, and when I stopped to get some gas the fellow at the station told me that the ranger, Ed Cazier, had said that the supervisor was trying to get ahold of me. So I got on the telephone and called him at his home, and he told me, "You are being transferred to the Challis National Forest. How soon can you leave?"

I said, "We'll be on our way tomorrow." I tried to drive back up to the timber survey camp, but the river had come up and I couldn't ford LaBarge Creek . . . so I hiked on up. The supervisor had told me, "You pick the man that you think is the best qualified to head up the crew . . . ." Andy McConkie was the best, and he was in camp, so I told him that from then on he was chief of party of the timber survey crew.

Jane and I got our stuff together–back then we could haul everything we owned in the back of our car. We moved around so much we'd never accumulated any furniture or anything like that. The next day we got squared around, and I had to go into the supervisor's office and turn in my time for the Wyoming Forest. We took off and made it to Mackay, Idaho, about fifty miles south of Challis, stayed overnight, and the next morning went on in to Challis.

Interim appointments   Mac McKee was supervisor of the Challis National Forest. I was fortunate throughout my entire Forest Service career to work with some darned good Forest supervisors–in fact, I can't say that I've ever worked on a Forest that I didn't have a good supervisor–but Mac was probably the best of them all. He told me when I got into the supervisor's office that I was to take over the Rapid River Ranger District, which was a fire district. (The Challis had two districts that they classed as fire districts–one was the Rapid River, and the other one the Loon Creek district.) The ranger that was on the district, a man named Hinkley, had requested a transfer to the Soil Conservation Service. For some reason there was a delay in his transfer, but Jane and Jimmie and I went on up to the station. The Hinkleys were living in the house, but there was a detached ranger's office with an extra room in the back that was big enough for us. It had no water or toilet facilities so we used a two-holer out a ways from the office, and if we wanted to take a bath we went up to the main house and took it in their bathroom. But it was a lot better than some places that we've lived in! [laughter] So we didn't object to it.

I felt kind of awkward, in that Hinkley was still the ranger, but at least the situation gave me an opportunity to get out and really get acquainted with the district before I would have to take it over. But for some reason Hinkley's appointment was delayed and delayed, and by the first of September it still hadn't come through. Meanwhile, they had decided to create the position of fire chief on the Challis Forest, and they moved the ranger on the Stanley district into Challis to fill this position. That left the Stanley district without a ranger, so they moved me down to Stanley, and I ran the Stanley district until the middle of October. Hinkley's appointment finally went through, so then I ran both the Stanley and Rapid River districts. Of course, the Rapid River was about ready to shut down for the year. The first good snowstorm would close the summit, and you'd be through for the winter as far as that district was concerned. But I ran both districts until the snow snowed me out.

About the middle of February they had a lot of work in the supervisor's office in Challis that wasn't getting done, so they decided to move me into it. As soon as the Loon Creek Summit closed, the Loon Creek ranger had also moved into the Challis office. This Loon Creek ranger, Jay Hahn, was the same ranger who was on the Evanston district that I took over for about three weeks in 1935. And I eventually replaced him again on the Loon Creek district . . . . [laughter] I stayed in the supervisor's office that winter, and in the early spring they got a ranger, Charlie Langer, for the Stanley Basin district. Then I went back to the Rapid River district as its ranger in the late spring of 1937.

Seafoam: the end of the road Of the two fire districts on the Challis, Rapid River was quite a bit the smaller. It didn't have near the fire activity that Loon Creek did, and it didn't have many fire personnel. There were four or five bands of sheep that grazed on it during the summer, and there was a fair amount of recreational use–a lot of deer hunting in the fall, and a lot of fishing on the streams. Then down on the Middle Fork of the Salmon River there were quite a few fishing parties that packed in from the end of the road over near the Seafoam ranger station on the Rapid River where the trail heads started. They would generally pack down on the Middle Fork for a week or ten days, but once in a while there were parties that stayed all summer down there. The packer would pack supplies in to them periodically if they stayed longer than a week or ten days, and pack them out when they got ready to come out. As soon as the heavy rains came in the fall and the fire season ended, all your lookouts, patrolmen, fire crews, trail crews, packer . . . everybody took off, and you were there all by yourself.

On both the Rapid River and Loon Creek districts, the only means of travel was afoot or horseback. The Rapid River district had one full-time packer during the summer, who had about ten mules and packhorses. The government also had around three horses, and the ranger himself had to supply three horses, at least one of which could be packed; the other two could be just saddle horses. Every bit of supplies–whether it was drill steel, powder, groceries for the lookouts; you name it–was packed. We even packed in the lookout buildings themselves. They were pre-cut so that they could be packed on a mule or horse.

The wives of back-country rangers had a special problem. The Seafoam ranger station on the Rapid River district was, you might as well say, the end of the road. People would come in–livestock people, miners, and others—and there was no place else for them to eat, so you fed them. Jane and I would sit down to supper, and before we were through we might have four or five other people around the table. She'd get up and open some more cans and pour some more water in the tea kettle, and it might not be the best of grub, but it filled them up. A lot of people were good about it and paid for the meal, but some of them didn't. That kind of bothered us, because we weren't making all that much money back in those days. Of course, Forest Service people, that was different; we were tickled to death to have them drop in for a meal there. At Loon Creek the dude ranch was just down the road about an eighth of a mile or so, and they could take care of anybody that came in, but at both places crews would come in and wouldn't have time to set up a camp, so they'd come in to the ranger station and maybe have a meal and then take off down country going to a lookout or some place. We were tickled to feed them, and during the summer I'd eat at the lookout . . . but I always carried grub with me. When I stopped overnight at a lookout, I'd put out part of the grub, and they'd put out part of the grub, and it'd come out more or less even.

One time a rancher came in to the Seafoam station, and it wasn't eating time, but he was very resentful that I didn't invite him to eat. I said, "It isn't noon yet, but if you'll stay we'll be glad to fix something to eat for you." Finally, I went up and told Jane to have dinner a little bit early, so we fed him. That didn't stand too well with me, but that was part of the job when you were living at the end of the road, and Jane was a good sport and it didn't bother her too much. I couldn't have married a better girl. She took to the isolation like a duck to water, and the happiest times in our lives were when we were isolated in that back country. Neither one of us ever needed outside entertainment very much; it's just in our makeup, and we were fortunate to be that compatible when we got married.

Dutch Charlie's vegetarian fur farm   As you go through life, once in a great while you run onto a person who is so different from the ordinary run of people that the only classification you can give him is that he's a character. That's what Dutch Charlie was. I met Dutch Charlie shortly after I became ranger on the Rapid River district in 1936, and I continued to visit him after I took over at Loon Creek. If any man was purely an individual, he was. He owned this fur farm . . . well, he didn't really own it: he had a special-use permit from the Forest Service for about fifteen acres. In years past the land had

Dutch Charlie operated a fur farm under a special-use permit on the Rapid River district from the mid-1930s to the late 1940s. "A character. That's what Dutch Charlie was."

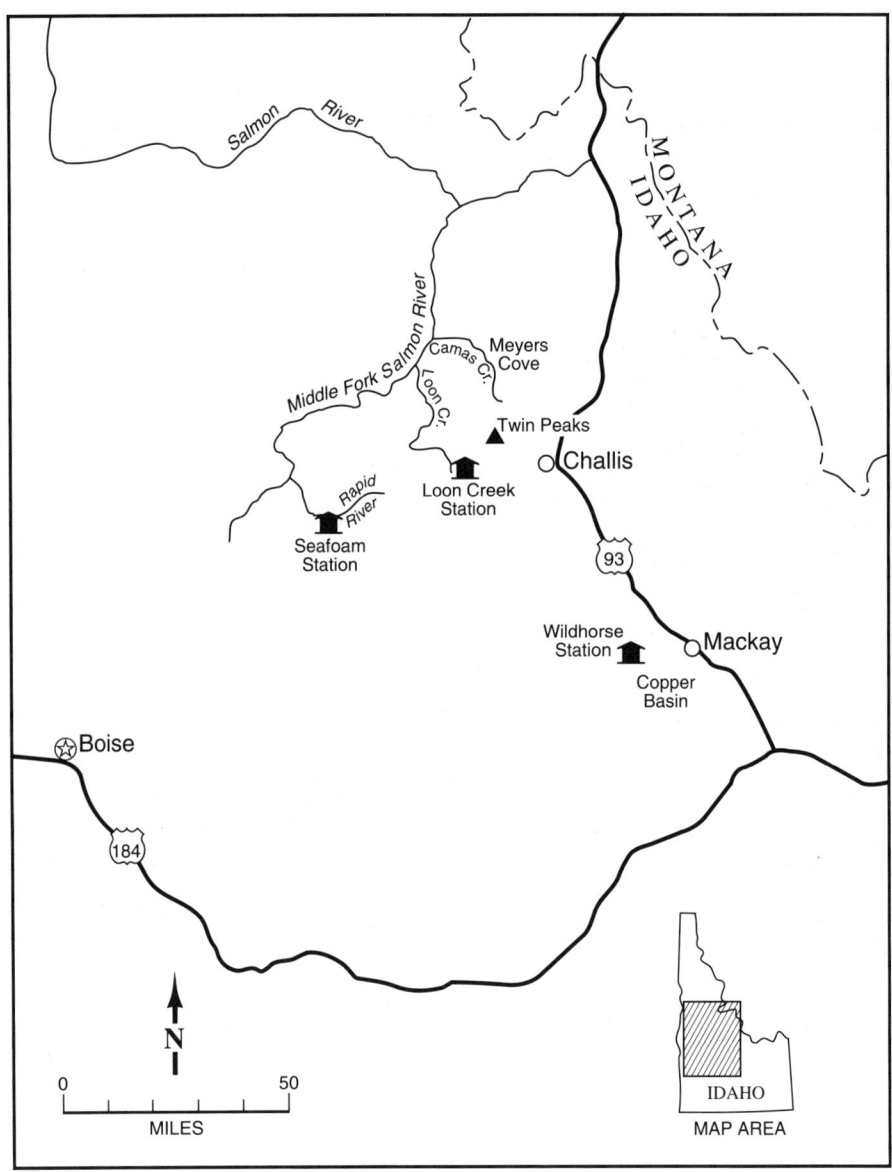

Map 3. From 1936 to 1937 Archie Murchie served as ranger on several districts of the Challis National Forest in central Idaho.

been straddled by a series of beaver dams, which had eventually filled up with silt, and the soil was extremely rich and deep. There he raised mink, silver fox and marten and the vegetables he fed them. All his initial breeding stock was native stock that he trapped himself, but later on, once in a great while, he'd buy a male mink and bring him in so that his animals wouldn't interbreed. He also did quite a bit of trapping, and replaced adult males and breeders with wild animals that he'd trap.

Dutch Charlie didn't have too good luck with the martens, so he went out of the marten business fairly soon, but he kept a female; she was just as tame as a cat. She'd climb up his pants leg and up his shirt and sit on his shoulder. He called her Suki. Suki Marten. [laughter] She finally got so old that he knew she was going to die. He couldn't stand to have her die right there at the place, but he just couldn't kill her, so he took her a little way out in the woods, in the mountains, and turned her loose. The next morning when he got up, Suki was lying outside his cabin door. She'd come back and died on his steps.

Although he didn't do well with the martens, Dutch Charlie was very successful in his mink and fox raising. In fact, he was averaging around five pups a litter with his mink; and your professional mink ranches on the outside, three was about their average. Men came in from all over, trying to find out what Charlie was doing that was so much better than what they were doing. Basically, if there was a difference, it was that he had changed his animals to practically straight vegetarians. He raised tons of carrots and turnips on that rich silt, and mixed with some goat milk, that was about all they were fed. Once in a while, if he had extra eggs, he'd feed some to them; and when he'd kill a goat to eat, he'd feed the liver and entrails–probably more to the fox than to the mink, but the fox were also vegetarians. The fox, he'd feed some whole wheat. He'd order his grain and have it come to the ranger station, and some weekends when we didn't have anything to do I'd load it in the car and haul it out to the end of the road at Indian Springs. Charlie had two burros–a black one and a white one–and he'd come up and load the grain on the burros and take it back down to his ranch.

The carrots and turnips were all boiled, of course, before Charlie fed his animals–he made kind of a mush out of them. Boy, they'd go after that just like it was the best piece of meat you could find! For some reason, on the vegetable diet their furs were a lot darker than their parents', and they grew to a lot bigger size. I'd say some of those young ones, especially the males, were a good third bigger than their parents who had been born in the wild.

Dutch Charlie was quite a guy. He was always experimenting with one thing or another. He took a number of fur magazines, and someplace he either read or he'd concocted the idea that arsenic would darken mink pelts. So he took two mink and put them in a pen by themselves, and he was feeding them a little bit of arsenic. I stopped in there one day, and I says, "Charlie, how is your experiment going?"

"Oh, it's going fine." He says, "Come on, I'll show you." He showed me, and they looked good. Boy, they had nice, dark pelts!

It went on for maybe a month or so, and I stopped by again and said, "Charlie, how is your experiment going?"

"Oh," he said, "not so good."

I said, "What happened?"

"Oh," he said, "I gave them too much arsenic. Killed both of them!" [laughter]

I think he wormed his animals twice a year. With the mink, he just put the medicine in their feed and they'd eat it. For the foxes, he used a capsule just like you use on a dog. He could worm the bitches, but with the dog foxes–especially one–he just couldn't do it alone, and he'd wait till I stopped by. The first time that I ever helped him . . . boy, that dog fox was mean! The only way Charlie could catch him was to get in the pen with him, get him in a corner, and then throw a blanket over him. Before the fox could get out from under it, he'd straddle him and grab him right at the back of the neck. When he pulled the blanket back over his head, the fox's mouth would be open and snarling. Charlie had a hardwood paddle about an inch-and-a-quarter wide and about a quarter of an inch thick; and my job, when the fox opened his mouth to snarl at me, was to run that paddle in and turn it upright so he couldn't close his mouth. Then Charlie would take one of these capsules and shove it down the fox's throat.

Since Dutch Charlie had a special-use permit, every year I had to make out a report. His was a rare special use, but it wasn't one that would be denied, because he was making excellent use of this little piece of government land–a far higher use than you could ever make of it for grazing or anything else. The special use was written out for the fur farm, but I guess that if he wanted to raise turnips, he could raise turnips . . .

or do anything else as long as he didn't damage the land. He did kind of violate things a little, though. He had these goats, and he had a pretty good-sized herd; he must have had twenty to twenty-five head. He had one billy, but the rest were either kids or milking goats. If he got a surplus of kids, he'd feed those to the foxes, and he ate quite a few of them himself . . . in fact, about the only meat that he ate was goat meat, and it was good. Those young kids, you couldn't ask for better eating. Sometimes he'd run short of feed for the goats on the fifteen acres, and he'd turn them out. They were really in trespass, but they didn't do much damage. Everybody knew about it–the Forest Service knew about it–and we never called him on it.

Charlie was also one of the best fire guards we had. Any fire, any lightning strike that was in old Charlie's range of vision, he was on it. We kept a smokechaser's outfit there at his ranch for Charlie's use, and I don't care whether it was day or night when a fire started, he was on it. He was the best unpaid fire guard that you could ever have asked for, so in the end we received from him a lot more than we ever gave. If he trespassed a little with his goat herd, it really didn't make all that much difference.

I really appreciated Charlie for his desire to protect the forest, and, in a certain way, his desire to protect the animals. The only traps he had were metal traps, and some of these animals that he had trapped had broken legs, but he made splints and set their legs and preserved those animals. Of course, he preserved them for breeding stock, but anybody else would have killed them and taken the fur off of them and forgot about it.

Charlie also had a sawmill, and we actually bought quite a little lumber off him. The mill was all run by water power. For a while he had a horse, and then the horse died, and then he had the two burros that were broke to work, and he'd skid his logs in with these two burros. He'd hooked up a set of Swedish gang saws that were set in a frame, and that sawed up and down. For inch lumber, he'd set them an inch apart; or he could set them at two-inch or four-inch or whatever you wanted. He had a big water wheel that ran the operation, and he'd fixed it so he could start a log through and then go off and do his other work, and when he heard the noise of the saw stop, he'd know that it had sawed through the length of the log. Then he'd go back and start another one through. He had a lot of tension on his saws, but he had them set so accurately that those planks or boards just about looked like they were planed. You could hardly see a saw mark on those boards. After he'd run a bunch of these boards through, he'd put all the boards of about the same width in a pile, then he'd run them through and saw off the bark edges. So when you bought boards from him, you never knew what width you were going to get. [laughter] They could be six inches, eight inches, ten inches, eleven and a half inches . . . ! You never knew what width you were going to get.

We were building a lot of catwalks around the lookouts, because at certain times during the day the glare on the windows made it awful hard for the lookouts to see through them. So we built these catwalks so they could go outside, and we used four-foot planking. Charlie cut out all the lumber for the catwalks and the supports and the rails and everything. We bought a few thousand board feet–it didn't amount to an awful lot, but it was good, solid lumber, and most of it was practically free of knots, because he'd only use the lower part of the logs. We gave him a timber permit for the trees, but we didn't check too close on the utilization. [laughter] There was all kinds of timber in there, so why worry about it if he left a few tops that should have been taken?

Charlie was running his fur farm when I came to Seafoam (I don't think he'd operated it too many years before that) and he was still operating it when I left the Challis Forest in 1947. But he was getting on in years; when I left he must have been in his sixties. Merle Markle and Charlie were good friends, and Merle talked him into moving out. I've been told that he and Charlie went into business and set up a fur farm on down the Salmon River below Challis. He ran that for a while; then Merle died of a heart attack. A big fur farm down around Idaho falls hired Charlie strictly as a consultant, and I was told afterwards that he used to drive up to Challis in a nice car, dressed to kill, with his mustache all trimmed and his hair trimmed, and they said you'd never know Charlie when you saw him! [laughter] But I imagine that he's surely dead by now, and gone . . . .

Working in to the Loon Creek station

In the early spring of 1938 Jay Hahn, who was the ranger that I had replaced on the Evanston district, was transferred to Paris, Idaho. That left the Loon Creek district vacant, so I was transferred from the Rapid River to the Loon Creek, and I stayed on the district from 1938 to 1942. The Loon Creek district was a bigger district than the Rapid River, and had quite a bit more activity on it. I think it ran around one hundred and fifty head of cattle and probably four bands of sheep. That was the size of it as far as grazing went, and there was lots of range.

At Loon Creek you never worried if stock got off their allotments, because there was adequate range and very few range problems. However, there was quite a bit of loss among the sheep bands, and I sometimes wondered whether it was really practical for them to run sheep back in that country. There was just thousands of acres of dead, bug-killed lodgepole, and a lot of it was going down or had already gone down. Grazing sheep in country like that, you had a lot of snagging. A ewe would jump over a log and there would be a broken branch sticking out and gouge her in the side. They'd sometimes rip off sections of hide four and five inches wide, and eight to ten inches long. When a patch of skin would be torn loose, blow flies would lay eggs in the wound. Maggots hatching from the eggs would feed on the exposed flesh, and the wound would keep growing larger. Eventually, the sheep would die. (There was also pretty heavy coyote loss on the district, and quite a little bear loss, and once in a great while there would be a heavy cougar loss.)

To get into those two back-country districts (Rapid River and Loon Creek) by vehicle on a road you had to go over a high summit, over the divide between the Middle Fork of the Salmon River and the main Salmon. These two districts both laid on the Middle Fork, and you couldn't drive in to either ranger station until late into the spring or early in the summer when the snow and the snow slides on the roads going in had finally melted.

The only way into the districts before the snow melted was to go in on horseback by the Middle Fork trail. Generally we tried to get away around the last of March, depending on the winter. When we'd go in we'd drive our horses and mules to a ranch that sat on the upper part of Camas Creek, which flowed into the Middle Fork of the Salmon. Some called that area Three Forks, but most knew it as Meyers Cove–a nice big ranch there. We'd drive the stock into Meyers Cove, and then the truck would bring in all our grub and saddles and pack equipment and tools and everything we needed. We'd try to take enough grub in for at least a month and a half, and the lookouts would generally take enough for two months, because they'd be up on their peaks quite a while before the summit opened and they could get another supply of grub in. One year it was July before you could drive a truck in to Loon Creek.

On much of this country they'd had a big beetle epidemic in the lodgepole pines. Nothing had been done to control it, and there were just thousands of acres of dead lodgepoles. I came in right at the period that the roots were rotting and the trees were falling down, and it was one of our big jobs to open up the trails in the spring by cutting out all this dead lodgepole that had fallen across them. There was also a big problem in maintaining the telephone lines, which were tree lines. A small tree or two could fall across them and just pull slack, but if too many trees fell across, then it'd break the line. Sometimes you'd have two or three breaks in a distance of not over four or five miles, and you'd have to splice them. You'd also have insulators pull off the trees, and in those places the line could saw into a green tree, which would ground it out; it wouldn't work.

We'd go in and the men from both districts would start working the Middle Fork trail as one crew. We'd just move on up the trail until we came to Loon Creek, and then the crews would split. The Rapid River crew would go on up the Middle Fork, and we went on up Loon Creek. From the mouth of the Camas to Loon Creek the trail was entirely on the Challis side of the river, but a little bit above Loon Creek it crossed a suspension bridge, and then it followed the Payette Forest side as far as the trail went. Rapid River was the next big drainage that flowed in to the Middle Fork from the Challis side up above Loon Creek, and the Seafoam ranger station was located in the upper Rapid River. To get to the ranger station, the Rapid River crew would cross back over the Middle Fork on the Pistol Creek suspension bridge just below the mouth of Rapid River, and work their way up the Rapid River trail.

I'd generally have anywhere from five to seven men in my crew, and the Rapid River ranger would have the same. (These were all the main lookouts and the packer for each district, and some of the patrolmen.) We had pretty-good sized crews, but there was an awful lot of work on the Middle Fork, Rapid River and Loon Creek trails. It was through rough country, and there was a lot of rockslide. Some few stretches of rockslide were huge: they would bury a section of trail hundreds of feet long. During the winter I guess the action of snow freezing and melting would roll rocks down onto the trail, and sometimes you'd have a rock as big as a table and no way you could move it. You'd have to drill it and blast it to get it out.

These trails that were built across the slide rock were all surfaced; it would be difficult to walk a mule across that slide rock without having it surfaced. We'd pack in small rock and fill in between the bigger rocks until we got it reasonably tight with rock. Then we'd pack decomposed granite in (it was all decomposed granite country) and surface the tread of the trail. During the winter and spring, water from the melting snow would wash a lot of that surfacing down into the rocks, and there were stretches that we'd have to resurface again every spring when we came in. It was hard work.

We used to try to bring a wheelbarrow in, but it was difficult to wheel one. If you put on any kind of a load in that decomposed granite, you couldn't push it. So we improvised: we'd take powder boxes and wire a pole at each end of the powder box. They're made good and heavy, but the bottoms are just nailed on, so we'd have to wrap wire around to keep the weight from pulling the nails out. Then if the decomposed granite was dry and you had two husky men, you could sometimes pack two powder boxes loaded with granite between these poles. You'd pack it in and dump it sideways and spread the surfacing out. It was hard work; it was tedious work. Two of us would pack surfacing for, say, an hour or so, and then two more would take a turn.

There were level stretches of trail, but much of the way it ran along the sides of fairly steep hills. The decomposed granite on slopes is constantly moving, and there was always slough coming down . . . there was places where you'd have enough vegetation to hold it, though. To deal with the slough we had a trail grader. It was just a blade that you could set to any angle you wanted, two handles and a good heavy workhorse on the grader. That saved a lot of hand work, a lot of shoveling. Most places we could use that trail grader to grade the slough off the sides, but in places with too much rock we just couldn't use it. Then we'd have to do all that shoveling by hand. Of course, there was a lot of places we didn't have to do anything, where the tread was just as good as it was the year before.

After we started up Loon Creek, and after the Rapid River ranger started up Rapid River, the telephone line to all the patrol stations, lookouts, and the ranger station pretty-well followed the trail. Even though this meant it was strung for a longer distance than if it had gone cross-country, it was easier to maintain–as you moved up a trail you checked your line and maintained it. At first we didn't have radios (later on we got radios on some of the peaks along with the telephone), and we had to get the telephone lines repaired or we didn't have communication when we went into the fire season. As a rule we got lots of trail money, but sometimes our telephone money was pretty skimpy, so we considered the job one job; and if we ran out of telephone maintenance money, we maintained the telephone with trail money. These were entirely different funds, and we really shouldn't have done it. [laughter] But the supervisor knew we were doing it, and the office knew we were doing it. It was the only way to get the work done.

Tracking a dude   Very little timber was sold on the Loon Creek district, but there was a bit of mining, most of it lead-silver, and there were quite a few prospectors around. Recreational use was also heavy. Fishing was good in all the streams and lakes, and a lot of fishermen would go clear down to the Middle Fork of the Salmon and camp there for days, or maybe all summer. On down below the station about fifteen miles was the Falconberry ranch, owned by an old man named Len Falconberry. In his latter years he sold out to a group of doctors from Idaho Falls that converted the ranch house to a fishing and hunting lodge. Then, just below the Loon Creek Station was Boyle's Dude Ranch, a combination dude and cattle ranch. They owned about twenty-five horses, and ran around twenty head of cattle. Boyle's did pretty good business, and we had quite a lot of contact with these dudes, as they called them when they came in.

When you're a forest ranger, you get involved in a lot of activity that in a way is connected with the Forest Service, and in a way it isn't. Like, for example, if a child gets lost—and they don't necessarily have to be lost on the Forest—one of the first people that's contacted to go out and hunt is the ranger. Or somebody steals a car and they think it's hid someplace on the Forest, the first person they contact is a ranger. (I got involved in one of those deals over to Ely, and we found the car.) One of the jobs that the ranger had in those back-country districts like Loon Creek was to plant fish. If the state couldn't drive a truck to the stream or lake, they didn't plant them. Well, a back-country ranger had a pack string, so the Forest Service cooperated with the state, and we'd pack the fish back to all these lakes. A lot of them were virgin lakes and had never had a fish in them.

One fall on the Loon Creek I was going to make a pack trip back and visit some virgin lakes and some that fishermen had told me were so fished out that they should be replanted. A lot of those lakes would support fish, but due to the lateness of spring at the high elevation, and the earliness in the fall that ice formed again, the fish wouldn't spawn, so you had to replant them periodically if you were going to have any kind of fishing. I was going to check these lakes over and see which ones should be planted and which ones shouldn't.

Joe Boyle, who owned the dude ranch, told me he had a fellow and his wife that wanted to go out on a pack trip, and they didn't really care too much where they went. She was a photographer and had more cameras than you could shake a stick at, and she was good, too–she showed me some of her work. They wanted to go out and take some pictures of game or whatever they could see. Joe hadn't decided where he wanted to go, and he asked me if it would be all right if the three of them went with me on my fish-planting expedition. I said sure, so we took off and went first to Horseshoe Lake and spent the night there. I checked it out, and then we crossed over into the Stanley district and came around the back side of my district.

There was a lake on the Stanley district that I was going to plant if it was barren, because the district lines didn't make any difference as far as the Fish and Game Commission was concerned. The lake was off the trail, and we had to climb up the side of the mountain quite a ways. We camped there, and the next morning I got up early while everybody else was still asleep and went out and wrangled my stock, but I couldn't hear Joe Boyle's horses' bells. I looked around some and finally hit his horses' tracks. During the night, they'd pulled out for the ranch. I came back and told Joe, "Your horses are gone." So I loaned him my saddle horse, and he took off. He trailed them clear back to the ranch before he caught up with them.

After Joe left, I got busy making breakfast. It was pretty cool up there, and it was on pretty late in the fall. The woman got up first, and she was wearing a heavy red blazer jacket. While I was getting breakfast she asked me if it was all right if she walked down to the lower end of the lake to see if she could find some deer and take some pictures. I said, "OK, but when I yell, you be sure and come back." So I saw her go down around the end of the lake just a short distance; I saw her going up the other side. On the far side of the lake was a big red bluff, and between the lake and this bluff was fairly heavy timber. Finally she disappeared, and I thought, "Well, she just turned back up into the trees."

Breakfast was about ready–the sourdough batter was about ready to cook and the bacon was about done, and so I called. It was a fairly small lake, and my voice would easily carry across. No answer. I thought maybe she was taking a picture of a deer and she didn't want to spook it so I waited a little while, and then I called again. Still no answer. Then I started getting worried, and I really bellowed.

My yelling had awakened her husband, and I told him what had happened. "Now," I said, "everything is probably OK, but I'm going looking for her." And I said, "Everything is about ready to eat. All you have to do is throw some batter in the dutch oven and fry some hotcakes, and the bacon and everything is there . . . if you want some eggs, fry some up. But whatever you do, don't leave this camp. Even if I don't come back until evening, don't you leave this camp." (I wasn't going to try to look for *two* people.)

I took off and went around the lake, and I could track her. She'd gone up and turned and gone straight back into the timber, gradually climbing. She'd stop, and you could see where she was looking around to see something to take a shot at. This must have all happened before I yelled, because she'd gone up and made a circle. Then she got her directions haywire, and instead of coming straight back she came off at an angle. Well, when she came back down she couldn't see the lake; she had hit the creek below the lake, and then she didn't know where she was.

Before she left, just kidding, I had told her, "Now if you get lost, all you do is head down country until you come to a trail, and then stay there." She'd wandered back and forth, and it was slow tracking her out because there were quite a few needles on the ground. Finally I could see her tracks heading down, and I figured now she had remembered what I had told her. I took off then, not tracking but trying to cut her off at an angle. I wanted to be sure that I hit the trail below where she would hit it. When I reached it I started coming up the trail, and sure enough, I saw where she'd come out of the timber and had hit the trail and sat down. She must have sat there for quite a little while, but then for some reason she started walking *up* the trail. Of course, the further she walked the further she got away from camp. I started walking up the trail, and I'd call. No answer. I went further, and I'd call. Finally, I thought I heard a voice, and I stopped while it came closer. To get to the lake we'd come over a pass and dropped into this canyon, and finally she'd walked up high enough that she could look back and see the red bluff above the lake. Then she knew where the lake was, so she started coming back down the trail. The creek made a curve, and, of course, the trail swung around on this curve and finally I heard her crashing through the brush along the creek. She was coming towards my voice, and she didn't trust the trail to come around the bend, so she come straight across. [laughter] Where she could, she was running, and she was just completely out of her gourd. When she saw me she let out a scream, and she ran up to me and threw her arms around me and started crying. Then she just slid down on her knees and sat there and cried. I didn't try to talk to her or anything; I just stood there. Finally she cried herself out and was just sobbing, and I told her, "Everything's all right."

This was getting on into the morning and it was starting to get pretty warm, but she was still wearing her heavy wool blazer. And sweat . . . she was just *wringing* wet. Even with that creek running next to the trail she hadn't drunk any water, and she was thirsty. I took her down to the creek and showed her how to dip the water up in her hands and drink it, but I told her the water was cold. I said, "Drink just a little." So I went back and sat down, and I told her to take her jacket off; then I could see how wet she was. Her clothes were just soaking, so I told her she better put the jacket back on, but to leave it unzipped. She went back down to the creek a number of times and had water and we sat there for quite a while. Then I said, "Well, no use waiting here. Let's head back for camp."

When we'd gone down a ways, due to the gradient of the trail it was easier to cut off cross-country to the lake than to go clear down to where the creek came in and then climb all the way back up. So I told her, "Now, we're going to leave the trail and cut across country," but she wouldn't go. I talked and talked to her, and I told her, "I know exactly where camp is, and I can take you right there." Finally we started out and went about fifty or sixty feet; you could still look back down and see the trail. She balked again. She said, "No. We're going to stay right here until Joe comes with the horses."

I finally convinced her, but I had to take it awful slow. She'd go just a short distance before she'd have to rest. Finally we got back to the camp about noon, and her husband was worried; he was just about frantic. I was going to get her something to eat, but she didn't want anything. She just went into the tent and laid down, and she slept until evening.

Joe got in around four o'clock or later. We were supposed to be gone a whole week on this expedition, but the next morning they decided that was it. [laughter] Joe was kind of fed up with the whole thing, too. So, I thought, "Well, if they're going in, I might as well go with them." So we went back to the station.

A ranger is always prepared

The Middle Fork of the Salmon River was where all the winter deer range on Loon Creek district was. It was low elevation and lots of years there was very little snow, but it was very badly overgrazed, while there was unlimited summer range. The deer, for some reason, would start leaving their summer range way early, even before there was any reason for them to leave it. Snow didn't drive them out, and there was adequate feed, yet they had developed the habit of moving down onto their winter range early. Some of them would get there a month before it was really necessary, and that meant that there was a month's use of forage taken off the Middle Fork that really shouldn't have been. In the fall of 1939 the Idaho Fish and Game Department wanted to put out salt fairly late in the fall to try to hold deer back from going down on to their winter range. They knew you could control cattle fairly well by salting, and they got the idea that maybe they could control the deer the same way. So that fall they brought out I don't know how many ton of salt. It was what they called sheep salt (which is kind of ground rock salt) in fifty-pound bags. We were cooperating with the state Fish and Game, and my job was to scatter this salt.

Shortly after I got started the Challis fire chief and the supervisor on the Sawtooth National Forest and a couple of others wanted to go deer hunting. Every fall this group would come out. They'd bring all their own grub and everything, and I would act more or less as a guide and pack them where they wanted to go and do the cooking. I told them, "There's all kinds of deer at Cougar. You can shoot a dozen four-point bucks every day of the hunting season if you want to." I said, "I'll scatter my deer salt while you hunt, and when you fellows get some deer I'll pack them out." So they hunted and they all got their deer and I packed them out. But the last night they were there it snowed; laid down right there at the lookout about a foot and a half or better. In fact, they had a hard time getting their pickups out, and I had to hook onto them with my lariat and help them in places.

I had one more load of salt to take up, but I got to thinking that there was no use scattering it up that high because since the first snow had come, the deer had already gone. Still, I had a whole pack string of it, so I thought, "Well, I'll load it up and I'll go down Cougar Creek and keep going until I start running into deer. When I run into deer, then I'll put my salt out."

It had already snowed about a foot and a half with no wind, and the trees were just covered with snow. An experienced ranger, when he leaves his station, he's always prepared for different kinds of weather, different conditions. It isn't too often that he's caught without necessary clothing or proper tools or whatever he needs. But this time I was. When I left the lookout in the morning it was fairly cool, quite a ways below freezing, but as I dropped down towards the Middle Fork it got warmer—there was probably a difference of at least three thousand feet in elevation from Cougar to the Middle Fork. On during the day it started warming up, and by the time I got my deer salt all scattered (by noon or shortly after) it was really thawing. The water was just dripping off the trees, and every time you got under one you just got soaked. I was wearing chaps, but I didn't have my slicker with me; I did have a good heavy wool shirt and a heavy jacket, which normally would have been plenty. (Late in the fall you'd never wear a slicker, because it would be snow instead of rain.) Well, I really got soaked; I got soaked from the skin out.

I could take two bucks and I had shot only one; I hadn't got my second buck of the season yet. So I thought, "Well, as long as I'm down here I'll just knock over a deer and bring it out." I got my buck and gutted him out and loaded him on a pack mule and started back, and it was quite a climb to get back to the lookout. On to about four o'clock in the afternoon it started turning cold, and in just a matter of three quarters of an hour the rigging on the mules was frozen solid. My jacket had been soaked and it was froze. I got so cold that I had to get off and walk . . . but after I walked so long, I had to get back on; I couldn't continue to walk. My gloves were just ordinary leather gloves, and I got so cold and my hands got so stiff inside the gloves that I could hardly hang onto the reins, but I didn't dare get off, because I'd never have gotten back on again. I knew it.

Towards the last I was beginning to get drowsy, and I knew what was happening–hypothermia was setting in. Once I thought, "I'll get off and just try to make it on foot." But then I thought, "Well, suppose I can't make it? Then what am I going to do?" So I just sat in the saddle and swung my arms and beat them against my body as much as I could, and I finally made it back to the lookout.

I wish I knew how cold it was that evening when I pulled into that lookout station. Man, it was cold! The Cougar lookout was on a tower–they'd built a log tower with the idea of someday extending the road and having a garage for a car underneath the tower. Well, instead they had created a primitive area up there, so they never could extend the road. The garage wasn't finished, but they had cut the door opening in the tower, so I took all my pack mules and my saddle horse and put them inside there. Then I was going to get a fire going in the cabin, get *myself* warmed up. Why I had locked the cabin door that morning . . . just force of habit, I guess; there was nobody around in miles. But I had locked the door, and now my hands were so numb that I couldn't feel anything. I finally got my keys out, but the only way I knew I had ahold of the keys was by looking–I couldn't feel. I got the key in the lock and finally got the door open and went in, and I tried to whittle some shavings to start a fire and I couldn't even hang on to the knife.

I didn't drink much of anything then, any liquor of any kind. But Merle Markle, the fire chief on the district, had brought up a jug of wine when he was up there hunting. There was about two inches of it left in the bottom, and it was sitting under the bed when he got ready to leave. I had pulled it out and said, "Merle, you better take your wine."

"No," he said, "keep it. You may need it."

I remembered this wine, and I sat down and got the top off, and I took a couple of good, big jolts of that wine and then just sat there. I could feel the warmth start coming through my body, and I got a fire going and got warmed up a little, and I went back down and gave the stock some hay. They were so cold they wouldn't even look at hay; they wouldn't even touch it. They just stood there with their backs humped up. Everything was froze–I couldn't undo the latigos; there was no way to get the saddles off; and all the lash ropes were froze solid. So I just left them with plenty of hay, went up and got myself something to eat and went to bed. Sometime during the night I woke up and I could hear the animals chewing hay, so I knew they were all right. The next morning was just as cold, but I packed up what little I had and headed for the station.

Fire up Pioneer Creek   On the Loon Creek district, the primary job was fire. Especially during the fire months, your time was pretty much 100 percent fire related. (You weren't always *fighting* fire, but it was fire related.) We had then, I feel, one of the best pre-suppression and detection systems that the Forest Service has ever come up with. To me it was far better than the fire patrols and whatnot that they have today, because we had, you might say, 100 percent detection of visible areas. Those men were there on duty twenty-four hours a day, and if a lightning storm went over at two o'clock in the morning, they were up. Some of them were married, and when they were gone their wives were just as good lookouts as they were, although they never got paid a penny. There's been wives that's been the lookout twenty-four hours a day for sometimes two days straight when their husbands were out on a smokechaser fire.

We had lots of fires. We had lots of smokechasing fires, lightning fires, but we had very few man-caused fires. On the Rapid River district I didn't have a project fire of any kind, and on the Loon Creek district I had only one project fire, and that was the summer of 1940. That was in a blind area, and we didn't even know we had a fire in there until it blew up. Of course, when it blew up, *everybody* knew we had one!

I was making a pack trip, and that morning before I left the station I talked to the fire chief, who told me that fire conditions were bad–it was dry and the humidity was low. We'd been having quite a bit of wind and I asked him, "Do you think I'd better take off?"

"Yes," he said, "go ahead. If we can't handle a fire that starts with you gone, we'd better be fired."

So I took off, and about two o'clock or so in the afternoon here came my patrolman up the trail. He had a white horse, and he was running that horse, coming on as hard as the horse would go. I knew something was up, so I stopped and got off and cinched up the cinch on my pack mule and cinched up the cinch on my saddle horse. Before he even got to me he started yelling, "You've got a fire!"

And I said, "Where is it?"

He said, "Up Pioneer Creek. They want you to get back to the station as soon as you can."

So I turned around. I could turn my little pack mule loose and she'd follow me any place I went, even afoot. (I'd used her a lot planting fish, and when I got into real bad country I'd empty about half the water out of the fish cans and take off afoot up through the rocks, and she'd follow me, and we'd dump the fish.) Anyway, I turned her loose and I was riding a single-footer; it's faster than a trot and a lot easier riding. I shoved him into this single foot, and boy, we took off and got to the station and I got on the telephone. They told me they had this fire up Pioneer Creek, and my administrative guard had gone to it. The fire chief told me he'd already dispatched a twenty-five-man CCC crew to the fire: they had had to hike in because there was no trail.

I put my radio on my back and took off. When I got to the fire I set up my radio and told them that it looked a bit more than a twenty-five-man crew could handle, but that I was going around it, and when I came back to my radio I'd call in and tell them what I needed. So I started around the fire and I ran on to my administrative or station guard. He and I went around the fire together, and by the time we got around I knew twenty-five men wasn't near enough. But when I got back to my radio, it had gone out! And this shows a good fire chief: when he didn't hear from me, he automatically started another twenty-five-man crew in. They got there probably about four o'clock. They had fire rations—that's all they had to eat—and nobody had any bedrolls or anything.

We worked on the fire the rest of that day and on into the night a bit. It was too dangerous to work men all night, but the crew was pretty-well played out anyway. So we came down to a meadow and built some pretty-good sized camp fires, and all the boys curled up around the fires, and I was sleeping out away from the fire.

Next morning, just breaking daylight, here came an aeroplane over to drop a camp in to us. It was flown by an old pilot name of Bennett, who lived in Idaho Falls. He'd been a World War I pilot and brush-hopped up in Alaska for a good many years after the war. At the time that I knew him he was getting on in years, but he was still a good pilot. Bennett had an old Waco biplane that he'd cut the wings down on, and then he'd put in a motor about double the horsepower of what the plane originally had. He dropped everything on parachutes that we rigged from burlap bags that were used for packing wool. (We would open the seam and tie a shroud line at each corner.) These chutes would support sixty or seventy pounds, I'd guess, but you didn't want to drop anything fragile with them.

Bennett dropped the camp by parachutes, except for the sleeping bags. The sleeping bags were done up in bundles of five, rolled real tight and then tied with two ropes, and they dropped them free fall, no chutes or anything. Boy, when they hit the ground they'd bounce about twenty or thirty feet in the air! It was smoky, and the pilot was dropping on this meadow and he couldn't see those boys. It's a wonder he didn't kill some of them, because those beds were dropping all around them. They also dropped another radio to me by a chute, and I had communication with the outside again.

There was no trail in to the fire, so it was supplied by air for three days until a trail could be built to it. Only one other fire that I was ever on had the fire camp entirely supplied by air: that was the Hells Canyon fire on the Payette National Forest in 1949. They even dropped drinking water for our camps on the fire line.

I decided that the fifty men we now had on the fire still weren't enough; that we'd need at least another twenty-five. So they got them started in. When we got the full crew of seventy-five men on it, we had it licked; it didn't go any place after that. One boy came down with appendicitis, and we had to pack him out, but other than that we had good luck on that Pioneer Creek fire. But it was an extra-period fire. This time I had kept very accurate notes. Boy, I put everything down in detail!

That winter they had their board of fire review and I was called in, so the Forest supervisor and the fire chief and I went over to Boise. C. N. Woods, the regional forester, who was very critical on board of fire reviews, was there. Also the assistant regional forester in charge of fire was there. As I remember, his name was Henry Shank. He was just as critical as Woods was. They were tough on that board.

There was a young ranger that had been transferred from a southern Utah district to one of the fire districts . . . I don't know whether it was the Payette or the Idaho. I guess he'd had a pretty bad fire, and in his telling what happened on the fire you could see right from the start that a lot of it was inexperience–that he did the best he could, but he just didn't have the experience to handle a fire like that. (On the Stanley district a year or so after, there was a good-sized fire. The ranger on it did not have much experience on fires, so as soon as I could get to it, the instructions from the Forest supervisor were that I would take over. And the ranger was tickled to death to see me when I got there.

[laughter] That's what should have been done with this inexperienced ranger who was before the board of fire in 1940. As soon as that fire started he should have been replaced by an experienced ranger, but they left him there and he had a *lot* of problems.)

There were other rangers from all over the region that were called in. Some of them did a pretty good job, and some had difficulties. Finally, towards the last they called me up, and I started going over the Pioneer Creek fire. They'd ask a question and I'd pull out my notebook and thumb through it, and tell them exactly why this decision was made and so on. I went through fine; I didn't have any trouble at all. They couldn't really chew me out for anything. In fact, the assistant regional forester in charge of fire accused the Forest supervisor and the fire chief on the Challis of coaching me before I had to go up. Merle Markle, the fire chief, was teed off and he said, "They didn't have anything to do with it. Those are exact notes that Archie kept on the fire." [laughter] But that all went back to the fire that I had had on the Kamas district on the Wasatch National Forest. That lesson stayed with me through my entire Forest Service career.

We finally got through, and the regional forester stood and said, "Is there anybody else who wants to say a few words before we conclude this meeting?" A supervisor on the Boise by the name of Guy B. Mains got up. (At that time you always had to have your hair cut; you couldn't wear a moustache; you couldn't wear sideburns; you couldn't wear whiskers of any kind–chin whiskers or beard, or anything. But Guy B. wore a Vandyke beard. I never knew for sure, but I was told that at one time he'd injured his chin and it was kind of deformed, and that's why he wore that Vandyke. Nobody ever told Guy B. to cut it off, either. [laughter]) Old Guy B. got up and he lit into them! He said, "You send young, inexperienced rangers up to these fire Forests and expect them to handle one of these big fires, and you don't give them any help. This ranger that you just raked over the coals is no more to blame for what happened than you are!"

He went through a lot of reasons why he disliked this board of fire review and the way they were handling their personnel. Now, he was talking to the regional forester and the fire chief and the personnel officer, but when Guy B. sat down they didn't have a word to say, except the regional forester said, "This concludes our meeting." Well, that ended our board of fire review, and we never had another one. As far as I know, Region Four had been the only region that had a board of fire review, and I'm just guessing, but it probably started when C.N. Woods became regional forester.

We decide to winter over

On those back-country districts you wanted to be sure you had your family out in the fall before the Loon Creek Summit closed or you didn't get them out. That could be any time between the first of October and the last, but generally by the twentieth of October you'd better have your family out of there. I'd stay in after the summit closed, and sometimes I'd work clear up to Thanksgiving, but then I'd have to take the stock out by trail.

The Rapid River ranger generally left a lot earlier than the Loon Creek ranger in the late fall or early winter, and both rangers would move into the Forest supervisor's office in Challis and stay there until work started the next spring. The Challis district ranger also had his headquarters there at the supervisor's office, so that meant that in the wintertime there were three rangers on the site. Lots of times it was kind of hard to find enough work for three men to do during the entire winter. (Of course, when we were making the seen-and-unseen-area maps and travel maps, that took a lot of time. We also worked with the CCC camps, and I taught some classes and lectured. It took up part of my time.) In the fall of 1941, we had a ranger meeting and the Forest supervisor brought up the fact that he didn't know what he was going to do to keep everybody busy that winter.

There was a lot of work that needed to be done around the Loon Creek station. We needed a cabin to house crews when they came in, instead of having to pitch a tent: we needed a place where they could stay and cook their own meals. Up until then the guard and his wife had lived in a tent all summer, which for a road crew of all men may not be too bad, but for a fellow and his wife it wasn't all that good. So I told the supervisor, "There's a lot of work at my station that I can do." I told him if Jane was agreeable we would stay at the Loon Creek station all winter.

That evening I asked Jane. We had four children by then; the youngest born in March of 1940. I wanted to be sure that Jane was agreeable, so I asked her what she thought of staying up there. Of course, she'd spent one winter in isolation at the tie camps, so it was nothing new to her. She said, "That would be fun; let's plan on it."

I told the Forest supervisor that it was OK with Jane, and if he wanted me to I'd stay in there all winter and go as far on building the log cabin at the station as I could working alone. We also needed to complete a guard cabin down at the Falconberry ranch that we had started in the summer of 1940, and I said I would go as far as I could on finishing it. It was agreed that we'd stay in.

Jane and I made out a grub list of everything that we would need for the entire winter, and then we made out a bid that we took to different stores in Twin Falls. It was written into the bid that the store that got the order had to deliver the groceries to the Loon Creek station. For the six of us we bought enough grub to last from October to . . . well, it turned out we bought practically enough to last us from October to the third of July. (During the winter we did run short of bacon and a few other things, but I could backpack in what we needed on snowshoes over the summit.) This whole list of grub cost a little over three hundred dollars delivered to the Loon Creek ranger station.

The summit snowed in about the very last of October, but the nearby Boyle ranch had kind of a landing strip–it was just a meadow; it wasn't developed in any manner, shape or form, but a plane could land on it if it had a good pilot, so we never worried about sickness or anything. We thought, "Well, if somebody gets seriously ill or gets hurt, a doctor can fly in." The telephone went out not too long after it started snowing, but we had a radio and we had plenty of batteries. We found out when we came out the next year that the doctor wouldn't fly. Of course, that really don't mean too much, because there was probably doctors in Boise or Pocatello or someplace else that would have come in.

"Worlds of time" to build a cabin

Before the guard cabin at Loon Creek could be built, logs had to be cut and brought to the site. There was a good place to cut cabin logs just up the road about a short half mile. The trees were lodgepole pines: nice, tall and straight, and the right size. Lodgepole are the best for cabin logs, because they have little taper, they grow straight, and they have thin bark that is easy to peel. Toward the end of October, as soon as we had snow on the ground so I could skid logs, I got started.

To help me with this I had the big old plow mare that pulled our trail grader. She had been bought for fire use, mainly to pull a fire plow, and we did use her some on fires; but that was awful rocky country, and where you have fire you have trees, and there was always rocks and tree roots. The plow was hung up most of the time on rocks or roots, and I never thought it was much of a success. If you were down in sagebrush country, it'd probably work all right if it wasn't too rocky, but most of the places we tried it, it didn't work too good. But we used the mare for everything else. She was a big, powerful, capable animal. You could put a pack saddle on her, and she could pack the fire plow or the trail grader or anything else you wanted.

I'd go up to the stand of lodgepole and drop the trees, limb them and cut them to the length that I wanted, and hook the old plow mare on them. The first few times, I'd lead her down to the building site and park the logs where I wanted them–some longer logs on one side of where the cabin was going to be and some on the other side, with the shorter logs parked at either end. Finally, the mare got so that I could hook her to a log, turn her loose, and she'd go right on down to the other logs and stop. Jane would come out and unhook the chain, step the mare up a little bit so she could pull it out from under the log, then undo one tug and turn her around and slap her on the rump. She'd come back up to where I was working. During the forenoon she'd make a number of trips that way, but when it got pretty close to noon, she wouldn't come back! [laughter] So when she wouldn't come back, then I'd go on down and have dinner. Same way in the evening, and when it came about five o'clock–she could tell time just as good as I could–she'd head for the barn instead of coming back. [laughter]

We soon skidded all the logs down. All the knowledge I had acquired in the tie camp on how to peel a tie came in handy, and I made me a spud, and it didn't take me long to peel the logs. I tried to get not only all the bark but the cambium, too, because the cambium, when it dried, would peel off and leave a strip on the log.

I had all the logs that I needed, and I got poles for the rafters and about everything necessary to build the cabin; but it was too cold to run any kind of foundation to build the cabin *on*. So I drug down three timbers that were considerably longer than the cabin would be, and I set them up on blocks. With an adz I leveled the timbers off, perfectly smooth-leveled each way, so I could build the cabin on top of these timbers. The following summer, when the fire crews and trail crews came in, I would borrow a cement mixer and run the foundation ... just pour it in under the elevated cabin.

Although I had never built a log cabin before, and had no manual on how to do it, I'd seen a lot of them. Up there at the tie camp there were *dozens* of log cabins. So even though I'd never built one, there was no problem . . . except figuring out how deep to cut the notches so the logs would fit down together just right. In theory it'd be nice if you could get your logs to barely touch, but you want just a little gap between them. If you took out too much, then you had a big gap that looked awful. (On the tie camp, if they accidentally did that they filled the gap with rags or moss. But I had lots of time; that was the beauty of it–I had *worlds* of time. I could just do it as careful and easy as I wanted to.)

Bob Melville was kind of a hermit who lived on the district, and who most years would stay through the winter in the Loon Creek ranger cabin as sort of a caretaker. This winter of 1941-1942 he was acting as caretaker on Boyle's dude ranch. He had built log cabins, so I asked him, "How do you figure out how deep to cut that notch?"

He says, "Simple: I'll show you." And he got a piece of quite small rod; it would hold its shape, but it wasn't too hard to bend. He used it like you'd use a set of calipers, I guess you'd call them. The only secret was to hold the thing vertical, so it would fit tight and true on the log. First, you'd put your log up and measure it. Say you had to drop your log four inches–OK, you would spread the ends of this bent rod four inches apart. Then you'd hold it vertical just on the edge of the log, and bring it up around and score the log that you had to cut. Then you'd go to the other side and do exactly the same thing. I'd take a carpenter's pencil and mark it good, so I could see it, and then I would start in with my double-bitted axe to chop the wood out. When I got down close to the line, I used a hatchet. And when I got *really* down close to the line, I had a real heavy wood rasp, and I'd rasp it out to the line. Boy, that log would fit just as snug as could be!

Once in a while you'd have a tree that had a little bit of a wow in it–not too much, but a little bit. Sometimes it'd have such a small amount that looking up it, you couldn't see it. When you used those, you always turned them with the wow up. You'd set your corners, then before you put on your next log you'd take a cross-cut saw and cut down right where the bend was. You'd cut down until you'd weakened that log enough so it would start settling. Sometimes one cut wouldn't quite do it, so you'd settle it down as far as it'd go and then you'd start another cut close to the same place. You cut in until you got enough slack to bring it down perfectly level, then you'd take shingles and drive them into that cut just as hard as you could–just fill it, so it wouldn't settle any more than what you wanted. And once in a while, if I'd made a pretty deep cut, I'd take a big spike and drive it in from the top, through each side of the cut to hold it a little bit better.

I had thought it was going to be an awful lot of trouble building the cabin alone, because those were pretty heavy logs; but I found that I could put slides up to the top log on a wall, and then, using the old mare, put a rolling hitch on a log and cross-haul it. I could roll that log up there just as nice as could be, and after I got it up there I could do my hewing and fitting and everything I wanted to do. As I laid the logs, I used drift pins in every one of them. I made my own drift pins from a lot of round iron that was left over from the CCC camp days. One log, I drifted in the center. The next log, I would put two drift pins in it about a third of the distance from either end. So you would have one pinned, then two, then one, and two . . . so that your whole wall was plumb fastened together.

When I started building, I was going to build like a regular log cabin with about a foot of log sticking out past each corner. I got about halfway through, and I got to looking, and I thought, "Gosh, with all these nice trimmed corners on the other buildings at the ranger station, someone is sure to make a point of it when they come in and see this cabin." So I cut all the ends off. On the new buildings where they used shevelen* siding they finished the corners with two two-by-fours set vertically with a piece of quarter-round set in between them. The only lumber I had was three-inch rough-sawed bridge plank. For each corner I ripped off two pieces of this a little wider than the four inches used on the other buildings, then I planed them. To make the quarter-round I cut a piece of bridge plank three-by-three inches and planed off one corner.

Mules, horses, and treacherous ice

For about a month I made good progress on the cabin, but then the weather turned so bad that I had to do something about the stock we were wintering over. The Rapid River district had very little pasture. Timber came right down to the buildings at Seafoam, and the ranger had just enough pasture to last him through the summer, so when the fire season was over and he was through with his stock, he'd

---

* This is a phonetic representation of Mr. Murchie's pronunciation of the term. The term could not be located in standard reference works.

bring them over to Loon Creek. We had worlds of good pasture. We'd just turn the animals out in the mountains: it was steep, but it was good range–open grassland and sagebrush. Then when I got ready to take our stock out, I'd take the whole bunch. Some years, if the snow wasn't too deep, I could take them out over the Loon Creek Summit, and be met at Bonanza guard station by a truck with hay so I could feed them hay that night. We would load all the saddles and pack gear on the trucks, and then the next morning I'd turn the horses and mules loose on the road going to Challis. They knew the way, and you'd have to ride like everything to keep up with them. (Without saddles or anything, they'd really take off.) If I couldn't get over the summit, I'd take the stock up a fork of Loon Creek. The town of Challis was on Challis Creek, and this trail would drop into the head of Challis Creek and eventually hit the road. I'd bring them right through town, on down to the supervisor's office.

Well, that winter of 1941-1942 we had decided that we'd keep the stock at Loon Creek. My family and I were going to stay through the winter anyway, and the year before, Joe Boyle at the dude ranch had grazed his horses out all winter, so we thought we'd save a lot of money on hay and pasture by running our whole bunch out all winter. The total number of stock for the two districts was thirty-three head, or perhaps a few more: the Loon Creek station packer had a full string of nine mules, plus two extra that I sometimes used when he wasn't using them; I had a short string of seven horses and mules, and then there was one crippled mule we could pack sometimes, and I also had three saddle horses; and the Seafoam station had a string of nine horses and mules and two Forest Service saddle horses.

There was plenty of feed for the stock, and there wouldn't have been any problem if the snow conditions had stayed right. On into the winter we had quite a lot of snow, but the horses could paw through it, and on the south slopes the snow melted enough that they still had good feed. Then at the end of November we had a warm spell and it thawed pretty well before freezing again. This formed a heavy ice crust on top of the snow. The horses could walk on it, but it was difficult for them to paw through. If they kept to the north slopes, they could still paw down to good grass, but if they got over onto the south slopes where there was glazed crust.... One mule lost his footing and went to the bottom and it killed him. A day or two afterwards another mule lost her footing, plowed into a patch of burnt lodgepole about two inches in diameter, and knocked a bunch of them down. (It stopped her before she got to the real steep part. She was skinned up pretty bad, but it didn't really hurt her.) I decided that if I kept the stock in there through the winter I'd lose a bunch. I'd have to take them out, and with the summit snowed in the only way out was down Loon Creek to the Middle Fork and on out by Camas Creek to Meyers Cove. Only the old plow mare would be kept back to help me complete the cabin.

This was all granitic country, but as soon as the fire season was over you'd jerk all the shoes off the stock. Any shoes that was still serviceable, we kept; so I tied in and shod that whole outfit, every one of them, with used shoes. Since this was only meant to be a temporary shoeing (I intended to pull them off as soon as we got to Challis) I did very little leveling of hooves or anything, and I attached the shoes with only four nails instead of eight. Still, it took me the best part of two days to complete the job. The following morning, about the first of December, we started out for the mouth of Loon Creek, where a fellow by the name of Sam Lovell and his wife had a little ranch. They kept a few cattle and horses, but they didn't really have any more hay than what they needed, so I packed enough hay from the station to feed my bunch that first night. The next night I figured I'd make it to Meyers Cove, where they had plenty of hay; I could buy hay there.

I wish I had a picture of what we looked like on the trail: I had that whole string of stock strung out; the mules had their saddles on, and, of course, the horses were tied head to tail. I got to Lovell's ranch OK, but Sam told me that there was a bad place on the Middle Fork trail where some little springs came to the surface just up the slope from it. In the summertime there'd be thin patches of water, not much more than would keep the trail wet, but in the wintertime this started freezing. Sam said the trail was covered with ice there, and that he had already lost two packhorses on that stretch this winter. He says, "You'll never make it through."

And I says, "There's no other way I can take them out. I've got to go through. If I don't, I'm going to lose a bunch of them up at the station."

"Well," he says, "I wish you luck."

I started out the next morning, and when we were getting close to those springs I pulled the string off the trail on to a little flat and secured them there. Then I went forward on my saddle horse to inspect the trail.

Sure enough, when I came to the place, it was bad. This was a one-horse trail, maybe three feet wide, cut into the slope a steep sixty or seventy feet above the rocky bank of the Middle Fork. The two hundred foot stretch where the spring-fed ice flowed across the trail was fairly level, but the ice was not: it was thickest to the inside of the trail, its surface declining toward the drop-off. I had difficulty walking on it. Clearly, there was no way that the stock could keep their footing across this place, because shoed horses and mules will slide all over ice even when it is flat.

Returning to where I had left the string, I broke the stock up into groups of four or five, and tied each group to a tree. Then I got the pincers out of my shoeing outfit and jerked the shoes off all the stock. This took quite a bit of time, and towards the end I was getting pretty tired. I hung the pulled shoes in the branches of a tree. [laughter] I hid them so nobody could find them, so that when we came back in the spring we could pick them up and take them back to the station.

On my saddle I always carried a cruiser axe in case I had to cut out a windfall or something. It was double-bitted, but it had a head about half the size of a regular double-bitted axe and a shorter handle. With the axe I chopped out footholds for the stock in the ice. I made each one about twelve inches across and maybe four inches deep, so a horse or mule could put its foot in it. Mules are careful, but horses just go along and they don't care where they're going, so I led each horse across individually. Then I turned all the mules loose, because I knew the only way a lot of them would come across would be to come alone. They crossed the ice, and not one of them slipped and pitched down into the river.

I put my pack string back together again and took off. There was still quite a jaunt down the river to the mouth of Camas Creek, and then I had to go up Camas Creek to the end of the road, which was a ranch at Meyers Cove. It must have been at least ten o'clock in the evening when I pulled into Meyers Cove. The rancher had already gone to bed, but the string of horses and mules made quite a bit of noise, and he got up. I told him I had over thirty head of stock and asked if I could turn them into the pasture. "Go ahead," he says, "There's quite a bit of snow, so just throw them out some hay." There was a stack of hay right close by, so I did.

They had a bunkhouse with six or seven bunks in it, equipped with a good kitchen stove and gasoline lantern and dishes and everything I needed. (I had grub with me.) I spent the night there, and the next morning the rancher came over and invited me to breakfast at the ranch house. After breakfast I called the supervisor's office and told him that I was at Meyers Cove, and asked him to send a truck out for all the saddles, halters and everything. Then I just turned the stock loose. Of course, the mules knew where they were, and they took off, and it was all I could do to keep up with them. It was quite a ways in to Challis, but with the mules loose, it didn't take them all that long. About halfway in I roped one of my other saddle horses and changed horses. We were in a good swinging trot most of the way, and we still couldn't keep up with the mules. Of course, it was way late when I finally got to the supervisor's office. The mules were waiting for me at the gate going into the pasture when I got there.

I worked in the supervisor's office for a day or two, and I bought a few groceries, and then the fire chief hauled me as far up the road to the Loon Creek ranger station as he could go and let me out. I'd brought my snowshoes out with me, and I snowshoed over the Loon Creek summit and back to the ranger station. I had been gone about six days.

Snowshoeing over the summit

After I got back to the station I resumed building the guard cabin. By the last of December I had the walls up about two or three logs shy of finishing the square to the proper height where I could start the roof, but I couldn't go any further on my own. So I radioed the Forest supervisor and asked if there was any chance of getting Herb Freece, the ranger from the Rapid River district, to give me some help. "Yes," he said, "there isn't enough here to keep him busy anyway, so I will send him in."

By this time we'd had a lot of snow, and there was danger of snowslides, so I didn't dare to cross the summit during the day. At night after it froze, the danger of snowslides was practically nil. As a half-way station for my periodic treks out to the end of the cleared car road from Challis, I had packed a tepee up to the foot of the summit and put it up underneath a big spruce where snow wouldn't break it down. Late in the afternoon I'd snowshoe up to the tepee, have something to eat, and get to bed very shortly afterwards. I'd sleep until probably ten o'clock in the evening and have breakfast, I guess you'd call it, and then snowshoe on over the summit and down as far as where Merle Markle could come with a pickup, and I'd meet him there. (Of course, as the winter went on and the snow got deeper, he wouldn't be able to get as far up the road.) Merle would usually bring some steaks out, and we'd have something to eat . . . and generally there would be a few groceries for me to take back. Jane and I thought we had bought enough of everything we would need for the winter, but we ran short of bacon, and about every time out I would pack a slab of bacon in. Also, before Christmas there were a lot of packages to pack in.

When I'd start back would depend on if it was a cold day. If it was, I'd start back earlier, because there would be less danger of snowslide; but if it was a warm day, then it'd be six o'clock or so before I started back, and I'd try to cross the summit after nine o'clock, something like that. Then I'd stay overnight at the tepee and go on in to Loon Creek the next morning. This system worked good; it worked good.

It was about the third or fourth of January, 1942, when Herb Freece was sent out to help me finish the cabin. When Merle trucked him out he also brought out a fellow by the name of Hank Halverson. During the summer Halverson was one of my lookouts, and he had a cabin down on Loon Creek where he spent his winters. The idea was that they'd meet me on the road, and then we would follow my procedure and come up and over the summit, and snowshoe on in to the station. I found them as far up the road as the fire chief could bring them. Herb was getting over a drunk, and Hank was coming down with the flu, so Herb says, "Well, I think we'll have to take our chances with the snowslides and start out early," and they left.

Merle and I visited and ate our steaks and whatnot before I started back, and by the time I got up pretty close to the summit it had been dark for quite a while. I could look up and see a fire burning on top of the summit, and I thought, "What in the world? Hank and Herb should have made it to the ranger station by now." I got up there, and both of them had shot their wad. They had halted on the summit and built a fire and not gone on. In fact, I didn't know for sure if I was even going to be able to get them down to the tepee. But they'd rested, and I always packed some coffee with me, and I made some coffee, and we started out. I got them down to the tepee, where I had a double sleeping bag—one inside another. If they kept their clothes on with that bag over them, they could make it—they'd be warm enough . . . but there was no room for me. Well, I'd been snowshoeing back and forth over that summit for a couple of months, and I was probably in the best shape that I'd ever been in my life, and I wasn't packing too much, so I took part of their load and the rest of mine and took off. I couldn't travel too fast, and it was about nine o'clock in the morning before I made it in to the station.

Cold . . . oh, it was one of the coldest nights I have experienced! When I was a kid my dad always got us moccasins from Canada. When I knew that I was going to be doing all this snowshoeing, I wrote my dad and asked him if there was any chance that he could get me a pair of those moccasins, because for snowshoeing there was nothing better. So Dad inquired around among some of his friends up in Canada, and sure enough they got me a pair. I normally take a nine and a half, but I had him get them large so I could fit three or four pairs of socks inside of them. Of course I was wearing these, and when I pulled in to the station that morning and went to take my moccasins off, the outside socks had frozen to the moccasins . . . but my feet weren't cold. No frostbite.

It was afternoon when Herb and Hank finally pulled in. We put Hank right to bed over in the guard station, and doctored him for four or five days before he finally got well enough that he could go on down the river to his cabin. Herb stayed with us for a couple of months, returning to Challis about the first of March.

Cabin fever   With Herb to help me I soon laid the last two logs at the tops of the cabin walls. The top log on each of the long sides extended eight feet beyond the front of the cabin to support the roof out over the porch. As we laid the logs at either end of the roof, we placed four poles on each side parallel to the eaves and finished with a heavy ridgepole. These parallel poles supported the rafters. One-by-twelve sheathing was put over the rafters. We had done as much as we could, but the windows, door, porch and shingles would have to wait until later in the year. In the summer we would also have to pour the foundation and lower the cabin down onto it.

Once we set the structure down off its blocks, it would probably have been heavy enough that it wouldn't have shifted, but I wanted to have it fastened to the foundation. So in the sill logs I drilled a series of holes–four on each side and two on the end sill logs. Then when we poured the foundation that summer, I set pins in it that would go up these holes. And, boy, I did some calculating to get those pins exactly straight with the holes. We took jacks and jacked the cabin up a little bit, and pulled the timbers out. Then we lowered the cabin down onto the foundation. Every pin fit the holes exactly, except in the southeast corner–that one was off *just* a little, and there was a crack ran down through the foundation, but it didn't hurt anything. I was surprised; I didn't think that we could lower it and have all but one of them fit that close, but they did. Of course, we put them in when the concrete was still soft, and with the wet concrete that one pin might have shifted just a little. Anyway, I'm going to say that was the reason it missed; it wasn't my calculation that was off. [laughter]

After we had done as much as we could with the cabin at Loon Creek, Herb and I took the old plow mare and loaded her up with what camp stuff we needed and tools and whatnot, and led her the fifteen miles down to Falconberry's and started work on that cabin. Guard cabins are for fire guards, but the Falconberry guard was really a fire *patrolman* due to the nature of the district. There were a lot of fishermen–some of them afoot and some of them on horseback–going up and down Loon Creek. We only had one fire started by a fisherman while I was there, but there was always the hazard of somebody dropping a cigarette or building a fire for a cup of coffee and not putting it out. The fire guard would patrol down Loon Creek to its mouth one day, and the next day he'd patrol up the river to the end of the road; and that was his beat. We tried, if we could, to get a married man for that job, so when he was patrolling somebody would still be there to answer the phone so we could get a message to him.

Down to Falconberry we didn't have any lodgepole, so we had to build the cabin out of Doug fir* and the logs were a lot bigger. They were a good fourteen to fifteen inches in diameter, and maybe some of them a little bit bigger than that. Also, they had quite a bit more taper to them than lodgepole had, and they're a lot harder wood; they were harder to work. I guess we laid up about seven or eight logs, and even with the old plow mare that was about as high as we could go. We'd have gone higher, but there was a danger of getting hurt. If we'd lost a log and it had fallen on one of us, it would've been curtains, so we left the cabin to be finished up in the summer by the guard and the trail crew foreman.

The family and I spent the rest of the winter at the Loon Creek station, and in the spring, when it came time for trail work, I snowshoed

---

* Douglas fir.

out over the summit, because my horses and riding gear and everything was in Challis. Merle met me and took me on into Challis, and we got the crews and the pack outfit organized, and we came back in to the district through Meyers Cove and down the Camas. I wish I could have taken the family out that way, too, but it was impossible. We had four children ranging in age from eighteen months to eight years, and we would have had to take them out on saddle horses. It would have been a three-day ride at least, camping out two nights. We waited until the road over the summit was open so I could drive the family out.

When Hank had come in with the flu, none of my family caught it. We never even had a cold all winter. But when I went out this time, one of the fellows on the trail crew had a cold, and I caught it. I should have stayed down there until I got over it, but I came home, and the first cold that the family had all winter was the one I brought in from the trail crew. By then, we had four children. When we went in there, the youngest one was nine months old.

The family was very anxious to get out. Jane was patching our clothing, but the kids had outgrown their clothes, and we'd run out of clothes. We had a big order from Sears & Roebuck waiting for us in Challis, but I wasn't going to pack it all over the summit! Even though it was getting on into the summer we could not leave, however, because a big snowslide had come down across the road. The snow was all gone on the summit on both sides, but this great big snowslide closed the road–it was a dandy. It had taken down trees two foot in diameter.

One day I called the Forest supervisor and said, "Can't you figure some way to get us out of here?"

"No," he said. "The Forest is short of road money. You got to wait until the snowslide melts." It got up to the first of July, believe it or not, and finally Mac says, "Well, I'm sending a Cat in with a crew." They plowed the rest of the snow out, and they had to cut all that timber out that was across the road, and we finally got out. When we got to Challis, we took the kids to the warehouse and dressed them from the skin out in the clothes we had ordered from Sears & Roebuck before we dared to take them out in public! [laughter]

A serious injury   That fall, after Jane and the kids had all moved back into town, I continued working on finishing up the Loon Creek guard station. I was building a porch for it. I kind of liked a guard cabin to have a porch on the front where the men could sit out in the evening; and it just looked better too. The last log on either side, when we came to the square, ran out eight feet further, and across that there was another log laid. There were to be three pillars to support the front, and the roof went out and covered the whole thing. When we built it I wanted a concrete floor for the porch, but we couldn't because it was cold; so I just put supports under the front end, with the idea that after we ran the foundation and the floor of the porch, I'd cut pillars for the front, and everything'd look natural. In the fall, after the fire season was over, I finally had time to put these pillars in, and put in the windows and door and so forth.

I wanted pillars that were as close to exact in size as I could get. They only had to be about eight feet long, so I hunted around and found a tree that I could get all three pillars out of by cutting about two feet off the bottom to get away from the buttswell. I drug the sections down, and was peeling them. I had two sawhorses, and I picked up the first pillar and carried it over, laid it on the sawhorse, peeled it. The second one, I did the same thing; only it was a little bit heavier. When I came down to the last one that had been cut from the bottom of the log, it was too heavy for me. So I stood it on end and took a bear hold on it and took it over; and with *all* the first-aid knowledge that I had and taught my men and CCC boys, I still took that thing and bent over with it sideways when it began to get away from me. If I had been smart and dropped it, I would have been all right; but I hung onto it. I later learned that trying to hold on to it, I tore two ligaments loose on the side of my back where they fasten onto the top of the vertebrae. When they went they took a little piece of the cushion with them. Well, that was bad enough; that was bad enough. Initially there was very little pain, but eventually the inflammation settled in the sciatic nerve, and boy, after several days I was having a problem.

The old cabins had been chinked with moss; and up at the tie camp they had chinked them with moss. But for a Forest Service building, I didn't think moss chinking would look too good. [laughter] So I chinked between the logs with mortar cement. It was practically insect-tight. I didn't cut the door and windows out until I had the cabin almost completely built. Alone, I was able to put the windows and door in, and I was lucky that there was an extra door the CCCs had left there at the station. The Forest Service had given me sixty dollars for materials, and that had been spent for shingles because we had none in stock. But when the ranger station had been built, I guess they had windows left over, because stored up in the attic of the warehouse there were enough to do the cabin. (The cabin was to have a double window on each side, a single window in the front, and a single window in the back.) However, there was no trim for the windows, not even a windowsill. To make windowsills, I took a three-inch bridge plank and planed it. You know, a windowsill is flat on the top where the window slides, but on the outside they have a good slope so the water will drain off. So I had to take a hand plane and bevel the outside part down to about an inch-and-a-quarter at the edge. That gave it the slope to drain.

I wanted to put the last window in before the snows came, so I kept at it, and one day I was smoothing the log for the windowsill to set on. I was standing on a plank on some blocks, working away with my axe, and . . . . Well, this darned leg: because of my back injury sometimes it'd just buckle. I went to take a step, and it buckled, and I tossed the axe as I fell. The handle hit the side of the window opening, and it flew over and cut me across my palm. [laughter] Torn ligaments and a sliced hand. Everything. And at night I couldn't sleep in a bed. The only way I could sleep was to get right next to the wall so I could lay kind of against the wall; that's the only way I could get relief from the pain in my back. But to make a long story short, I got the window in.

The cabin still had to be shingled, which I just could not do in my condition, so I called the supervisor and finally told him that I had hurt myself. He wanted me to come out, but I said, "No, I got to get this thing finished."

He said, "I'll send in Charlie Langer and Art Cusick," the ranger at Clayton. So they came in, and it didn't take them long to shingle it. But I hadn't checked my property, see. You had to check your property every year and account for every knife, every hammer, screwdriver–everything. So they loaded me in a wheelbarrow, [laughter] and I knew where all the property was, and they'd wheel me around to the different places, and we'd check the property. It started to rain, and it rained pretty good, and they kind of worried about getting over the summit if it was snowing up there. I says, "There's no use worrying about the summit. If you look up there at Pinyon Peak, that's a lot higher than the summit. If there's no snow on Pinyon, you don't have to worry about the summit."

Well, it came Saturday, and we'd had breakfast and were sitting around, and someone said, "Let's go out and see our wives."

"Good idea. Let's go." So we jumped in the pickup and took off and started up to the summit, and got up a good distance when we hit snow.

I tried to help push the pickup, but could do very little because of the pain, and there was very little shoveling I could do. But they shoveled and bucked snow, and shoveled and bucked snow. We were still another half mile at least to the top, and the snow was deep. I was in such pain and misery that I was just about ready to scream, but we finally got to a place where we could get the pickup turned around, and we came back to the station.

Art called his wife, Cassie, there at Clayton and told her to come up the other side of the summit the next morning as far as she could come–that I was going to try to hike over to her. So we went to where we'd turned around, and we took snowshoes. Art went with me and helped me. It was tough traveling on snowshoes. We finally made it over, and I never saw a woman that was as welcome as she was! [laughter] She loaded me in her car and took me into Challis, and I was in bed for a week or so. But every time I started getting over it, I'd do something and hurt it again. I went to a doctor in Pocatello, and he kind of got me squared around, but I hurt it again; and I went to a doctor in Boise, and he shot me so full of B-12 that you could smell me for a mile. I told him, "You're not doing me a bit of good. Is there any chance a belt would help?"

"Well," he says, "a belt might help. I've got a friend that makes belts." So I was measured and fitted up with a belt, and it worked. I felt the best I'd felt in a long time. But I'd get to feeling good, and I'd take the thing off, and then I'd hurt myself again. So I went to a doctor in Idaho Falls, and he told me, "We can do one of three things. We can put you in the hospital and strap you to a board and let you lay there for a month or so until those ligaments grow back; or we can open up your back and perform an operation; or you can put the belt on and *keep* it on." He scared the daylights out of me when he started talking about this operation and what it consisted of. So I decided I'd put my belt on, and I wore it continuous for five years, and I was still wearing it when I transferred to the Ely district in 1947. I'd worn one belt out and bought another one. After that, any time I had heavy work to do–like going out on a fire or doing heavy lifting–I'd go get the belt and put it on.

That was the only serious injury I suffered during the time I was in the Forest Service. Looking back now at the horses I've worked around, the pack strings I've pulled, the places I've gone alone, and traveling at night as much as I did, I marvel that I was never seriously hurt. I never even got injured on a fire. Oh, once I froze a toe on a snow course and they were going to take it off, but that really wasn't . . . .

8

ON THE CHALLIS FOREST: WILDHORSE, 1943-1947

It was fairly early in the winter of 1942 that I got back in to Challis. Naturally, I supposed that I'd be the ranger on the Loon Creek district again the next year, but the Forest was having some problems on the Wildhorse district. A lot of sheep ran there, and a lot of cattle. It was an odd situation, in that the cattle grazed the low country and the sheep grazed the high range, and there was no definite line that had ever been established between the two. They were having quite a little conflict between the sheep permittees and the cattle permittees as to just where the boundary was. The ranger there, Sterling Justice, was an old-time ranger who had come in on the old Ranger Exam. Justice was pretty much a cattleman, and he didn't care too much for sheep.

McKee, the supervisor, came into my office one day and asked me, "Would you be interested in taking over the Wildhorse Ranger District?" I'd had about all the experience that I wanted on a straight fire district, and the Wildhorse district was primarily grazing, so I told him sure; I'd be glad to take the Wildhorse district.

An easy district to administer   Early in 1943 they transferred Sterling Justice to the Paradise Valley district on the Humboldt National Forest out of Winnemucca, and that's where he retired. I moved down to Mackay and became ranger of the Wildhorse district. Wildhorse was primarily grazing, but we also had two campgrounds. There was quite a lot of fishermen (most of them daily fishermen, just driving up and back out), very few fires, and a little timber. There was one portable sawmill that ran during the summer all the time I was there, and sometimes when fall came the sawyer would move his mill down below Mackay and cut cottonwood for the ranchers around there. Another portable sawmill operated for a short time over on Antelope, and those two mills were the only timber activity on the district other than a few Christmas tree sales. So my work, you might say, was entirely range . . . which, being technically trained in range, I really enjoyed.

After we got the problems settled between the sheepmen and the cattlemen, Wildhorse was an easy district to administer. Most of it was very accessible: by hauling a horse out to the end of the road, I could hit any part of it in a day's ride, which was a lot more than you could say for Loon Creek or Rapid River or even the district there at Kamas. I had only one area that I'd have to camp out at night to reach.

Much of my work was in Copper Basin, where thirty-two hundred head of cattle ran on a single allotment, and that didn't include anything under six months of age when they went on. In Copper Basin there was a nice stopover cabin, with a stove and all the accommodations, and it had a good pasture for my horses. Then over in Antelope, an area just south of Copper Basin, there was another cabin where a guard and his wife lived during the summer. Although he was hired as a fire guard, his work was 90 percent range. He was a good range man, but at least part of the summer he worked as a fire guard.

Probably the thing that most made the district easy to handle was that each cattle allotment had a cattle association, and they were very active associations. They had their president and vice president and secretary and treasurer and the whole works, and you could work directly with the association. Oh, you worked with some of the permittees individually, but if you had a problem on the allotment, you didn't go to an individual; you went to the association. I attended all their meetings, and if they had a problem, they told me their problem; and if I had a problem, I told them our problem, and we'd sit down right there and work it out. If it meant that we had to take a ride out on the range, the president of the association would appoint a rancher to ride with me: we'd go out on the range; we'd work out our problem; that'd be it. The sheepmen also had an association.

A dispute and a killing   The conflict between the sheepmen and the cattlemen on the Wildhorse district went back a good many years. The fault wasn't entirely with the cattlemen–the sheepmen were as much to blame, probably. One of the problems was that the sheep allotments rimmed the high country around the basin, and in order to get to these allotments the sheepmen had to bring their bands right through the center of the Copper Basin cattle allotment. In the spring when the lambs are real small you don't travel too fast–some days you're lucky if you make six miles with those small lambs. Even so, they were taking far more time going across the basin to get to their allotments than they really should have. Of course, in the fall there was no excuse for the sheep to stop in the basin at all, because if they left their own allotment in the morning most of them could be out of the basin by evening. But even with their older lambs, they were still taking plenty of time crossing the basin, and that was one of the big problems between the two groups of livestock.

I was told that some years before I came there, some young fellows from the different ranches were riding up in the basin one day. They ran onto this sheep camp, stopped right in the middle of the basin, and indications were that he'd been there for quite a little while. They ordered him to round up his sheep and take off. (There's different stories about exactly what happened. Of course, the only side that you could hear from the cattlemen was the side of the boys that were involved.) Anyway, they claimed that the sheepherder stepped into his camp wagon and reached for a rifle . . . at least they thought he reached for his rifle. Most of the boys had revolvers on them, and one young fellow just pulled out his gun and shot the sheepherder. Well, that created quite a stink, but the boy was from an important family there around Mackay. I was told that it was fought through the courts, and that the young man served a little time in prison, but not too much. I believe he eventually became involved in politics for most of his adult life. I could tell his name (his brother was running cattle in Copper Basin when I was there), but I will not name him, because I am not absolutely certain that the story I heard was accurate.

This problem, you see, wasn't just of short duration. They'd had problems there for a long, long time. And Sterling Justice, the ranger prior to me, was pretty much a cattleman at heart. If a problem arose he would favor the cattlemen rather than the sheep owners, so the sheepmen were upset that they didn't get due consideration. My first job was to settle that problem. This is how I resolved the conflict: First I went to the cattlemen's association meeting and told them that I realized there was a dispute as to the boundary between the cattle allotments and the sheep allotments, and that I intended to settle it, one way or the other. Then I went to the sheepmen and told them the same thing. They picked three of their own to meet with the officers of the Copper Basin Cattle Association, and I got everybody together and we sat down and talked it over. It was finally decided that we would determine a boundary line between the classes of livestock, and that each group would abide by it.

We rode out and took each allotment one at a time, and we decided a line that was agreeable between the two groups. Then we put signs up to mark it, so there was no question that the herders and cattle riders could see where the boundary was. Most of the canyons were quite wide where they entered Copper Basin, and they had big meadows that ran quite a ways up the canyons. That is what had caused much of the trouble. The sheep allotment line, we finally decided, took in part of those big meadows, and the cattlemen really didn't like that, because they'd been used to going up on the sheep allotment to graze the meadows. On these meadows we had to fence the line–it wasn't sheep-tight, but it was cattle-tight–because there was no other way the cattle herders could keep their cattle down. Without the fence they would run them down and take off, and the next day they'd be right back up there. The Forest Service did the fencing: we bought all the material and built the fences ourselves. Finally, we got this line established, and after that there was no problem.

The sheepmen and the cattlemen all lived more or less together, ranches adjoined, in what they called Barton Flat up above Mackay. This had caused a conflict between neighbors even when they were off the range. (It wasn't real bitter, but it wasn't the best for the community.) But after this boundary problem was settled, then everything went smooth between the permittees off the range and on the range, and we had no more problems.

Roundup   The cattle association ran thirty-two hundred head of permitted cattle on its allotment. They had two riders that handled the cattle, and generally in September they would have a big beef roundup. They handled their cattle just like they handled them way back in the 1800s. They had what they called a holding ground, and they'd bring what cattle they could gather in a day onto this holding ground. Riders would hold the cattle in a loose herd while the owners went in and cut out their beef and kicked them out to one side, where pick-up men would pick them up and form a beef herd. On the other side of the holding ground they'd kick out the cows with unbranded calves. Branding fires would already be going, and they'd rope the calves and brand them and earmark them and castrate them, and then they'd be turned loose. They'd take the beef to a beef pasture, and when all the beef had been gathered they'd have a big drive from the basin out over the top of the mountain to Barton Flat, where they had a holding pasture. Then the different owners would each cut out their own beef and take them down to their ranches.

Each spring the association would pick a roundup boss to be fully in charge of the roundup. He ran it; that was his job: and boy, you did what he told you to, or you were in trouble! Two or three days before the roundup, the boss and one of his riders would take off in a pickup and drive around until they found a good-sized steer–it didn't make any difference what brand was on it. They'd shoot it in the head and cut its throat and dress it out, and that was the meat for the roundup. They had a screened-in cage right close to the creek–in fact, part of it was over the creek–where they hung beef so it kept nice and cool.

Most of the time everybody'd show up the afternoon before the roundup. (I'd show up too, but just to help as a rider. During the day I was just another rider; then in the evening I could be a forest ranger again.) [laughter] We'd always have a real early breakfast, and we started riding shortly after daylight. Every rider would bring at least two horses, and sometimes three. We would turn them out in the pasture, and there was always a bunch of teenage boys that'd come along to be the horse wranglers. They would have our horses in the corral when we got ready to ride in the morning, and you'd go out and rope whatever horse you wanted that day.

If the ranchers had any colts–unbroken horses–they'd always bring them in, too, because these teenagers . . . some of them weren't very old, but those ranch boys, now, they could ride! I don't know what they'd do in one of these rodeos, but they could *break* a horse. They'd get some of these colts in and top them off in the corral. Then when we were assigned where we were supposed to ride, some of the teenagers would each take a bronc and go with one of the older men–his dad or somebody else. For a while he'd lead the bronc until he got some of the energy taken out of him, and then he'd ride him, and when they would come in at the end of the day the horse was practically broken. The next day they'd take some more broncs out and break them, and by the time the roundup was over they had some good riding horses out of those colts.

In the morning we'd all be lined up, and roundup boss'd say, "You, you and you take this area; you take this area; you take this area," and he'd divide us all up, and we'd take off. We'd ride the area as thorough as we could. Once in a while we'd miss a few cattle, I'm sure, but we'd take our time and we'd clean the area pretty good. The first bunch down would bring their cattle to the holding ground, and then as others came down they'd put their herds in with the rest until everybody was in. By the time all the cattle were on the holding ground other teenagers had brought out a big lunch with lots of coffee from the rider cabin, and we'd have a good feed, and then we'd go and cut out the beef and the unbranded calves.

I had a good cutting horse, a roan that was born and raised on a ranch. He was a darn good cutting horse, and he was a good catch horse too. When they'd kick a steer out, you had to catch and turn him before he could break back into the herd. And boy, if you didn't get him and he broke back into the herd, you were told off! If that happened they weren't too happy, because they had to work pretty hard to get some of those steers out. That's what my horse liked to do: he was in the height of his glory when he was a catch horse. He'd stand there sometimes watching that steer, and he'd just pat his feet on the ground, and you never had to speak to him. The minute that steer cleared the herd, he was gone, and all you did was hang on! He'd get so tired sometimes that his flanks'd just be shaking, and you'd have to rest him. Once in a while one of the ranchers would want him for a cutting horse, and he was good at that too. I'd ride my other horses in the morning, but I always saved the roan for the afternoon. When the boys brought out the lunch, some would bring extra horses out for us—we'd tell them before we left in the morning what horses we wanted—and I'd always leave the roan to the afternoon for a catch horse.

Helping with the roundup was interesting and enjoyable, and the permittees appreciated that you could take off your ranger hat and go out and work and sweat and eat dust right along with the rest of them. The fall roundup was something I wanted to be involved in very much, but it also was really, I guess, part of my job, because I could work with those ranchers then; I could talk to them. When the day was done and we were sitting around having a shot of whiskey after we got through eating, if they had problems they'd come out with them. And if I had problems, I could come out with them. We could sit there and talk things over, and there was no bitterness. It worked.

The ranchers that ran in Copper Basin were a fairly close-knit group. One time a fellow by the name of Dale Parsons broke his leg in the spring right after calving. I was sitting in the office, and a rancher stuck his head through the door and said, "Arch, they are castrating calves out at Parsons's tomorrow." That's all he said, and he left. Well, next morning bright and early I got in my pickup and drove out. Most of the ranchers were there, and we castrated and branded and wattled all his calves. If any Copper Basin rancher was hurt or too old to get out on the roundup, they'd cut his beef out and brand and take care of his calves just the same as their own. And the wives from surrounding ranches would all gather at the ranch, and if it was nice weather we had a big dinner out on the lawn and passed the bottle around a few times, and that was it.

I haven't found that type of cooperation and support on other districts. When I transferred from Wildhorse to Ely in 1947 there was such a great difference that at first I just couldn't believe it. There was actual animosity between adjoining ranches and ranchers with adjoining allotments.

Trespass is trespass We had a little trespass on the Wildhorse, but most of it wasn't intentional... still, trespass is trespass, and I had to trespass them. And some of the ranchers had more cattle than they had a Forest permit for, so the extra ones were turned loose on the BLM, and they ran there until the ranchers could get all their hay up.* In that country they all had to feed during the wintertime. They had great big meadows, and they put up tons and tons of hay. This was ordinary slough hay, and there's only one cutting, but after the cutting there was quite a lot of regrowth. When they'd bring them in, the cattle would feed on that regrowth until the snow covered it; then they'd have to start feeding hay.

They would run their extra cattle on the BLM. The Forest boundary was fenced, but there were a number of gates in it—only one road went clear over the top and down into Copper Basin, but there were a lot of other roads went up a little ways into the Forest. I don't believe any permittee deliberately went out and opened a gate, but, like people everywhere, they would sometimes open a gate and not close it. Well, the extra cattle out on the BLM may have run in the basin on Forest land the year before, and they sure didn't want to eat that dry feed out on the BLM when they could get nice green feed in the basin; so if they got a chance, they'd get through the gates.

The CCCs had also built a fence just on the edge of the basin between what we called the summer range and the spring range, and the gates on that fence were always closed until the middle of the summer. Around the first of July, after the cattle had fully utilized the forage on the spring range, these gates were opened and the cattle were allowed on to the summer range in the basin. A few days after all the cattle had gone through, the gates were closed again. Trespass cattle would often pile up along this fence, so I always insisted that the riders keep those gates closed. They watched the gates pretty close and kept them closed, so if cattle came into the basin from BLM land, eventually I'd find some of them along that fence dividing the spring and summer ranges. But not all trespassing cattle would come down. There was plenty of water in the higher country, and some of the spring range would remain fairly green, so some cattle would always stay back on the spring range.

---

*Under the Taylor Grazing Act of 1934 grazing lands on the public domain were managed by the General Land Office (GLO), which underwent significant evolution during the Roosevelt administration. In 1946 the GLO was merged with the Grazing Service to create the Bureau of Land Management (BLM). When Mr. Murchie says BLM, he is referring to public domain grazing lands in general.

One of the beauties of that country was that when they were on the feedlot I could count their cattle and know exactly how many head they had. Suppose a fellow had a permit for 500 head, but when I counted them in the feedlot he had 560 head. Well, when he turned them out, that 60 head over had to be put on the BLM someplace, not on the Forest. And I could kind of keep track of the quantity of cattle out on the BLM, because it was flat country, and you could see a bunch of cattle for quite a ways off.

Most ranchers always cut down to about the right number before putting them on the Forest. In the fall they would always hold over a few, because they'd lose some during the winter, and during calving they'd lose a few head; but when they got ready to go on the Forest they were usually just about right on the permitted numbers. This one summer it sure didn't look like there was many cattle out on the BLM; some of the ranchers clearly had a lot more cattle on the Forest than they had a permit for. But I found a way to count them.

A young fellow who had an airplane had just come into Mackay. He wanted to have commercial flights, and he wanted to build a hangar. I got acquainted with him, and one day he came into the office and said, "Archie, do you know anything about laying up cinder block?" I told him no, but I'd laid up an *awful* lot of brick. He says, "Well, will you come out and help me?"

I said, "OK. Can't be much difference between laying up cinder block and laying brick."

So I went out every Saturday for a number of Saturdays, and we built his hangar. We got to talking, and he says, "Do you have any flying to do in the Forest Service?"

"Well," I said, "we hire planes to survey for fires once in a while after a bad lightning storm. That's about the only time that we rent planes."

"Any time you need a plane," he said, "you let me know."

He was in the office one day, and I just happened to mention my suspicions about trespass, and he says, "Why don't you go with me and I'll fly you over that BLM country? It won't cost you a penny."

I says, "Fine! That's just right!"

He said, "Just after daylight tomorrow morning would be a good time, because the air will be heavy, and we can get right down over their backs so you can even read their brands." [laughter] He didn't go down quite that close, but he flew me low, and I located all the cattle on the BLM. As soon as I got back I jumped in my pickup, and I could drive to practically every one of them and check brands. Then I knew right to the number whose cattle was out on the BLM, and therefore who had too many on the Forest. I went to each of the three ranches, and there wasn't a one of the ranchers that disagreed with me. But it ticked me off, all the same. I trespassed them, and they paid up. That's the only time I had a trespass while I was on the district.

If a permittee had ten or fifteen head more than his permit allowed, I'd tell him, "OK, you can turn them on," providing the range was good. (If the range was poor, I wouldn't do it.) "You can turn them on and I'll figure out the animal months extra that you're taking by turning this number on, and then at the beef roundup you got to cut out enough cows to make up the difference, so that at the end of the season you won't have exceeded what range you've paid for." Most of them would be tickled to death to do that. It worked out fine. A little bunch like that, it's hard for the rancher to handle–if they just turned them out on the BLM they'd have to ride out and check on them every once in a while, even that little herd. Allowing some flexibility helped them, and it didn't hurt the Forest Service any. When they had a big number, though, that was different.

Wartime use of the Wildhorse The United States was fighting the Second World War when I was on the Wildhorse district, and this had some effect on Forest Service operations. During the war there was a lot of mineral exploration going on, and miners could apply for mine-to-market roads. The money would be allotted to us, and the Forest Service would build the roads. We had a little trouble with these roads, because sometimes miners or prospectors would apply for these funds but not get them until late in the fall. One fellow got his allotment after freeze-up, in the winter of 1943. Of course, he insisted that we had to go in and build the road, frozen conditions or not. He said that he had a bunch of ore that he had to get out because the war effort needed all this copper . . . which I imagine it probably did. The conditions that we had to build that road under were just unbelievable! Before we could even *start* the road, we had to dynamite the frozen ground to loosen it up enough so a Cat could get in and start moving the earth, and it was dangerous. One time we had a little D6 Cat pioneering, and then the big D8 was coming behind. He got out too far on a frozen slope, and over it went–rolled it clear down to the bottom of the canyon. As luck would have it the driver felt it slipping, and he jumped and got clear of it. We finally salvaged the tractor, but it took a lot of money to put it back in shape again.

Every morning the ground would be frozen. You wouldn't have to blast it again, but you'd have frozen ground to work. Then we'd get a big snowstorm, and we'd have that snow to fight. On the latter part of the project we had to remove two or three feet of snow from the ground before we could even start blasting so the Cat could work. We finally got the road built, and the mine owner got out quite a bit of copper, but it wasn't an amount that I would say justified what the road cost, by any means. [laughter]

Other than building mine roads, World War II didn't affect the Wildhorse district very much. As far as maintaining our equipment went, we had plenty of tires, plenty of batteries and plenty of gas. What we were doing was considered important to the war effort. Of course, back then there was very little rehab; in fact I would say there was no rehabilitation of fire-damaged forest like they had later. By rehabilitation, I mean reseeding and planting ground cover and trees, and trenching to prevent erosion, and things like that. None of that was routinely occurring back then anyway, but if it were, we probably wouldn't have gotten any money for it.

During the fire season we still had plenty of funds to hire men to fight the fires . . . but the men we would hire for lookouts or patrolmen or trail crews were older groups than before or after the war, because all the young men had either enlisted or were drafted. A lot of men that we hired were married men, so we had more women in patrol camps and on lookouts–with their husbands, of course. On one lookout on the Wildhorse, I had a woman who manned it . . . or womanned it. [laughter] She was a darned good lookout, too, although she wouldn't have been too good at fighting a fire, because she was just a little bit of a girl; I don't think she weighed over a hundred pounds. But as far as detection, it didn't make any difference whether it was day or night, if a lightning storm went over, she was up. I was very satisfied with her work, and I gave her a real good recommendation when she finally left.

There was another part of my Forest work that was affected by the war: during that period we received no money for range improvements of any kind, be they fences or water developments or whatever. That was completely eliminated. We had a little money to maintain what we had, but we didn't receive any construction money to build anything new. Sometimes when we couldn't maintain a fence or water development, the permittees did it, which really helped. Another thing was that we were allowed to accumulate our annual leave up to ninety days. I believe it was forty-eight days we were allowed to accumulate normally, but during the war they wanted us to put in as much time as we could without taking vacation. Most years I didn't use all my annual leave, and I lost what I didn't use. But I kept my wartime ninety days, and when I retired I got paid for this ninety days of annual leave, which helped.

I was fortunate that on the Wildhorse district, except in exceptionally dry years, every allotment would carry its permitted livestock without any problems, and not even for one season did we have any overgrazing. At that time there were probably very few districts in Region Four where the allotments were as close to being properly stocked as they were on the Wildhorse. In fact, in some *good* years—especially on the Antelope cattle allotment—there was feed left over. The meadows on the Copper Basin allotment were sometimes grazed down a little closer than I'd like to have them, but the only way you could control grazing on your meadows was by salting away from them. When the herder saw a meadow was getting pretty-well used, then he'd salt away from it. You couldn't keep all the cattle off the meadow that way, but you could keep a big part of them off.

Even during World War II we didn't raise the permitted numbers of any allotment, because the numbers were already about what the allotment would carry, but we did make a change to the non-use option for grazing permittees. On some of your allotments—and especially on a big one like the Copper Basin—there'd generally be a rancher who for one reason or another wouldn't be able to fill up his permit; then he'd fill out a form, and he'd take non-use. (This wouldn't affect the status of his permit—all that it'd affect was the fees that he paid for that particular year.) Or sometimes a ranch sold out, and the rancher sold all his cattle. The new owner would want to bring in his own cattle, but maybe he would go for a whole season and not fill up the permit. During the war we tried to fill this non-use. If on the Copper Basin allotment the total non-use that was taken was, say, 150 head, then we'd let ranchers that had extra cattle put them on the allotment. One rancher couldn't gobble up the whole 150 head—we'd more or less divide it up—but we'd fill up all our non-use. We never did exceed the total preference on the allotment as far as I can remember, and in later years, after the war was over, we generally used non-use as a cushion to let the range build up a little bit more.

During the war, especially in the northwest and California–and even Region One and Region Four–there was a lot of concern about the possibility that the Japanese could set our Forests afire with balloon-carried incendiary devices. One day we were notified of a meeting that was to be held at Salmon City on the Salmon River (I believe this was in 1943). Even the supervisor didn't know what the meeting was for. They just said there was going to be a meeting–bring all your rangers. And we went, and it was strictly about these incendiary balloons that Japan was sending over. One of the balloons had already killed a teacher and some children in either Oregon or Washington. The Forest Service had diagrams of exactly how the balloons were built, how the ballast was dropped, how the incendiary bombs were dropped . . . and of the altimeters and whatnot that they had on them so they wouldn't go too high or come down too low. We were very well drilled in all that. If we saw one of them, we were not supposed to touch it. As I remember, what happened with this teacher and the children was they saw this device and they didn't know what it was. They went over and pulled on it, and the thing exploded.

Poison larkspur and warbles

The Copper Basin allotment was the largest cattle allotment on the Wildhorse district, but it was only one of four: there were the Copper Basin, North Fork, Alder Creek and Antelope cattle allotments. Copper Basin had very little larkspur on it, but just about every other allotment had a fair amount, which created quite a problem at times. The plant is poisonous to cattle–it is a deadly heart stimulant. During the CCC days, when the Forest Service had lots of manpower, they thought they could grub the larkspur out and get rid of it. But they worked and they worked, and I think if anything they probably scattered it more than they grubbed it out, because it sure wasn't successful. About the only thing they could do was fence the poison areas off and keep the cattle out of there. In doing that they inadvertently grew real thrifty stands of poison larkspur, and the larkspur eventually moved down the canyons outside of the fences, and they were in trouble again.

One time I was riding with a herder whose bunch of cattle had gotten into the head of a canyon that contained the plant. For some reason cattle really *like* the larkspur—cows especially—and this one cow was just going down on it like everything. The herder swore at her, and he said, "You won't do that very long." And sure enough, when we came back down the canyon she was dead.

Another time on the Alder Creek allotment the herder and I were riding up in the head of a draw—somebody had left the gate open, and a bunch of cattle had gone up it, and they were feeding on larkspur. There was a lot of young stuff, so we didn't bother them, but we went around to see if there was any more cattle above. Coming back down we were discussing whether to try to move them out or not. (Excitement and energetic activity intensify the effect of larkspur on cattle's hearts.) "Well," he said, "there isn't very many of them. There's only a couple of cows, but there are four or five yearlings in the bunch, and I don't want to ride back up here to keep checking on them. Let's try to move them out." So we started real slow. We didn't have very far to move them down the canyon, but these darned yearlings got in front, and you know how frisky calves are! They kicked up their heels and started running, and the whole bunch took off. Well, first one of the yearlings went down from the poison, then another one; and then one of the cows went down. Somebody had told the herder that if you bled them you could probably save them, so he jumped off his horse and lifted the cow's tail and cut her right at the root of the tail, and boy she bled! He cut an artery, and she really bled. She died, but I don't think the bleeding had much to do with it. And two calves died, but the rest of them apparently didn't have enough poison in them to kill them.

They had more larkspur problems on that allotment than they had on others. They also had a bad lice problem on the Antelope allotment. The ranchers had built a big dipping vat, and every spring they would round up all their cattle and dip them before they went on the Forest. They did this before calving, because calves were too small to put through the dipping vat, and they had to be awful careful running those pregnant cows through that vat.

There was a problem with warbles on the district. Warbles is caused by a fly—I don't know whether it's related to the heel fly or not, but they'll sting a cow generally in the heels, and other places, too. Apparently, they'll lay their eggs and then sting the animal, and the cow will reach down to lick the sting and she will get the eggs in her mouth, and she'll swallow them. They go down into the stomach and hatch. They're extremely small larvae at first, but they start working up towards her back, going up through the tissue and organs. When they get ready to pupate, they form lumps right under the cow's hide, and they chew holes through her skin where the larvae get air. As I get it, the larvae mature; they crawl out and fall to the ground and pupate and then come out as mature flies and repeat the cycle again. Well, in that country I won't say *every* critter had a lump or two on its back in the spring, but at the Antelope allotment some of their backs were just solid with knobs with the pupae underneath. These warbles would just completely ruin a hide. When they came out they'd leave a hole, and the darned magpies would start working on the wound. I've seen holes half the size of my fist that they pecked out of a cow's back.

When most of these warbles had first opened up the little hole on the cow's back, that's when the ranchers would treat them. I don't know whether they might have added a special pesticide for the warbles or whether it was the same one that they used on the lice, but it would get in there and kill the warble. Then the next year they wouldn't have so many warbles to deal with. How much it actually controlled the warbles, I don't know; it seems like *every* year they had a lot of warbles in their cattle. Every spring they treated them. That was one of the jobs of the association: when they'd meet they'd decide how much insecticide they were going to have to buy and when they were going to treat.

Frostbitten on a snow course   Years ago rangers measured the water content of snow packs on all the Forests. Snow courses were set up for this purpose by the SCS, but I guess they just didn't have enough manpower or experience in taking long snowshoe trips alone to do the measuring, so they asked the Forest Service if rangers couldn't measure the snow courses.

Dr. J. E. Church at the University of Nevada had developed the snow course concept, and he must have been a pretty smart man or did a lot of studying, because this was an exceptional idea which worked perfectly. Dr. Church's system used a tube made of an alloy that was very resistant to snow freezing to it, and with it you would take a core sample of snow. Especially at higher elevations, you'll get snow, and then you'll get a few warm days and freezing nights and you get an icy crust, and then you get snow again; so the bottom section of the tube had teeth around its lip, so you could grind down through the crust. Sometimes at the bottom you'd have an ice layer that you'd have to grind down through, and you'd always go down until you brought up a little bit of dirt. Before you measured the snow sample, you'd take your knife or a pencil and dig the dirt out. Included in the measuring kit was a scale to weigh the tube with the sample in it, and it read out in inches of water in the snow. This tube was in sections that you could unscrew, and there was a nice sack with shoulder straps, and an aluminum tatum holder with your measurement sheets in it.

I started measuring snow courses while I was at the Wildhorse ranger station. In February of 1945 the first measurement we were going to take was up in Copper Basin. I drove as far as I could up Lost River to some ranches up there, and the rest of the way I had to go on snowshoes. Two brothers lived up there: one of them had a pretty good-sized ranch, and the other one was a government predatory animal trapper. The SCS allowed me to hire one man to go with me, and I knew I could depend on John, the rancher brother, so I snowshoed into his ranch and asked him if he was interested in going.

John had skis, and for one man to have skis and one man snowshoes don't work too good. So he says, "I got an extra pair of skis. Why don't you take them?"

I said, "Sure." These were old-time, homemade skis. All they were was . . . they steamed the end of a board and curled it up–they didn't even have a groove down the middle–and then they'd take a strap and nail it to either side, and you'd slip your foot through it. They weren't very good going uphill, because your foot kept sliding out.

The next morning John and I took off. It was a good, long day's trip from his ranch into the Copper Basin snow course. There had been awful heavy snow that winter, and I remember we skied up on top of the roof of the rider's cabin and ate our lunch sitting on the ridge pole! [laughter] But anyway, we measured that snow course.

On the way back, the end of one of my ski straps tore off right where the nails go through. When snowshoeing I always carried a roll of rawhide thongs with me, because you never knew when you might break the webbing, and I had a roll of this with me. I tied my foot to the ski, but then I couldn't slide it; I had to just walk on it. I made it into the ranch OK, but I told John that I had better take my snowshoes the next morning. "No," he said, "I'll fix your ski." He had a shed out back, and he took it out there and fixed it.

The next snow course was up Pass Creek. We could have made it, but it would have been a long day if we'd gone from his cabin. He said, "Let's walk down the sleigh road to my brother's place and stay overnight." So we did. We had a good breakfast the next morning and took off on our skis. What he'd done to my ski was just cut the torn end from the strap and nail the strap back down, which shortened it and made it much tighter over my foot. We hadn't gone over a mile or so, and my foot started hurting. I told John, "I don't know—if my foot keeps hurting, I'm going to have to turn around and go back and get my snowshoes." We went on a little ways and it stopped hurting, so I thought, "Well, the leather has stretched." So we went on up Pass Creek.

There was a fellow that had a portable sawmill up there, and the course was just a little bit above the sawmill. We had coffee at the sawmill, measured the snow course, and turned around and came back. We got back to John's brother's place probably about seven o'clock or so, and his wife had waited supper for us. I had on fairly heavy boots, and was wearing a pair of light socks, then a pair of heavy red wool socks. I took my boots off, and we were sitting there eating supper, and all of a sudden my foot started hurting. I thought, "What in the world?" Finally, it really got to throbbing. I wasn't even through eating yet, but I slid back and took my sock off, and my big toe and part of the toe next to it were just as red as fire. I thought, "Uh-oh. I froze them." The reason I hadn't felt any pain was that the darn strap had cut the circulation off, and it had gone numb.

I didn't sleep much that night. *Oh, it hurt.* The next morning I could just hobble around, and could no more get my boot on than anything. John's brother was a fairly good-sized man, and he said, "I'll tell you what I'll do: I'll give you a couple more socks, and you put them on and then put on my overshoes." I did, and I had to loosen the strap way up on my snowshoe to get it on, but I made it out to where my pickup was at the end of the cleared road. We didn't have a doctor in Mackay, so I just hobbled around on that thing for a while. Finally it got so it didn't hurt too much, but I was only wearing a bedroom slipper on it, and that toe had turned just as black as coal for about an inch down.

A week afterwards the supervisor was down. He looked at my slipper and says, "Arch, what's wrong with your foot?" I told him, and he said,

"Let me see it." I pulled my sock off, and he says, "You get out to a doctor! Don't you wait a minute; you get out to a doctor."

There was a doctor at Arco, but I just had a feeling that I might have to go a lot further than Arco, so I packed a bag, kissed Jane good-bye and took off. The doctor at Arco took one look at it and said, "You get out of here and head for Pocatello."

Our family doctor at that time was a fellow named Doc Newton, who lived in Pocatello. I had gotten acquainted with him mountain goat hunting. It was evening by the time I got there, and I called Doc, and he says, "You go on over to Saint Mary's Hospital," which I did. He took a look at it and said, "It don't look good." It was black, and there was the start of a little red line around the edge of the black. He said, "If that continues moving up, it's a sign of gangrene, and that toe is going to have to come off." The next morning he looked at it again. There was no feeling at all in it; it was just as dead as a stick of wood. He shook his head, and he said he wanted to cut it off right then.

I said, "No, no. You're not going to take my toe off."

"Well," he says, "I'll give you until four o'clock this afternoon. If no feeling comes in it by then, it's coming off. I can't let you go any longer, because gangrene is going to set in just as sure as not."

They had been putting sulfa on the toe, and they kept putting this sulfa on it, and I *still* don't know . . . . I've wondered whether I *really* felt pain in my toe by four o'clock. [laughter] But anyway, I told him that the pain was starting to come back into it. He says, "OK, we won't cut it off." Well, all that black flesh (the whole end of the toe) came off eventually, anyway. It looked like the end of the bone with just some flesh on it. It finally scabbed over, but it took a long, long time for the toe to heal up. Finally, my nail came back too, but the toe never filled out as full as it was before, and to this day there's no feeling in it.

Thank goodness for sulfa! If the supervisor hadn't told me to go out and see a doctor . . . . Well, I wasn't too smart on that. It didn't look too bad. It was black, but I figured the frozen part would finally just drop off. Not the whole toe, but [laughter] the frozen part! Anyway, that took care of that.

Not an administrator In the late spring of 1947 they had a supervisors' meeting in Ogden. Mac McKee and I always got along exceptionally well, and he was going to the supervisors' meeting, and he stopped in the office. He told me, "Arch, if I can work it, would you take a job as my assistant supervisor on the

On the Challis: Wildhorse

Challis?" If it had been anybody else, I'd have turned him down, but with Mac it would have been all right, because he would have more or less just turned me loose. I thought it would be fun working as assistant supervisor with Mac, so I said yes. "When I come back," he said, "I'm pretty sure you're going to be assistant supervisor of the Challis Forest."

Well, Mac got to Ogden and brought it up, and they said, "No. We've got a problem down on the Ely district in the Nevada National Forest, and we're going to send Archie down there." When Mac came back he felt bad. He was really disappointed, but in a way I was kind of glad that I didn't get the appointment because I liked being a ranger; I liked the freedom of being my own boss. (I never had any desire to move into administration . . . except when they finally moved me over to Reno in 1959. Then they told me, "That's it; you're going." They gave me a choice of going to Reno or the Caribou Forest at Pocatello or someplace down in southern Utah. Of course, southern Utah was out!) [laughter]

One advantage of being a ranger at Mackay was that it was a small community and you got to know the people well, and you could work with the people. I had actually become part of the community, more than in a lot of other places I've been. I worked with the Boy Scouts and Lions Club, and was a member of the Masonic Lodge there. Women's clubs and different groups, if they wanted somebody to give a talk, they would call on me. Also, the community was a very nice community to live in. Our children were all in school then, except our youngest boy, John. They had good schools, and the classes weren't so big that they didn't get a good education.

I thought that everything was in pretty good order when I left the Wildhorse district in 1947. I even managed to get out of there without being charged for missing equipment. [laughter] I got out in good shape. In fact, the Copper Basin Cattle Association gave me two bronze horse statuette bookends as a gift for what I'd contributed to the grazing of their livestock.

9

Raising a Family on Remote Ranger Stations

YOU TAKE A RANGER and his wife, they had problems that people living in town never experience. The main problem was isolation in remote places. There's been lots of rangers' wives that just couldn't stand that, and it either came to a divorce or the ranger having to be transferred. Now I don't think you have a ranger station that has the isolation like what we had years ago. Back then you didn't have a choice, either. When they said go, you went. An old ranger once told us that every time a truck drove into his yard, his chickens rolled over on their backs to have their feet tied. [laughter] Lots of times you didn't get all that much notice, either. Jane and I have packed up and loaded the car and been gone the next day. One time when I was transferred without notice we had to can our peaches all night so we could leave the next day. Jane lost the diamond out of her engagement ring in the peaches, but she found it before we left.

We moved with one night's notice from the Wyoming Forest to the Rapid River district on the Challis. I was chief of party of a timber survey crew at the time, and I had part of my stuff up to the camp. One day they asked me how soon I could report to the Challis National Forest. Jane and Jimmie were in Pinedale, and the rest of our stuff was in Kemmerer, but at that time we could load everything we owned into the old Chevy, and that was it. We had no furniture, and not all that much clothing. We took off. The longest notice that we ever had that I was going to be transferred was when we moved from the Challis to Ely; then we knew better than a month ahead.

Back in those days the ranger station was often at the end of the road, and that was it. Many a time Jane would have a meal fixed for us, and we would be just about ready to sit down, and a bunch of people would come in. And by the time we got through eating, we'd have six or eight people at the table. It wasn't just ranchers and miners, either: a lot of Forest Service people came. You were kind of a ranger station and restaurant combined. This could be quite a burden, especially when you have children and sometimes you might have a child that's sick. Since we ate out of cans—at Loon Creek you got groceries delivered every two or three weeks; you didn't have fresh meat or vegetables—Jane would just open another can and fix it up. A lot of other rangers' wives resented the fact that they had to do this, but Jane knew it was part of the job, so she just did it and never complained. She was young and she accepted these things, and that was expected of rangers' wives by the Forest Service, because you were there at the end of the road . . . where else could people go for a meal?

There were two reasons why I went into the Forest Service: one was because I really liked that outdoor work; and second was the security. From 1933 until I retired I was never out of a job. You might as well say my entire life I was never out of a job, and there wasn't a day that I wasn't being paid . . . or that I didn't have to work. [laughter] As a result of that we had a good-sized family, but Jane did the best part of the raising of the kids because I was gone all the time. She was left alone a lot. In the spring on the Challis, when we went out on trail work, I could be gone a month or two at a stretch, and I worked from Monday morning till Saturday night during the war years in the 1940s. Even after that, I would sometimes be gone out on the Forest for ten days at a stretch. I was gone so much that it was difficult for our children to be born at home, and our youngest son was really the only one that we could say was born in the same town we lived in. When Jane was pregnant with Jean we were stationed at Challis, where I was the ranger on the Rapid River district. Jane's Aunt Belle lived in Salt Lake City, so we contacted her and she said she'd be glad to have Jane come down. Jean was born in the LDS Hospital in Salt Lake—she must have been

"Back in those days the ranger station was often at the end of the road, and that was it. Many a time Jane would have a meal fixed for us, and we would be just about ready to sit down, and a bunch of people would come in. And by the time we got through eating, we'd have six or eight people at the table."

Left to right: Johnny, Jean and Carol Ann, pictured with Jane and Archie in 1956. Missing is Jimmie Murchie.

close to a week old before I saw her. When we brought Jean home we were living in town, but as soon as the summit opened we went back in to Seafoam. She was probably three months old.

When Jane was carrying Carol I was ranger on the Loon Creek district, and we were still living in Challis. Around the first of September I took Jane and the two children to Salt Lake City, to her aunt again. This time she gave birth in St. Mary's Hospital. I was to Ranger School at Logan when Carol was born, and Jane's aunt called me up there to tell me of the birth. A lot of the rangers and people at the Ranger School were Mormons, but I still had to buy cigars for the whole bunch of them. [laughter] Then they let me take off for a couple of days to go and see my wife and new daughter.

When John came along we decided to have him born right there in Challis, because by then there was a pretty good doctor there. (Before that, we didn't have much of a doctor.) So, of all our four children, John was the only one whose birth I was present at.

The Seafoam ranger station was a modern, two-story house with running water. We didn't have any electricity, though; we used Coleman lanterns for light. At Loon Creek and Wildhorse we also had running water but no electricity. Jane was born in town and was afraid out there in the forest without electric lights, so she'd go to bed at dark. [laughter] At Seafoam Jane washed all of the diapers by hand, but when we were at the Loon Creek station I bought an electric washer and hooked a gasoline motor to it. In the wintertime, when we moved into Challis, I'd take the gasoline motor off and put the electric motor back on. (As I remember, the gas motor ran slower than the electric motor, but I used a smaller pulley when I hooked it up, and it worked fine.)

We didn't have driers of any kind, of course, and all the drying was done either in the house or on a clothesline outside. I think the best dryer there is is a clothesline in the sun if you want to have nice, bleached clothes. Of course, Jane did all her own ironing with flat irons. Flat irons were real heavy and had a handle that you put on. She heated them on top of the wood stove, and the ironing board was right by the stove, so she stayed nice and warm. [laughter] But they were good irons. Jane made most of our children's clothes and she would often buy fabric like plisse, that didn't have to be ironed. Even after we moved to Ely, Jane was still making clothes for the children. She was a good seamstress, too; she made excellent clothes on one of those treadle-driven Singer sewing machines. We bought our first Pfaff electric machine when we moved on to the Toiyabe Forest in 1959; up until then it was all the treadle sewing machine.

Up at Seafoam the timber was dense and it came down close around the house. Jimmie got lost one time at Seafoam. [laughter] He and the dog went on a trip:

It was spring or early summer of 1937 when the lookouts and the patrolmen and other men were coming in to go to work. This lookout pulled into the station, and he had a female dog with him that was in heat. Down the road below the station was this big black dog . . . just standing. So I asked the lookout, "Is that your dog?"

He said, "No."

I thought nothing more of it, and the next morning we went out to round up our stock (I was taking him out to the lookout), and this dog was with the horses. He saw us coming, and he took off into the timber. We packed up and left, and I took him up to the lookout, where I stayed overnight. The next morning I headed down, and where the trail came off the peak and made a switchback I looked back and saw the black dog following me. The lookout was watching from above, and I yelled up to him, "That dog is following me!"

He called back, "He isn't mine. If he wants to follow you I can't do anything about it. Just take him!" [laughter] So he followed me back to the station. We didn't feed the dog at first because I really didn't want him to stay. He was kind of unfriendly, too–he liked to be around the horses, but whenever you'd get close to him he was gone.

The dog made a couple of pack trips with me. When he came on a pack trip, whenever I had sourdough hotcakes I'd deliberately make a little bit more and I'd take them out and pour bacon grease over them to attract him. The first time I did that he was hungry, because he'd followed me all day and he hadn't had time to hunt. I took the hotcakes out and put them on a rock. He was standing out at the edge of the timber, watching me, and I spoke to him . . . but no recognition of any kind. He could smell the sourdough hotcakes and bacon grease, though, and he stood there and looked, and he began getting closer and getting closer, and he walked around and finally he grabbed those hotcakes and he was gone! [laughter] And so after that when I was out camping I'd always feed him a little. He got so that when he saw me do it, he'd come in and eat. He got used to the taste of sourdough hotcakes, I guess.

At the station the dog was able to hunt, and we never fed him. One morning we made sourdough hotcakes, and when Jimmie was practically through eating, Jane rolled up one more hotcake and gave it to him. I was in the office working and had the door open, and I heard Jimmie talking like he was talking to somebody. I looked out, and he had this hotcake in his hand and that darned dog had his mouth over his hand trying to get the hotcake. [laughter] I was going to yell, but I just thought, "No. If I yell he's going to clamp his jaws shut." So I just watched, and pretty soon Jimmie opened his hand and the dog ate the hotcake. [laughter] And that dog and Jimmie were practically inseparable after that. He finally tamed down, but he was never a tame dog. You could pet him, and he loved Jimmie; but with me or Jane, he could take us or leave us.

He was a Belgian shepherd–black, and he had just a little white strip. Sharp ears like a German shepherd. Pretty good-sized dog. We named him Spider, and one day a herder recognized him and told me the history of him. Another herder had bought this pup to try to make a sheep dog out of him, but he was a little bit too rough with the sheep, and the herder gave him quite a beating. The dog took off, and as far as anybody could figure out, he ran wild. (He'd been seen a few times.) I'd say he was at least two years old and maybe older than that when I got him.

One morning Jimmie was out playing, and we missed him. We called him, and no answer. Seafoam Creek came right down past the station off just a little ways. There was a footbridge across it (just a log; not a real bridge), and there was a tent on the far side where the men coming and going would sleep–we didn't have room for them anyplace else. A little rain had fallen during the night and it had settled the dust, and I could track Jimmie over to this footbridge. (Boy, I got scared–my son getting close to that footbridge. It was quite a swift stream, and we'd cautioned him time and time again to stay away from the creek.) The tracks stopped short of the creek, and I could see that the dog was following him, and they came back and went up past the barn. We had a two-rut road that went up to a powder house, and they went up this road. I'm sure Spider then was leading Jimmie; I *know* he was leading because I could tell by the tracks. They pulled out on the ridge to the south of the station. I could track them up and they went on up the ridge, and finally the timber got heavy enough–enough needles on the ground–that I couldn't track them. I just *couldn't* pick up a track. But they'd been holding steady, so I thought, "Well, maybe they'll continue to hold to the top of the ridge."

I went up another quarter of a mile and I was about ready to give it up. There was a CCC camp at the Bonanza guard station, and I thought, "I'll get in the car and go up there and get all those CCC boys out to hunt for Jimmie." But I went just a little bit further, and I thought I heard Jimmie's voice. I stopped, and sure enough I could hear him, and I went a little ways and looked down and just watched them a while. [laughter] They'd found a gopher hole, and Jimmie'd dig a little while and then he'd stop and the dog would commence digging a little while! I watched this for a little while, but finally I went down and picked Jimmie up and bawled him out. He started crying and Spider took off. I believe the dog thought I was going to whip him. He had already had one whipping in his life and he wasn't going to have another one, and I'm very sure that's the reason he ran. When I brought Jimmie down to the station, Spider was lying in front of the kitchen door, but when he saw me coming he got up and walked around the house, and he was gone.

The next morning, the ranger over to Loon Creek called me and says, "Where's your dog?"

"Well," I said, "he's out here someplace, probably out with the horses."

"No," he said, "he's over here." Across country, he had to go way over the top of a real high range of mountains. I'm just guessing, but it'd be a good twenty miles if I remember. I told Jay to catch him and I'd come over and get him. Well, Jay couldn't catch him, so Spider hung around there just a short time and then he disappeared and nobody knew where he was. Way over on the east side of the district a sheepherder complained about a big black dog killing some of his sheep. So I put two and two together, and Jay did too, and I says, "Well, go over there, and if you can catch him or get him to follow you, OK." And I said, "If not, shoot him, because we can't have my dog killing sheep." So Jay went over and the dog followed him back and Jay told me, "He's back to the station."

It wasn't too far from there to the Cougar lookout. I said, "You go up to Cougar lookout, and if he will follow you, I'll be there." Jay beat me there, and Spider was standing out at the edge of the timber. I rode up and he didn't wag his tail, he didn't . . . . He was watching me, but there was *no* recognition. Then I got off and walked toward him just a little ways and called to him, and here he came, and I threw my arms around him. [laughter] Everything was forgiven then, see, and he followed me back.

Spider was a great dog–courageous and tough and very protective of Jimmie. But he was a fighter and a little wild. After we moved into Challis, he was going to get me in some serious trouble, so I gave him to a rancher.

During my career the size of Forest Service staffs was much smaller than today. Sometimes, like on the Challis, the Forest supervisor, the administrative assistant, a truck driver, a warehouseman and a couple of stenographers were all the people that were in the supervisor's office. And on the districts, there were just seven or eight rangers. With very few people involved, we were just like a big family, real close-knit. If anybody needed help, all piled in. If someone rented another house and had to move, they'd be there to load up and take you over. At Ely when I put a full basement under my house, the whole Forest Service crew turned out, and we ran concrete till we were blue in the face. We used to have a lot of picnics, a lot of get-togethers. But I think as the numbers of people increased on districts and on Forests, this closeness gradually decreased some, and it's natural that that would take place.

When we were transferred from Challis to Mackay, we decided that we wanted to buy a house, but we had to rent at first because nothing was for sale. There was a mortician who owned a house, and he had the porch all enclosed where he kept his caskets and whatnot. He either went out of business or something, and the house came up for sale, but nobody wanted to live in a house where a bunch of dead people had been. It smelled of formaldehyde. We could taste it. [laughter] But we decided to buy it, and we could get it very cheap.

There was a Jewish banker down to Arco, and we went down and talked to him, and, "Sure," he said, "I'll loan you what money you need to buy it." So we bought it, and we did quite a little work on it. We tore all the porch closures off and we did some work inside. In 1947, when we got ready to leave, we better than doubled our money on the house. There was a rancher who was wanting to retire, and as soon as he knew we were moving out he wanted it, so we didn't have any trouble selling it.

I guess many young couples in the Forest Service might have had one problem that Jane and I had. We were at the age that we didn't fit in with the older people in a community; and we didn't really fit in with the younger people, because we had a better job and were making a little bit more money. (This was during the Depression, remember.) Among the younger people there may have been a little bit of envy of our prosperity, and perhaps even envy of the position we kind of had in the community, because back in those days the public had a lot of respect for the Forest Service. The Forest Service was it: they were the managers of our public lands. In the 1930s and 1940s, to be holding this prominent position and to be having a paycheck coming in regularly . . . the young people really didn't associate with us too much.

Also, as a ranger I was a man who was supposed to have the authority to tell a rancher he had to get his cattle off the Forest or that he wasn't supposed to come on yet. There were a lot of ranchers, especially the older ones, that kind of resented being told what to do by some young pup. I've actually been told that. They'd say, "You're just a young sprout. You can't come out here and tell us what to do!" That was a problem when we first started out, and it was a problem up in Montana, and it was a problem on the Challis. It was even a little bit of a problem down to Ely, but I was getting a few years behind me when I went down to Ely, so that it wasn't too bad. By the late 1950s and early 1960s, just the real old ranchers that I ran into—people like Fred Dressler or old Pop Mormon or men like that—would feel that there was that age difference.

I think one of the reasons that Jane and I succeeded so well as a team on those isolated ranger districts goes back to the fact that I cut my leg severely right after we married. When two people get married, the best thing that can happen to them is to really get acquainted, and when I was bound to the house with my cut leg we really got a chance to know each other. The times that we were isolated and snowbound in the Uintas at the guard station, and later on with the four children at Loon Creek, we not only got to know each other better, but we got to know our children and our children got to know us. A lot of problems with families now is just that the children don't know their parents and the parents don't really know their children. But that didn't happen to us, and I think that is one of the reasons why we've been married going on fifty-eight years.

Raising a Family on Remote Ranger Stations

Although a ranger station was often in a remote area, it could provide a very good environment in which to bring children up because so much of the time they had to use their own initiative, rely on their own resources. They couldn't go down to a movie; they couldn't go over to a neighbor's house and play with their kids; they had to create their own entertainment. When we were at Wildhorse our children would invite friends from Mackay out to spend a few days or a week, but most of the time they were on their own. Jimmie and Jane liked to fish, and they'd go catch their limit and make our other kids eat it. [laughter] When the kids got older we were always near good fishing streams, and they spent a lot of their time fishing. Even to this day, as old as some of our children are now, they still tell about how much fun it was growing up at these various ranger stations. I wonder how many children in a lot of towns can tell about how much fun they had growing up?

One disadvantage to our isolated life was that maybe the kids lacked enough association with other children to know how to accept them. When we moved into town is when our kids would get into trouble. There were twelve children on our block in Ely, and our oldest boy was the eldest one, so he sort of ran things; and there were a lot of things going on. It was kids getting into trouble playing–throwing a rock through somebody's window or filling a paper bag with water and getting on top of a building and dropping it on somebody's head as they went by, and things like that. You know how when you were a kid . . . . [laughter] After Christmas one time the neighborhood kids were burning the trees and a girl picked up a little tree and bopped Johnny right in the nose and rearranged his face. There were other incidents like that, but out in the wilderness they never seemed to get into trouble.

Only once was schooling a problem, and that was when we transferred from Challis to Mackay. Jimmie was in the second grade, and they were reviewing, and the Forest Service moved us right in the middle of the winter to Mackay. There was such a difference between the schools at Challis and Mackay that Jimmie was completely out of it for about half a year, and that really affected his health, too.

Jimmie was thirteen years old when we moved to Ely, and Johnny was in the first grade. So you might as well say that other than Jimmie all our children's education was obtained at Ely. For one of the children, that move to Ely was not so good. Mackay was a small school, and Carol Ann was fine there, but at Ely Jane would take her to school and Carol Ann would beat her back home. [laughter] She adjusted to that eventually.

She finally got her friends and it worked out, but it took her a little while.

When we were living in Mackay there was two vacant lots owned by the county. They were thought to be not cultivable, and they had sagebrush growing on them. But we wanted a garden, so one day I borrowed an Abney level from the Forest office, and Jimmie and I ran a bunch of levels on those lots. Now, everything in Mackay was watered with open ditches–the lawns, everything, was watered with open ditches–and Jimmie and I ran some levels on those lots and found that I could water one; I didn't have much of a head, but I could water it, and I could water about three-fourths of the other lot. So I bought them. I got the first for five dollars, and I developed it. It wasn't quite enough for what garden I wanted, and I thought, "I'll buy the other one." So I bought the other one, and they figured I must have something on my mind, so they charged me fifteen dollars for the second one. [laughter]

Everybody thought I was crazy when I started clearing the first lot; they thought I was *crazy*, and said, "Those lots have been sitting there since Mackay was established and nobody can get water on those." I didn't say anything. Finally I got them cleared and Jimmie and I put the ditches in and turned the water in . . . and then they accused me of being Mormon–said I was making water run uphill! [laughter] It really wasn't uphill; it just appeared to be. It was an optical illusion.

I had to have a good head in the ditch, and then it was just about a level grade through most of it, but it was adequate for irrigating a garden. We grew corn on them, and we had to go clear across town to take it home, and everybody would stop us along the way to buy corn. It was just beautiful corn. One time the kids started back with their little wagon full, and before they got home the wagon was empty. [laughter] I sold both of those lots for five hundred dollars apiece when we left Mackay.

In the summer after school was out, Jane and the kids would leave our house in Mackay and come up to the Wildhorse ranger station, where I would be. They brought the necessities with them, but as far as furniture and stuff like that, they never really *moved* up. Jane would just bring sleeping bags, and she wouldn't have to keep house. She could drive by then, and she'd take the kids and go up to Copper Basin or wherever they wanted to go on day trips. When their clothes got dirty they'd go on back into Mackay to our house and wash their clothes and come back out.

During those summers, while Jane and Jimmie fished in Copper Basin, the girls and Johnny had definite instructions not to leave the station. The girls had their dolls and different things, and they just played like kids play. There were enough things to do, and Wildhorse didn't have any timber around it that they could wander off and get lost in. They couldn't get into trouble, but they used to scare Johnny to death: they'd make footprints in the dust. [laughter] Other than that they got along fine, and we didn't worry about them. In fact, one of the advantages to living in a Forest Service family like ours over the years was that out in areas like that we didn't have to worry about the safety of our children.

Back then it was so different. You didn't have to lock your house and you didn't have perverts to worry about. There weren't many people up there at Wildhorse, anyway. Sometimes people would drive by, but there never was anybody that would hang around. Then too, there at Wildhorse we had a guard (Field Winn and his wife, Roma), who lived just across the road. Roma was home most of the time, and that made a little difference. The kids didn't bother Roma too much, and she was nice to Jean, especially–she'd help her learn to cook.

Living out on the ranger stations probably brought Jane and me closer together than if we had lived in a big town, and Jane recalls that she didn't mind my being gone so much. [laughter] I was always glad to get home, and she was always glad to see me pull in through the gates.

10

Nevada and Humboldt Forests: Ely District, 1947-1959

I WAS SUPPOSED TO TRANSFER to the Ely district of the old Nevada National Forest* the first of June, 1947, but Jane and I and the children had never really taken what you'd call a vacation, so I told Mac, "I'd like to take a little vacation beforehand."

Mac says, "Sure, go ahead; take all the time you want." Well, I built a little shell on a two-wheel trailer that we hauled behind our car. (We could sleep in it at night if we had to.) We took off, and we wanted to go over to the coast, but I thought, "As long as we're wandering around I'll drive down to Ely and just see what the situation is down there—what the country's like." Johnny Herbert was the new Forest supervisor, and when we got to Ely I called him up and introduced myself, and he says, "Where are you staying?" I told him in a motel, and he says, "Why don't you and your wife come over to the house this evening?" So we went over and had quite a long visit. I was really impressed with John; I liked him, and he turned out to be one of the better supervisors that I've ever worked for. He wanted to know when I was going to report, and I told

---

*Public Land Order 1487 abolished the Nevada National Forest, which was located along the eastern side of Nevada, transferring its lands to the Humboldt and Toiyabe National Forests, effective October 1, 1957.

him what Mac had said about our vacation. [laughter] He really wasn't all that happy, because the Ely district had been without a ranger for the best part of a month already. But we completed our trip and then wound up back in Mackay.

I arranged with Mayflower to haul our household goods and everything down to Ely when we were ready. We had bought a house when we first went to Mackay, and we were lucky that we were able to sell it before we left. We wanted to buy a house in Ely, but we just couldn't find a place, so we had to delay bringing the furniture down, and the people that we'd sold our house to couldn't move in until our furniture was gone. So we just lived in a motel in Ely for a while, Jane and I and the four kids. Finally, we bought a little house–it was the only thing that we could find. It had a front room and a small kitchen and one bedroom . . . for six people. (It was on a very narrow lot, so the only way I could enlarge it was to go down. I had to build a basement under it . . . but that's another story.) Finally, we got settled in Ely around the end of June. The livestock were all on their allotments, and everything was going full force when I took over, but it wasn't going well.

Bad relations with permittees

Before I had transferred to the Ely district, Mac had told me that there was a very bad situation on the Nevada National Forest, and they were cleaning the whole slate. In other words, they were replacing the supervisor and all his rangers . . . and when I got there they had a new administrative assistant, too. They replaced the supervisor with Johnny Herbert, who was tough; he wasn't very big, but he was tough.

The worst problem on the Ely district was that there was a conflict between the Forest Service and the livestock people, and the livestock people were very bitter. One of the biggest permittees was Bert Robison, who ran five bands of sheep. I hadn't been there just a very short time before I went to see him about his grazing permit. I stopped at his ranch about six-thirty in the evening. He wasn't there, but the foreman told me that he'd be in shortly, so I thought, "Well, I'll wait till he comes." In a little while Bert pulled up in his Cadillac, and he saw the Forest Service pickup parked there, and he came over. He wasn't too happy. He wanted to know who I was, and I told him, "I'm the new ranger."

He swore, and he says, "We finally got rid of that other so-and-so." Then he says, "Have you eaten?" I said no, and he said, "Well, I'm not going to talk until I have something to eat. Come over to the cook shack and we'll have something to eat."

So we went over, and I tried to talk to him. Bert's sheep were already on the allotment, and I wanted to get a count. I wanted to arrange with him–in case he wanted to be there–where and when would be the best time to count, but I could see by his attitude that I wasn't going to make any headway. I got up to go, and I asked him what I owed him for the meal, and he said, "You don't owe me anything." Well, I shouldn't have done it, but I had a dollar in my pocket (we had cartwheels then) and I just took it out and laid it on the table and started out. He was still sitting down, and I just got to the door when that dollar hit the door jamb! Man, if it'd hit me in the head, I don't know . . . . He threw that, and it hit the door jamb, and I went on out.

(Later on I found out that I could work a lot better with the herder or with the camp tender in counting Bert Robison's sheep. I counted his sheep time and time again, and he never knew anything about it. [laughter] It was far easier doing it that way than trying to work through a permittee, but I didn't know it then.)

Alfred Uhalde was a Basque who had a ranch and ran sheep. I believe Alfred was college-educated, and he was a prince of a guy as it turned out afterwards. I wanted to see him for the same reason I went to see Bert–to get a count on his sheep. While at a service station I saw this fellow standing there getting gas, and somebody called him Alfred. I asked, "Is that Alfred Uhalde?" They said yes, so I went over and introduced myself and told him that if he could come over to the office sometime I'd like to talk to him about counting his sheep and about his permit. He says, "I'll never set foot in that office as long as I live. I'm through with the Forest Service."

I asked Johnny Herbert, "What in the world's going on? Twice now I've tried to contact sheep permittees, and all I got was friction."

"Well," he says, "it's a bad situation. We're going to have to really work if we're going to get this thing solved."

According to the story that Bert and others later told me (this is all secondhand information, now) sometime before I got there the Nevada Forest supervisor was Alonzo Briggs–the same Briggs that took my place at Kamas–and the ranger was Foyer Olson. When Foyer had been on the district a fair length of time, according to the story, one day a Basque sheepherder of Bert's was coming down from his sheep camp. Apparently the herder was tired of eating mutton, and he saw a deer and shot it. He dressed it out and brought it down to the ranch where Bert had a big cooler where he hung his meat, and he took this deer and hung it up in there. I don't know why Briggs and Olson thought it was there, but they went into Bert's cooler and saw the deer . . . which they should never have done; they should never have entered his property. They went back and told the game warden, and the game warden came out, and when Bert came back to the ranch that night, the game warden was there and arrested him.

Well, poor Bert! [laughter] He didn't know anything about the darn deer. He didn't even know it was in his cooler. Bert was kind of a rough, blustery sort of individual, and I guess he blew up, which didn't help any. The warden finally took him over, and Bert went in the locker and looked, and sure enough the deer was there. Bert really didn't have anything to do with it; and this Basque, he didn't know there was a season on deer or anything else, see. When you come right down to it, nobody was at fault, but it really upset Bert. I don't know whether it was that single incident that affected all the permittees, but . . . .

We didn't have too many cattle permittees, but most were bitter towards the Forest Service, and *all* the sheep permittees were bitter towards us. Three or four years passed before we finally eliminated the bad feelings between the Forest Service and the livestock people. Eventually, Bert Robison turned out to be one of my best permittees, and I probably got along with him better than with any other permittee I had. One thing about Bert: he was hard to get to agree to something, but once he agreed, you didn't have to write it down in black and white–that was it. He'd bend over backwards to see that what he'd agreed to was done.

Rosendo, the sheepherder   I had to work with somebody on the range problems–if I couldn't work with the permittees, then I worked with the herders or the camp tenders. Most of the sheep camp tenders were easy to work with, and by working with them I gradually showed the owners that what I was trying to do paid off, because as they began using the range better, they brought out more lambs and they brought out bigger lambs.

Bert Robison had a herder by the name of Rosendo. Rosendo it was; that was his name. In fact I knew none of those herders' first names, and they didn't know my first *or* last name—I was just The Ranger. [laughter] Everybody thought Rosendo was a Basque, but according to him he was French. He was born in the Pyrenees, but for a Basque he'd speak good English and I could really work with him. He was an excellent herder, and Bert used to give him all the twin lambs. Of course, the twin lambs are always smaller than the singles, but by using his range right Rosendo was bringing out twin lambs that were as heavy or heavier than the singles that some of the other bands were bringing out. And it wasn't just my working with him that raised his lamb production: Rosendo was an exceptional herder. I could go visit him and say, "Rosendo, I'm going to be back in here two weeks from today. Where can I find you?" He'd tell me, and he'd be there. He knew for a week or two in advance *exactly* where he was going to have his sheep each day.

We were trying to get once-over grazing, and most of that range—even the steep country—could stand it. If they open-herded there would be very little trampling damage in once-over grazing, and most of the allotments, on most years, could adequately carry their sheep. But where you didn't use *all* of your allotment—and you went over a piece of range two or three times in some of that steeper country, like the head of Cleve Creek—the sheep did a lot of damage. Establishing a consistent policy was difficult, however, because herders changed quite often. There was a few like Rosendo (who worked for Bert for a number of years and finally died working for Bert), but on most of the bands you often had new herders every year.

To look at some of the range when I came, if they had been trying to practice once-over grazing, they weren't accomplishing it too well. Some of the range wasn't in very good shape, and some of the herders, when I'd try to talk them into once-over grazing, really didn't know what it was. At that time herders were either Basque or Mexican; most of them were Basque, and open herding was always a problem with new Basque herders, because apparently in the Pyrenees they close-herded. Of course, their bands weren't near as big in the Pyrenees—probably just several hundred head of sheep was a band there. And for some reason, on the Ely district they were afraid that they were going to lose their sheep if they open-herded. In open herding you *did* lose some—not to the extent that they were gone forever, but they would get over onto another allotment and mix with another band.

Every fall when they brought their sheep out, the owners would cut out the various strays and then they would notify the ranchers who owned them to come and get them. But when the owners came to get their strays, none of the ewes ever had their lambs with them. The herders lived on these stray lambs, and I wouldn't accuse anyone, but I'm sure that if the rancher wanted a couple of dressed lambs down on the ranch to feed his hay hands, that *they* were probably stray lambs, too. Of course, everybody did it, so it was tit for tat, and it was just an accepted thing.

A lot of times I'd camp with a herder. Sometimes I've camped a couple of nights with a herder, and that's one thing they *really* appreciated: if you spent a night in a sheepherder's camp, it wasn't long until every sheepherder in the country knew it. They liked that! You were a good ranger, boy, if you'd spend the night in a herder's camp! [laughter] Sometimes you could sleep in the herder's tepee, but most of the time you couldn't; you'd have to roll out outside someplace, because he'd have salt and all his grub and everything in his tepee. But I enjoyed spending a night, even with some of those Basques that couldn't speak too good English, and you bet they enjoyed your company!

Sometimes you'd run onto a herder who was short of grub if the camp tender hadn't delivered it in time, but I always carried quite a surplus. Of course, it'd be canned stuff–stuff that wouldn't perish–and I'd go into my outfit, give him some, and forget about it. One time Rosendo ran out. He was down to the nubbins, so I gave him some grub, and when I got back to Ely I called the ranch and got hold of the foreman. As had happened, a member of the camp tender's family had died, and he'd taken off to Oregon or Washington or someplace and been delayed getting back. So when I called the ranch, the foreman says, "We'll get somebody up there. If I have to take it myself, we'll get grub to him."

Well, maybe a week or two afterwards I was back up, and Rosendo was camped alongside the road. He was ready to move when they brought him his grub, but they had doubled up on it and he couldn't pack it all . . . well, he could have made two trips, but old Rosendo *wasn't* going to make two trips. [laughter] He was packing up his pack mules when I got there, and he picked up a can of peaches, and says, "Ranger, do you like peaches?"

I never wanted to take anything from him. If I was in the camp and ate with him, I would eat mutton or lamb, but I wouldn't take grub. So I said, "No, Rosendo, I don't need any peaches."

Down the mountain he'd throw it! And, "Do you like carrots?"

"No." . . . I don't know how many cans of stuff he threw down the mountain.

Politics on the range Occasionally, ranchers who tried to exert political pressure could make it difficult for you to do your job. On the Ely district a sheep permittee named Dan Clark was awful hard on the range. He was using his allotment a lot harder than he should by grazing too many animals on it and by staying on it too long; he was trespassing, and we trespassed him several times. Finally in 1949 Johnny Herbert and I talked it over and decided that we had to do something. Johnny suggested that we ask Clark to take non-use for five years for 10 percent of his permitted sheep. (This meant he would not turn 10 percent of his permitted livestock onto the Forest. In other words, we would reduce his permit by 10 percent, and then at the end of five years the 10 percent would be restored to him.)

Well, I tried to get hold of Dan Clark. I called his house and left word with his wife and family for him to call me, but he didn't respond. It was in the fall and the sheep had just come off the Forest, and he was shipping his lambs from the stockyards there at Ely, so I thought, "Well, I know I can catch him there." I went down . . . and I probably picked a bad time, because there were a lot of other sheepmen around. I went over and spoke to the bunch–I knew everybody–and I told Dan there was something I would like to talk over with him, and I asked him if we could walk off a little ways and talk. So he came with me, and I told him exactly what the Forest supervisor and I had decided on. "No," he said, "I'm not going to take non-use."

I did everything to encourage him, but he absolutely refused to take non-use. Then when we came back to the group he started laughing, and he turned to the other ranchers. He said, "You know what that ranger is trying to do? He's trying to talk me into taking 10 percent non-use on that Success Summit allotment." He was making quite a joke of it.

Well, it didn't bother me too much, and I just walked off and went on back to the office. The supervisor asked me, "Did you get your non-use?" And I told him no. I told him what had taken place, and so he says, "OK, we're taking a 10 percent *reduction*." And that was permanent, you see—that 10 percent reduction was permanent. So I went back to my office and wrote up a letter to Clark and told him inasmuch as he refused to take non-use for 10 percent, we were taking a 10 percent reduction on his permit . . . period!

It went on for a while, and he finally came in one day and said, "I thought it over, and I decided I'll go for the 10 percent non-use."

I told him, "It's too late now. We're taking a 10 percent reduction." He really got teed off!

In the Ely district there were a number of ranchers whose families had come from Utah right shortly after Salt Lake Valley was settled. They settled Baker, and the Snake Valley was practically all Utah people; Spring Valley was heavy with Utah people. Over on the west side of the district—White River, Lund, Preston—they were all originally Utah people. So it was a big early group of Utah people that came in there, and most of the ranchers in the district were descendants of these early pioneer families. Dan Clark's people were originally from Utah, and he and his wife knew a Utah senator, so Clark had his wife write to the senator. They told him about how the Forest Service was persecuting them, and how the ranger wasn't even able to count sheep, and a whole bunch of other stuff. The senator didn't know what was going on, so he takes it over to the secretary of agriculture, who sent it over to the chief of the Forest Service, who fired it back to the regional office in Ogden. And the regional forester fired it down to the supervisor of the Nevada National Forest. The supervisor brings it in and lays it on my desk and says, "Archie, would you answer this?" [laughter]

I dug out all the violations that we had against Clark, and there were a lot of them—trespasses and excess numbers and staying on when the season was over, which, of course, was a trespass. And I told them exactly how the 10 percent reduction in his preference came about: that we gave him an opportunity to take 10 percent non-use and he refused, and the whole thing. I had it typed up and took it in and had the supervisor look it over and OK it. He sent it into the regional office, and they sent it to the chief of the Forest Service, who sent it over the secretary of agriculture. [laughter] And eventually it wound up on the senator's desk. Well, that was the last we heard of it.

Deception on Success Summit

In the letter that Mrs. Clark wrote to the senator, she had mentioned my inability to count sheep. This is what happened: Up in Mosier Canyon, a part of Dan Clark's allotment, there were two water developments that needed a little work, and I was trying to get them in shape for when the sheep came on. I had talked to Dan a few days before; his sheep were over on the other side of Steptoe Valley, and I asked him when he was coming on. He says, "Well, I'm coming on opening day."

Four or five days before opening day I was up there working on the water troughs. I got through pretty late in the afternoon, and when I came down I saw where a band of sheep had gone up the canyon. I knew whose they were. I went up the canyon a little way afoot, and I could hear the sheep up above, so I went and got ahold of the herder and told him that Dan had told me that the sheep were coming on the Forest on opening day. "No," he said, "Dan came out this morning and told me to bring the sheep on now."

I said, "Well, I want to count them." So he rounded them up.

The best time to count sheep is first thing in the morning, just when they're breaking the bed ground. The second best is to count them off of water; the third best is to count them after they have finished shading up and started moving out. And of course, the worst time is when you have to round them up and count them. But I wanted to get a count, and I discovered that he was two hundred and some head over. (I don't think that I was off one way or another by more than ten or fifteen head.)

As soon as I got back to town I called Clark. He wasn't home, but his wife was there, and I told her to tell Dan that he was in trouble. Not only had he come on early, but he was also more than two hundred head over his permit. I told Dan's wife to ask him to come into the office. Dan came in, and he told me, "Your count was completely off. You couldn't count sheep out like that. Your count was way off."

I said, "OK. Fine. You pick the place early in the morning where we can count them off the bed ground, and I'll be there."

He said, "On Success Summit."

When I got out there he had two men from BLM with him, and he turned to me and said, "I don't trust your count. I want these BLM men to count, too." It was a good place to count, and we didn't have any trouble, and it came right out on the allotted number for his permit. So I told Dan, "Well, I goofed. Some way or another I got two hundred head over." So, that was that.

Dan had another herder with the sheep, a Mexican by the name of Garcia who herded for him that summer and then left, and I didn't see him for a couple of years. Finally one day I was driving up Timber Creek and I saw a sheep wagon down by the creek with a fellow standing out in front of it. I thought, "That looks like Garcia." When he saw the pickup he waved to me, and I went down. Sure enough, it was Garcia. It was about noon, and he wanted me to come in and have a bite to eat. We got to talking and he started laughing. I asked him, What's so funny?"

He said, "You remember that morning that you were counting sheep for Dan up there on Success Summit?"

And I said, "Yes."

"Well," he said, "*your* count was right. He came up there early and took two hundred head or so sheep out of the herd and drove them over the top of the ridge, down to the other side." While we had been standing there talking after we'd counted the sheep, one lone sheep came back over the ridge and into the herd, but I never connected it with what Dan had done, and nobody else did . . . except Dan, of course, and his herder.

The next time I saw Dan I jumped on him and told him what a rotten deal that was. He laughed and said, "I sure pulled one on you that time!"

I said, "Well, you did, but I still don't think it's much of a deal–that is no way for a permittee to act towards a forest ranger." Anyway, that's what his wife was referring to in her letter to the senator–miscounting two hundred-and-some head in his band of sheep. [laughter]

A mile-wide sheep trail   On the Ely district the stockyards were just out of town on the railroad. Some permittees were trailing bands of sheep in to the stockyards from close to forty miles away on one trail that started at the north boundary of Siegel Basin. The first band that went on the Forest would go on south of Schellbourne Pass, and then the trail ran along the ridge of the Schell Creek Range out to Duck Creek. In the Schell Creek Range it more or less followed the ridge, but on top of the ridge there isn't much feed, so the herders fed their sheep down the sides. Of course, by doing this they were taking feed that the permitted livestock were supposed to have been feeding on in the heads of those basins, but weren't getting.

The trail came off the Forest out of the narrows there in Duck Creek and crossed BLM land on out past McGill to the stockyards east of Ely. About ten bands of sheep had made and used that one trail, both coming in and then going back out after the lambs were shipped. Bert Robison alone had five bands. There were generally around a thousand head of ewes to a band; that's what they tried to run to. Of course, with the ewes and lambs . . . you'd have better than a thousand lambs in a band, because you'd have a lot of twins in there. Bert always had one of what you called twin bands, made up entirely of twins. So in it there was a thousand head of ewes and *two* thousand head of lambs. And that probably didn't include all his twin lambs, because in his other band he had twins, too.

The first band over the trail always got along pretty good, but when the second band came along a lot of the feed was gone, so the herder would move the band out a little bit further. Each succeeding band either spread out or moved to one side or the other until this trail was a mile wide in places. This was causing *a lot* of damage. The only solution to the problem was to get the owners to truck their lambs out. The Forest Service had worked on this a little before I got there, but they hadn't had much success. We didn't make much headway after I arrived, either. Finally about 1951 a fellow named Cliff came out from the regional office's Range Management Division, and we told him about our trailing problem and took him out to show him the trail. He said, "You got to do something about it. We can't put up with this."

Bert Robison ran his five bands of sheep over in Spring Valley, and he was kind of a leader in the community among the sheepmen. He was the biggest permittee, and I had been working on him, but I didn't think I was making any headway. One day I was in the office and Bert came in. He was kind of a gruff guy. Just out of a clear sky, he says, "When are we going to build that loading corral?"

And I says, "Any time you want. You got a spot picked out for it?"

He said, "Yes, I got a spot picked out for it. Come out and get in the Cadillac, and I'll take you out and show it to you." He took me up to a canyon which was centrally located—he couldn't have picked a better spot—but there was no road into it. We had to leave the car and hike about half a mile up the canyon. He showed me the area, and he says, "You furnish the material and I'll build it."

I said, "OK." I was glad he would take care of it, because I didn't have any idea what number of corrals he wanted and where the cutting chute and loading chute would be and all that. So I told him, "You make out a list of all the material that you need and you get it to me, and I'll see if I can get a road punched out to this site. Then I'll blade the area off where you're going to have your corral so you don't have a lot of brush interfering with working the sheep."

So he says, "OK," and I went back and told the supervisor.

Johnny said, "I'm going to get ahold of the road crew and get them over there right away." They had to load their tractor on a trailer to get up there, but the next day they were up there. The foreman was a darn good tractor man, and he could blade a smooth grade. Bert had told me, "You get it plenty wide, because those big stock trucks have a hard time if the road's too narrow." So we made it wide, and we made a big turnaround up in front of the corral so they could turn the trucks around. In the next day or two he brought me the list of material needed to construct the pens, and I went right down to the lumberyard and got everything he needed. (He was going to furnish the cedar posts.) I hauled the stuff out, called Bert and told him, "Everything's on the ground. When you going to get started?"

He said, "I'm going to start Monday morning."

I didn't want to get out there right when they started–thought I'd wait until about noon or so–and when I got there they were going right to town. In about two days they had the whole thing built. Afterwards, whenever they shipped lambs I would go out to that loading corral, and old Bert would put me to work just like I was a sheepherder. He worked the socks off me and I ate a lot of dirt, too. [laughter]

Bert found that he could load his sheep on trucks at the corral in the afternoon, and by truckers driving all night he could have his sheep to market in California the next morning. If he shipped by rail from Ely the stockcars first had to be pulled clear up to the main line at Elko, where they would wait on a siding until they could be hooked onto a train, and then taken maybe to San Francisco. And if they wanted to sell them in Utah, sometimes it would be three or four days and longer before those lambs got to market, and they lost a lot of weight in the process.

After that first fall, you couldn't have talked Bert out of trucking if you'd wanted to. He was sold on it. Although that corral wasn't on Bert's allotment, the understanding when he built it was that he'd have first choice of it in the fall, but anybody else that wanted to ship from there could ship. Of course, the minute the other ranchers found out that it was such a good deal for Bert, then everybody went to it. Dan Clark continued to ship by railroad occasionally, but by the time I left the district everybody else trucked their lambs out.

Once Bert began trucking out his lambs there was quite a bit of improvement in the condition of the trail. The one place I noticed it most was in a fork of Duck Creek called the North Fork. Up in the head of North Fork there's quite a big basin, and a lot of the bands had been coming through .... You see, there was no place to graze them around Ely. They had had to cut the lambs out and get out as quickly as they could, because there was no range at all around the stockyards. Because of that they'd often held a band or two in the head of North Fork until they were ready to load them, and when I first came to Ely North Fork was pretty badly overgrazed; it looked bad.

We didn't even know until some years later that there was a lot of bitterbrush growing in this area–the heavy sheep use had kept it cropped so close to the ground that you never noticed it. But after they took the sheep off the driveway and this country could be protected, the bitterbrush started growing just like everything. At one time the trailing area was probably an excellent winter deer range, too. All that stuff started coming back, and if you'd get up high enough there were a lot of deer wintering in there. It started going back to a winter deer range.

Balancing sheep with cattle

Getting stockmen to close gates behind them could be difficult. Most of the Forest boundaries were fenced, and it was a stockman's responsibility to get up there when he turned his cattle out in the spring to see that the gates were closed. As much as I could I checked to see that they were, but invariably someone would leave a gate open, and you'd either have a bunch of cattle on their allotment too soon or you'd have a bunch of cattle up on some sheep allotment. And then you're in trouble. Of course, the stockman would always blame the open gate on somebody else.

If there was a problem between permittees you generally tried to straighten it out. I remember one time Bert Robison got a bunch of cattle up on a sheep allotment, and I could tell by their cow patties and tracks that they'd been there for quite a while. Of course I trespassed them, and he got them off. But when fall came and I told the sheep permittee that he had to get his sheep off his allotment by such-and-such a date, he said, "Well, how about all of that feed that Bert Robison fed off of me? I'm going to go over and take the rest of my time on his allotment." Well, you can't do that without getting yourself into an awful lot of trouble . . . but it isn't really right for a permittee to pay good money out for feed and then have a bunch of cows come in and eat part of it up on him.

We had sheep on one allotment, and it wasn't handling the sheep too well—there was too much use taken of it. So I talked to the permittee and asked him to ride the allotment with me. We rode and he agreed with me. He said, "It's been grazed too hard. How about changing to cattle?" They used to have a rule of thumb that five sheep-months on a range equalled one cow-month. It was a figure that the Forest Service came up with some time in years past, but I never believed in it, and I still don't believe in it. So I told the permittee, "Well, I don't know how many cattle we can allow on there. You set the date, and you and I will ride the allotment and figure out about what area the cattle are going to feed."

Way up on the high slopes and up on the ridges where sheep would go, the cattle wouldn't. You'd have to figure that it was your stream bottoms and, depending on the slope, the lower half or at the very best two-thirds of the slope that the cattle would graze. And, of course, the type of forage that cattle consume is different too: cattle eat a lot more grass, and sheep eat more forbs, which they prefer. Sheep will generally leave grass and feed on the forbs and brush if they have the opportunity.

The permittee and I rode his allotment, and we decided what it'd carry in number of cattle per season. We arrived at a number that was agreeable to me and it was agreeable to him, and so he started running cattle. The agreement was that he'd run cattle for five years and then go back and run sheep for five years; then he'd go back to cattle again—just keep rotating back and forth. By doing that, his allotment would carry the allotted sheep and also carry the number of cattle that we'd agreed on, which I feel to this day was a very excellent arrangement. But there were a lot of cattle or sheep outfits that weren't that flexible—they didn't have the balance between sheep and cattle to be able to change back and forth like that.

Reducing the deer population   Deer overpopulation was a big problem on the Ely district when I took over in 1947. Duck Creek and Murphy's Wash, especially, were way overstocked with deer, largely because of the unrealistic hunting regulations. Game administration was handled by Nevada counties rather than by the state, and other than the game wardens there was very little administration by the state Fish and Game Department. The counties set up the hunting seasons and they set up the numbers, and this recommendation was sent in to the state game department, which approved what a county wanted; and White Pine County was dead set against doe hunting. Of course, the only way you can reduce a herd is to reduce the does. Shooting off the bucks don't do any good. If you shot every four-point buck on the range, you still wouldn't influence the deer population other than by the number of bucks that you killed. You wouldn't slow the population growth.

Since the county wouldn't allow any killing of does, the deer population had gradually built up and built up. On much of the district summer range was unlimited, but winter range was pretty short and the deer suffered on it; that's where some of them starved to death. The winter range was the limiting factor on the growth and development of an animal. A fawn might live through the summer in good shape–the doe would give plenty of milk–but by the time he went to the winter range, he was about ready to be weaned; and even if he wasn't, the doe would start losing weight due to the limited feed and pretty soon she would quit giving milk. That fawn would gain but very little during the winter months, and, in fact, he would become stunted.

(We had found stunted growth among deer down on the Middle Fork of the Salmon, too, where the winter range was very limited. It got so bad there that hunters would kill mature four-point bucks that wouldn't go over 135 to 140 pounds, when normally a mature buck should go up to 200 pounds. The does finally started producing fewer and fewer fawns, but even after we'd solved the problem and had adequate winter and summer ranges for the remaining deer population, the does that had been born and grown up through these depression years never did regain the full ability to produce fawns.)

Before I transferred to the Ely district the Forest Service had sent out a fellow by the name of Doc Aldus and a younger man by the name of Les Robinette. They were biologists for the federal Fish and Wildlife

Service, and they had set up a number of continuing studies, including browse transects that went through bitterbrush stands. Two limbs on each bush were tagged, and in the fall before the deer went on the winter range you measured and recorded the current growth on every twig on that tagged limb. Then in the late spring, after all the deer had left the winter range, you went out and measured what was left. From that you could determine the utilization. Well, we were getting 100 percent, and sometimes they were going back a little on the old growth! Some of those bushes in the spring looked like you'd taken a shears and rounded them off, just like you'd trim a shrub around your home.

Another study included photo transects on which we had to take pictures of the browse. They'd have a certain bush that was tagged, and a photo point for it. You'd take your picture from *exactly* the same spot every time, and you could see just exactly what was happening to this one particular bitterbrush. We also conducted a pre-hunt sex count and a post-hunt sex count of the deer, and in the early summer we would conduct a doe-to-fawn ratio count, which gave us the number of fawns per hundred does. The following spring we would make another doe-to-fawn ratio count, and this would tell us how many fawns survived the winter. This number was always less than the early summer count, and sometimes the loss of fawns during the winter was very heavy. (Some of this fawn loss would be to cougar kills, as we had a lot of cougar at that time, and the cougar also killed a lot of adult deer.)

We conducted regular deer population counts in the spring on the low ground around Duck Creek and over in Spring Valley and on Murphy's Wash. The alluvial fans that came out into the valley were covered with cheatgrass, and in the spring that cheatgrass would be just as luxuriant as could be–it was probably the only time of year that deer would really go for grass. They would gather in the mornings on these alluvial fans to feed on the cheatgrass, and if you got out at daylight to count them there, you could sometimes count a hundred or more in a herd. It wasn't a complete count of the deer on the district, and there was no intent of it even indicating that; but we'd count one year, and we'd count exactly the same area at the same period of cheatgrass readiness the next year, and from this we could tell whether our herds were going up or going down. Each year the numbers were always just a little bit higher, a little bit higher. Then, after we started getting control of the deer problem, they started going back down.

Archie Murchie standing by the Success Summit Game Studies Exclosure fence on the Ely district, the site of an artifical revegetation experiment. Archie is standing on the sheep-proof side of the exclosure fence, while the other side is elk-proof. Notice how deer and elk have grazed the vegetation in the sheep-proof area.

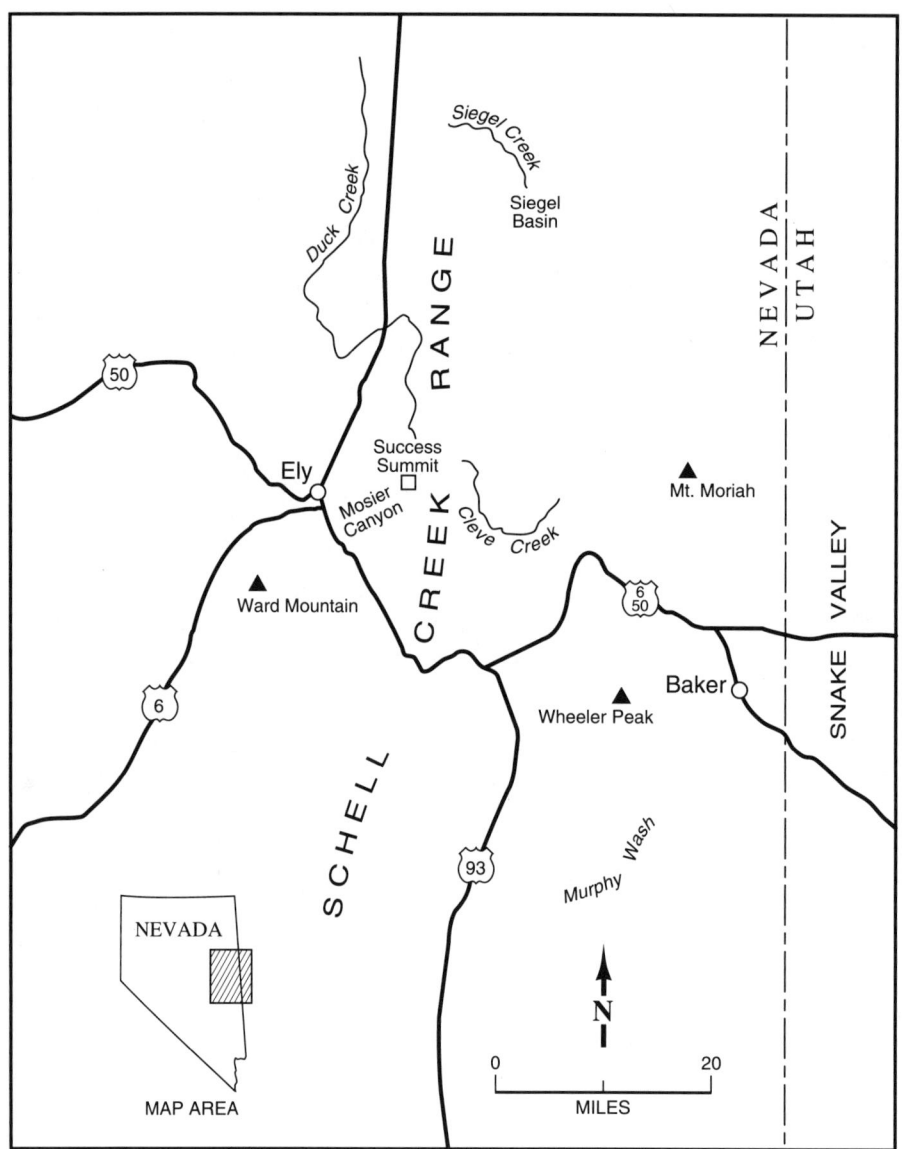

Map 4. The Ely district of the old Nevada (since 1957, Humboldt) National Forest was Archie Murchie's last ranger appointment. He served there from 1947 through 1959.

These spring deer counts had started before I arrived on the district. When I took over, I wanted to add a civilian sportsman to the group that did the counting. This was not easy to accomplish. The White Pine County Sportsmen's Association had an annual spring meeting, and in 1948 several of us attended and had to talk like everything to get them to send a representative. Even after we showed that first representative the conditions out on the range, he was reluctant to agree with us concerning the effects of deer overpopulation. When we got through with the counts we had a meeting of the counters and the Sportsmen and the two men from Fish and Wildlife, and we talked over what we had seen. The representative then agreed there were too many deer, but he didn't have much influence on the rest of the Sportsmen. The next year they sent a representative named Suzy Simes, who worked for Kennecott Copper.* After we'd had our meeting, he says, "I agree with you. I think that there's too many deer, but I can't go back and tell the Sportsmen that. They'd hang me if I went back and told them that," and he didn't, either.

The next year we had an up-and-coming young fellow by the name of Shirley Robison that went with us on our spring count. When he came back he laid the law down to the Sportsmen: they were going to have to do something to control the deer herds. I think that was probably the start of reducing the herds, because that fall the Sportsmen recommended a doe hunt. On the resulting permit you could kill a buck, and you could kill one doe. Well, a lot of the Sportsmen still wouldn't kill a doe, but there was a lot of people working for Kennecott and around that wanted meat, and they did take off a lot of does. Finally we got the size of the herds down, and then they went to permits on which you could take either a buck *or* a doe, but not both.

When I left the Ely district in 1959 permits were still being issued for either sex, but they soon restricted the hunt to straight bucks. Yet now, even after they've had only buck permits for twenty-five years or better, deer overpopulation hasn't occurred again, so it wasn't just a problem of

---

* Kennecott Copper Corporation operations dominated the economy of White Pine County. In the 1940s and 1950s Kennecott operated a large open-pit mine in Ruth, a railroad yard and hospital in East Ely, and an ore mill and related facilities in the company town of McGill. Kennecott's political influence was considerable, but apparently it was not exercised indiscriminately.

too many does. There were things about managing deer that even the biologists didn't understand.

On the Ely district, there was such an abundance of summer range that even when we had our overpopulation of deer they had little effect on it. Most of the winter range was on fairly level terrain, down in the lower elevations just off the valley bottoms where there was predominantly bitterbrush or cliff rose. But mixed with bitterbrush there was always pinyon and juniper and mahogany. The deer really high-lined the mahogany, but the most damaging effect of this over-browsing was that there was absolutely no reproduction of bitterbrush. Well, I won't say there was *no* reproduction–sometimes a bitterbrush'd start right in the center of a sagebrush and would be able to survive until it came up above the height of the sagebrush, but out in the open there was no regrowth. There was also a lot of bitterbrush–they only lived so long–that just died of old age. So on the winter ranges the bitterbrush was gradually decreasing, gradually decreasing, and I'm sure if the overgrazing had continued the deer actually would have eliminated bitterbrush from some of those winter ranges.

Fortunately, the deer were not competing directly with the cattle, and only to a certain extent did they compete with the sheep. Sheep are browse eaters–not to the extent that deer are, but they consume a fair amount of browse–but we never felt that there was a necessity to reduce sheep numbers or cattle numbers on the allotments due to the overgrazed winter deer range. Of course, any time that we tried to take action to reduce the deer numbers, the cattle and sheepmen supported it. They never attended any of our Sportsmen's meetings or anything like that, and they never came right out with a strong voice and said what they wanted done, but if we did anything, they supported us.

The White Pine County Sportsmen's Association had completely controlled deer hunting in that county. It wasn't just White Pine County; I think every county in the state of Nevada had the same kind of a setup–strictly county management of deer, and they just didn't want any interference from an outside agency. They didn't want anybody taking over any of their responsibility as they saw it. *They* were going to decide exactly what was going to be done and what wasn't going to be done, and they didn't care that the land that the deer were grazing was Forest

Service land. The Forest Service had no rights to come in and tell *them* what they could do or what they couldn't do with their deer.

Some of those Sportsmen's meetings . . . some of those got rough! I remember one meeting that Johnny Parker and I went to, and we almost walked out of it. They called us about everything they could lay their tongues to. But we stuck it out; we stayed in there. That was just an incident, and things finally got straightened out, and we got the deer population under control, but I don't believe that we'd have solved the game problem near as quick as we did if it hadn't been for the two Fish and Wildlife Service biologists, Les Robinette and Doc Aldus. They really helped us on the studies and through attending our meetings and expressing their views as to the excess number of deer and a possible solution. If it hadn't been for them, I doubt we'd have gone very far, because they were biologists, and we were just a bunch of Forest Service guys that didn't know anything about deer management according to the Sportsmen's Association. Actually, the Sportsmen were pretty rough on Aldus and Robinette, too, so I've got to take my hat off to those two men for really helping us.

The very severe winter of 1948-1949 had some effect on the deer population in White Pine County, but there were some herds, like the Murphy Wash herd, that only had to cross a valley to find adequate feed. Probably the herd that was hurt the worst was the Duck Creek herd. The ones that died were the big, old mature bucks and, of course, fawns. After breeding season, if the weather permitted, the bucks would move back up into the high country, and they were the ones that got caught in the snow. I went up afterwards in the spring and checked some of them out. You would see where a buck had finally made it down to a big juniper. (During the years of overpopulation the deer had eaten everything that they could reach on the junipers, so the trees were all high-lined.) These bucks'd make it to one of the trees, and that's as far as they would go. You could see where one had tramped a circle out underneath a tree, and there he'd lie–dead.

Elk poachers   I believe it was in 1935 that they brought a small herd of elk from Yellowstone Park and put them on the Ely district. There was two plantings up in what they call the Success Summit area in the head of Duck Creek. For quite a while the elk didn't do much, but finally the population started building up, and along about

1955 the county had a drawing for elk tags. I think the first time was for around forty tags, and they finally drew up to a maximum of sixty tags. That was for *either* sex, and why they ever did it, I don't know, because they killed everything. They killed cows and calves and yearlings and whatever they could see, because they wanted an elk. Boy, that was something if you went out there and shot an elk! I had a real problem both with the great number of tags that were issued and with the fact that they were for both sexes. The elk numbers diminished real fast, and it wasn't too many years until they had to eliminate the hunt entirely.

Probably as big a problem was poaching. There was *a lot* of elk that were poached on Forest Service land, but nobody was ever caught. One time I was riding up by Success Summit, and out on an open sagebrush hillside I saw a dead elk, a great big bull. I went over there, and found that all the poachers had taken was the elk's teeth–they'd knocked the teeth out and they'd left the rest. Well, you can get good money for elk teeth from a member of the Elks Lodge–the B.P.O.E. The Elks buy elk teeth and they hang them on a fob. Have you ever seen Elks with their elk tooth hanging on their watch chain? Maybe now they don't wear them, but years ago that was very common. You weren't an Elk unless you had an elk tooth hanging on your watch chain!

Once while riding I ran onto two sets of horse tracks. I followed them a little ways, and I got curious. The guy apparently had a packhorse–it wasn't anybody riding with him; you could tell by the tracks that it was a packhorse–and I just couldn't figure out why he'd be going that direction, unless he was a miner or something. I followed him, and we got into some pretty heavy timber, and there he'd stopped and he'd cut the legs off of this elk. There was four legs lying there. I followed him off the Forest, and I probably should have kept on, but I wasn't a game warden. The tracks were several days old, and it wouldn't have done any good to have followed him out, anyway. I would guess it was probably one of the people working for Kennecott. They poached deer, and there was supposedly a lot of cattle poached out of Duck Creek by workmen from Kennecott. Maybe I'm doing wrong in accusing them, because I don't have direct knowledge, just hearsay, but they'd be the closest people that would benefit from poaching the elk or the cattle.

Well, even with the poaching, the elk herd gradually increased and finally spread to Ward Mountain, and it spread south over into what they call the horse-camp, cow-camp country; and I have been told that they'd actually spread to Moriah and the Mount Wheeler area. Now, of course, there is an elk hunting season every year, and the state has planted elk in various places all over Nevada. I've seen elk range in Montana and Idaho and Wyoming, and to me that Ely area had some of the best elk range I've ever run onto. The biggest bull elk that I've ever seen, I saw there on the Ely district. Some biologists had thought that the Ely area wasn't really suitable for elk, but they're finding out now that either the range is better than they think, or the elk have adjusted to the range.

Controlling cougar and bobcats

We had a lot of cougar on the district, but they never bothered livestock–cattle or sheep–because their main diet is deer if deer are available. Of course, cougar will kill rabbits and smaller animals, but their primary diet is deer, and we had plenty of deer. During the time we were conducting all these deer studies in Duck Creek, one game biologist wondered what effect the cougar population had on the deer herd in Duck Creek. So they decided to hunt them out and kill every cougar in Duck Creek. They brought a hunter in by the name of Snap Palmer, who had a bunch of real good dogs that he had raised and trained. I used to go hunting with him every once in a while. It was fun cougar hunting with him, because he had some *awful* good dogs.

As I remember, Snap took something like seventeen cougar out of Duck Creek. One time he had five cougar hung up on the side of a log cabin at once. He probably took *all* the cougar out of Duck Creek, but like everything else with wildlife, when you do that you create a vacuum, and it was just a very short time until cougar started moving in from outside. So it had little effect that we could see on the deer population.

Cougar were no problem for the stockmen, but the sheepmen did have quite a problem with bobcat, especially in the spring with their young lambs. Bobcat killed a lot of young lambs. To combat this the federal Fish and Wildlife Service had a predatory animal control man who did trapping of coyotes, bobcat and cougar that were causing a problem for the ranchers. Also, 1080 poison was still being used, and this poison was used a lot, especially on coyotes.

Fish and Wildlife also used kind of a pipe that they drove into the ground, containing a .44 shell loaded with powder and cyanide, on top of which was a wad of cotton. The cotton was fastened to a trigger and a firing pin. They put a scent on the cotton wad, and when a coyote'd reach down and grab it, pull on it, it fired the .44 cartridge, and it'd shoot the cyanide right into the coyote's mouth. He was dead in just a short time. They called these things game-getters.

On the Ely district they also did an awful lot of trapping to control bobcats. Of course, if bobcats came into a 1080 station, they would die just the same as a coyote would, but for some reason members of the cat family could consume considerably more of that 1080 poison without dying than coyotes could. Members of the dog family were *very* susceptible to that 1080 poison. One cold morning late in the fall . . . . This predatory animal control man had bought an old horse from a rancher. He'd killed the horse the evening before, but he didn't dress it out or anything; he just cut it up into pieces. Then he shot the pieces full of 1080 poisoning and loaded them in the back of his pickup: he was going to poison coyotes with the meat. This morning he was ready to take off, and we were out there talking, and the blood from the poisoned horse meat had run out of the pickup bed and dripped down onto the bumper and froze. This rancher's dog came out and started licking it, and the trapper told him he'd better get that dog away from there, or it was going to kill him. The rancher said his dog only licked it a couple or three times before he called him off, but in a little while the dog was dead. It didn't take much of that 1080.

Years later, bobcat pelts were worth three hundred dollars, or maybe a good one would bring four hundred dollars. Incredible! I've driven up to the end of a road on the Ely district–going up into the Forest or close to the Forest–and found piles of bobcat carcasses. A trapper would run his trap line, and just take the bobcats, shoot them and throw them into his pickup, and when he got to the end of the road he'd dump the bodies out. Trappers only had to turn in what they called the scalp to collect their bounty. They'd cut off a piece of the scalp with the two ears attached, and that's what they turned in for the record of their kills. Sometimes there'd be eight, nine, ten bobcat piled up there in a heap at the end of a road. Think what their fur would have brought during the high price of bobcat hides!

Dynamiting beaver dams   We were bothered by beaver up at Cleve Creek. The creek and a fairly good road ran up a narrow canyon, and whenever the beaver would dam Cleve Creek the water would come right down our road and wash it out. When this happened I would notify the game warden. Sometimes the state trapper would already be in the area; if he wasn't, he was generally up around Elko. He'd come down, and he'd trap out all the beaver in Cleve Creek. Beavers would come down in early summer, but if he once trapped them out you weren't bothered with them until the next year at the same time.

One time they were bothering us and washing out the road, so I notified the game warden, who said, "I'll get the trapper out real soon." Well, he didn't come, and the road was impassable; you couldn't even walk it. So I told him *again;* I said, "Now, we can't maintain the road–nobody can go up and down it, and if you don't get him down here I'm going to blast the dam out."

"OK," he said, "I'll get him down."

Well, he didn't, and one day the Forest supervisor says, "Archie, why don't you go ahead and blast that dam out?" So I went up and blasted it out with a pole charge. A few fish got killed, and they drifted on down the stream. I had a trail crew camp just a little bit below the dam, and that evening when they went to get water for camp, here these dead fish laid. They knew that I'd blasted the dam, and they knew what had killed them and that they'd be good eating. So they gathered them up and were going to have them for supper, and I guess they were in the process of frying them when the game warden came up. He didn't arrest them, but he was sure teed off at me, and he went to the supervisor and told him he was going to arrest me. John told him that *he* was the one that told me to go up there and do it; so then he was going to arrest the supervisor. But he never did.

I think it was the next spring that we had the same problem, and I notified the game warden, and the trapper didn't come down. Finally one morning the game warden stopped at the office and said, "Archie, will you go up with me, and we can blow out that beaver dam? I don't have the powder and I don't know exactly how to do it, so I want you to get the powder, and any cost that you have, the state'll pay for it." So the next morning I had the powder and caps and fuse and everything all ready to go. The warden called me a little bit before eight o'clock and told me that he couldn't make it. He asked if I'd go out and get everything set up, and he says, "If I don't show up by such-and-such a time, just go ahead and blow it." I knew he wouldn't show up; I was *sure* he wouldn't show up. But I went out and got everything ready to go, and he wasn't there . . . so I blew it. After that, whenever we had a beaver problem I'd notify him and he'd say, "Archie, you go up and take care of it." Everything was all right. No more complaints about dead fish downstream.

Just about every spring we'd have to take the beaver dam out. I don't know where those beaver went after the blast. Now, whether I scared them so they went back up North Fork or not, I don't know, but they would never rebuild their dam that season. You only had to blast the dam just that one time.

Dealing with the mustang problem

What brought about the mustang problem on the Nevada Forest, and it was probably the same on all the Forests, was that ranchers would turn their horses out on what then was public domain and later became BLM land.

Then when they wanted some horses, they would just go out and round them up and bring them in. Some ranchers were good ranchers, and every spring they'd go out and round up all their horses, brand and castrate the colts, and turn them back out. But other ranchers weren't that energetic, and they wouldn't round up their colts. Pretty soon you had mature studs and mares running around unbranded, and these were the animals that eventually they called the mustang.

The Indians in the Ely area had horses with which they'd do the same thing. In fact, it was probably more common among the Indians than among ranchers not to round up their stock and castrate and brand. For example, an Indian by the name of Adams had a Forest permit for fifteen or sixteen head of horses that were running on Ward Mountain. For a while at the end of the grazing season he would faithfully round up his horses and bring them in, but finally he started getting on in years and he quit rounding them up. In the wintertime they didn't bother the Forest because they were running on BLM land, but in the summertime they were right back up on the Forest.

Adams finally decided that he wasn't going to pay grazing fees; that he didn't want a grazing permit. So we told him, "OK, get your horses off."

He said, "I can't. They're too wild. I disown them." There you had the start of another mustang band, and it built up to a fairly good size.

If you let them go, mustang can overgraze or damage a considerable area, probably more so than cattle. One of the big problems with them is that they are on the range twelve months of the year, and in the spring when the snow is melting they can do a lot of trampling damage. I have seen where mustangs have left tracks six inches deep in the mud on a hillside where they were grazing. They also are grazing the range a long time before it is ready to graze in the spring. The grass and other plants never get a start under this kind of use, and they soon die.

Mustang will trail long distances to water. They'll water, and they'll hang around water for maybe an hour or two; then they will drink again, and then they'll move out. For some distance around those water holes they clean up just about all the vegetation there is–a lot more so than cattle will do. After a big drink like that, mustang can go for two or three days without water. They like to graze the high elevation where it's cooler, and during the day they like to get on a ridge top where the wind's blowing one way or the other and it's easier for them to fight flies–sometimes if it's cool enough, there won't be any flies. Then on towards evening they go down and start grazing again, or go in to water.

You'd have a fairly small spring where deer would come in and drink, and maybe it was big enough that there'd be water running down out of it and a few head of cattle or sheep could come in and water. But mustangs have a bad habit when it comes to a little spring: if they can't get enough water to drink, they start pawing, and pretty soon they have that spring all tore up and so muddy that how they can ever drink that water, I don't know. Cattle wouldn't drink it, and I don't think sheep would drink it, and I think even deer might have a problem drinking it. Mustangs can really ruin a spring. This is one of the reasons why ranchers were very much in favor of closing orders. Some of them had horse permits in the Forest, but they never questioned it at all when we refused to issue a horse permit for a particular year.

In the twelve years that I was at Ely we had to have closing orders two different times to bring the mustang population down. A closing order was issued when mustangs in a certain area built up to the point where they were damaging the range. It wasn't that they were taking feed from sheep or feed from cattle, but that they were actually damaging the range–they were cropping it too short, and their trails in and out to water were causing erosion. When it got too extreme we'd go to the county commissioners and ask for a closing order for one year. They would issue one, and during that time we could go out and shoot any horse that we found on the Forest.

The closing order wouldn't go into effect for three or four months. We would notify all the livestock people, whether they ran on the Forest or not, that the order was going to go into effect on such-and-such a day, and that if they had any horses out in trespass on the Forest, to get them off. (If they didn't get them off before that closing order went into effect, they'd probably be shot along with the mustangs.) Notices were published in the paper for maybe four or five issues, and signs were put up on the road so that anybody that had a horse in trespass had adequate time to get out and round it up and bring it in. Most of them did that. The only two branded animals that I know of that were shot in a closing were a horse that got away from a hunter and went with the wild bunch, and a mule that a sheep permittee owned that had got into the wild bunch. The permittee had captured that mule a couple of times, but the first chance that she could get away, she was gone to this wild bunch. He'd actually told us, "If you run on to her, you shoot her." He didn't want to have anything more to do with her. And she was shot.

They issued closing orders and hunted mustangs on the Nevada National Forest a year or two before I got there, and during the twelve years I was there we hunted them twice. We hunted mustang over the entire Forest, but the two main areas where we had the biggest problem were Murphy Wash on the Baker district and the Success Summit area on the Ely district. On the last hunt there had also developed a bad mustang problem over in Siegel Basin on the Ely district.

The Army had checked out a Springfield .30-'06 rifle to the previous ranger, Foyer Olson. When I got the rifle, its stock was fastened together and wrapped with rawhide; Olson must have broken it. Also, someone had knocked the front sight blade out, and taken a penny and flattened it and put it in the blade slot and pinned it: that was the front sight. After I got the gun, I lost the improvised front sight and I did the same thing. I put it in a rest and started shooting, and just kept filing the penny down until I got it hitting right on target. But bar none, that was the most accurate rifle that I've ever shot in all my life.

The rifle was specifically for killing mustangs, and the district had acquired a lot of surplus Army ammunition for it; there were even some tracer bullets in the bunch. Why they ever acquired or even used that steel-jacketed Army ammunition to hunt mustangs.... You could have put half a dozen of those bullets through a mustang, and it wouldn't have stopped him. They should have used soft-nosed, expanding bullets. You shoot at a mustang for only one reason, and that's to bring that horse down as quick as you can–kill the horse as quick as you can, and have the horse suffer as little as possible. That steel Army ammunition was made to shoot at humans, and it would whistle right through a horse causing very little damage. If you use an expanding bullet, a lead bullet, it may not go all the way through the horse, but it'll create enough damage to kill the horse immediately, so you don't have any wounded horses walking around.

I would never think of using our Army ammunition to try to kill a horse. When I got to Ely we had a new supervisor who told me they had a thousand or better rounds of Army ammunition down in the basement. I told him what I thought of it, and he said, "Let's get rid of it." So we dumped the whole works; we didn't save any of it. (Why they had tracer bullets in the lot is more than I could understand.) After that we'd buy the heaviest load of soft-nose .30-'06 hunting ammunition that we could get, and that's what we used to hunt mustangs.

The way I approached it, hunting mustangs was pretty much incidental to my other work. If I was riding and came over a ridge and saw a bunch of mustangs a couple of ridges over, maybe the next day I'd go mustang hunting and shoot as many as I could out of that band. Oh, once in a great while I would make it a project. One time the ranger from one of the other districts came over and wanted to go mustang hunting with me, and we made that a two-day project. And one time the administrative assistant wanted to go, and we made that strictly a two-day mustang hunting trip. But most of it was pretty much incidental to my other work.

The first time we hunted mustang was probably 1950, and the next time was 1955. We were just about due for another hunt when I left the district in 1959, but I doubt very much if they ever hunted mustang again, because by 1955 there was beginning to be some talk against it–it wasn't very strong, but it was beginning to show up before our last hunt. The county commissioners were kind of reluctant to give us a closing order, and when they did they pretty much told us that this would be the last one that they would issue–that we would have to come up with some means of getting rid of the mustang other than just going out and shooting them.

We had an extraordinarily severe winter in 1948-1949 that was tough on mustangs, and there were quite a lot that perished. I've seen places where mustangs had finally worked their way down into a creek bottom, where they'd eaten the willows down to the size of your thumb, and the sagebrush the same way. They'd just be nubbins before they finally starved to death. If we ever get another real hard winter like we had in 1948-1949 and these people that are so much for the mustangs now finally see some that have starved to death . . . . They had eaten off all the hair on the manes and tails of each other; they'd eaten such coarse . . . I'll call it food, like sagebrush or limbs (they didn't eat pinyon too much, but they ate the limbs of juniper), and the sticks were so big (and maybe the mustangs were so weak that they couldn't chew adequately) that they passed through, and in the rectum where they had been trying to pass them, the sticks had gouged the colon so bad that there'd be blood. Before they died, there'd be these sticks pushing out of their rectums and you could see where blood had run onto the ground.

Now, if these people who are so strongly supporting mustangs want to see them in that condition when they do have a real tough winter . . . . And one is coming. I don't know how far apart they are, but we're bound to get them. Of course, maybe the mustang enthusiasts will put pressure on the government to spend millions of dollars to feed them hay; I don't know. But with these overpopulations of mustangs, some time down the line they're going to have to pay.

Winter disaster: the Forest Service responds

The extreme winter of 1948-1949 didn't do any lasting damage to the range. Actually, as far as the temperature went, it wasn't a bad winter—it was just the amount of snow we had and the wind. Due to the continual drifting and blowing of snow, which covered the vegetation, quite a few deer and mustang perished, and there was also quite a few cattle that died, and some sheep. The weather started getting bad, as I remember, around the early part of December in 1948.

In that country, the small operators around Preston and Lund brought their cattle in and fed, but the big ranchers all grazed their animals out. Most of the sheep herds went south down towards Pioche to winter on BLM land. A fellow by the name of Pete Johansen, who had a ranch over in Spring Valley, wintered his sheep around Cow Camp or Horse Camp, down in that pinyon-juniper-sage country. When these winds came, they came pretty much out of the southwest, and the south and southwest-facing slopes were blown bare. So Pete just broke his band up into little bunches of sixty or seventy head to a bunch, and had his herder put them on these snow-free slopes that were scattered around.

Normally you'd be bothered some with bobcats and coyote, but they were smart; they'd taken off for someplace else. He didn't lose any sheep, and he didn't buy a spear of hay. He wintered them out on the range, and everybody thought the guy was crazy, but he came out the best of any of them.

When conditions reached a certain point the ranchers contacted the Forest Service for help. They probably contacted the regional office first, and it came on down to us. After that we were able to take our Cats anyplace within the area–it didn't make any difference whether it was private land or BLM or where it was–to open up roads. We would try to punch a road out to a herd of cattle that were snowbound so the rancher could get feed out to them or get the cattle back to his ranch to feed them.

One time another ranger and I took a big Cat over to Lund. They had lots of hay, but we had to plow a road so the ranchers could get it in to their animals. They'd get a couple of loads of hay in and they'd say, "Well, that's enough for today," and they'd quit. The next morning their roads'd be snowed in again, blown in. We cleaned their roads out twice, and then we told them, "Now, this is the last time. You spend all day hauling your hay in, so that you can get enough to last you awhile, because we're not coming back. There's a demand for the Cat in a lot of other places."

Once I started out with a helper for Butte Valley with a Cat and a four-wheel-drive Jeep pickup. *Oh,* the wind was blowing! I started plowing snow at the junction of the highway and the road to Ruth. The pickup was right behind me, and when it came noon I told the fellow driving it to stop and have lunch, then catch up with me where I was plowing and take the Cat and I'd have lunch. Well, I kept on going, and he didn't come up and relieve me. It got on about three o'clock and I thought, "I'd better stop and see what's going on." I figured he must be right behind me pretty close, so I just shoved the snow off the side of the highway and parked the Cat and started walking back. I'd only walked a couple of hundred yards before where I'd plowed was practically full of snow. So I hurried back and cranked the Cat up before it cooled off, and turned around, and I had to plow all the way back–the same amount of snow as I plowed coming. Finally when I found my helper's pickup all you could see was the upper part of the cab. The hood was completely covered over. And worried—you never saw such a worried man in all your life!

I plowed him out; then I turned around and hooked onto him, drug him out, and we got as much of the snow off of the pickup as we could. I started back to the junction going to Ruth where we had begun. By pulling it, we finally got the pickup started, and we left the Cat parked there at the junction and went on in to Ely.

It was way after dark when we got into Ely. I called the Forest supervisor, and he says (brusquely), "What are you doing back in town?" I told him what had happened, and I said, "You're not going to get to Butte Valley unless you take a house trailer and a truck with plenty of grease and gas and oil, and just make an expedition of it. Let it snow in behind you if it wants to, but keep the units all together." So that's what they did clear in to Butte Valley. Bert Parish was the one that was in trouble, and they plowed his cattle out and got them into the ranch, but some of them had already died. There was quite a few more that died after they got them in and on hay.

Eventually the Air Force began flying baled alfalfa hay out from Fallon to feed the starving animals in the eastern part of the state. They used what they called Flying Boxcars with doors that opened in the back end. They'd carry, I guess, at least twenty-five tons of hay. Most times they'd fly right from Fallon to a particular rancher's livestock where the hay was supposed to be dropped. After they had dropped their load, they'd land at Ely to pick up a guide who would go with them to show the pilot where the next drop would be. Then they would fly back to Fallon, pick up another load and repeat the exercise. Sometimes they'd fly directly in to the Ely airport and stockpile the hay, but when they did that they couldn't carry as many tons, because they couldn't land with that heavy a load. They'd stockpile hay; then maybe there'd be a demand come in. They'd generally try to keep at least one of the Flying Boxcars on the runway there at Ely, and they'd load whatever hay the rancher wanted, take it out and drop it to the sheep or cattle.

All the drops were either at a ranch or on the BLM. I know a rancher just north of Ely where they dropped on his land, and he complained that they were dropping it too far from the ranch house. The next day they came over, kicked the bale out, and it came down and flattened a shed back of the house that his wife had her washing machine in. I don't think this was intentional. [laughter]

I went on a number of the flights. They'd circle the drop area while the pilot decided where he wanted to make his run, and while the plane was circling we'd shove maybe fifteen or twenty bales to the very back. Then the pilot would take the plane down real low on a long dive. Two Air Force men with long safety belts on them so they couldn't fall out the big, open door would kick the bales out, and the plane would go up on a pretty steep climb. If you were walking back towards the front of the plane when it came back up, it'd just drive you right to the floor! Your legs wouldn't hold you; you'd just go right to the floor! [laughter] Then they'd go aloft and circle, till you got enough hay moved to the back for another drop. Finally you'd be rid of all of it, and they'd fly back to Ely and let you off or put on another load and take off for someplace else.

When we really started handling hay, the cattlemen asked the Forest Service if we would keep track of the amount that was sent to various ranches. We had no way of weighing it, so we weighed up a few bales to see what they averaged, and then we just counted bales. Most often, these planes took off from Fallon and flew directly to the ranches, and the only count we could get on the number of bales dropped to a ranch was from the guides that went with the planes. They counted the number of bales that were loaded, and turned that over to us. When the airlift was finished the ranchers paid up their bills. There was no squabbling: there was no criticism or saying that they didn't get all the hay that they were charged for; and it was surprising that there was no problem, because some of the bills they had to pay to the hay farmers in Fallon were pretty good-sized bills. The price of hay went up *considerably* during this operation.

The Air Force would fly the hay in, but for some reason—I can't tell you why—they didn't drop hay where the sheep were located. But the Army came in with trucks, and a road was plowed out to the sheep, and the Army hauled the hay out. We must have had twenty-five Army trucks there, and when they would leave Ely going down to Horse Camp or Cattle Camp, a distance of sixty or seventy miles, they were completely out of communication. If they got part way down and got snowed in (which happened), there was no way they could get word out for help, so they asked the Forest Service if there was any way that we could set up radios so they could communicate.

Well, the radios had to be where they could be manned. You couldn't just set them up and forget about them; they had to be manned and maintained. There was a highway maintenance station at what they call Lake Valley, right where the road left the highway and went back into the foothills. We went out and set up a radio station there, to be manned by the maintenance man and his wife. It was close enough that when the trucks got snowbound one of the men could snowshoe to the maintenance station, then radio in to our set at Ely for help. We could then get word to the Army, and they could haul tractors out and plow them out. That happened a number of times.

The morning we decided that we'd set up the radio at the Lake Valley maintenance station it was a nice day—the wind wasn't blowing much. A young fellow who worked for the state Fish and Game happened to be over at the Forest office when I was ready to go, and he said, "Do you care if I go with you?"

I said, "No, come along. I enjoy company."

So we took off and never had any trouble, and we got to the maintenance station and started setting up the radio. We put up a big aerial, and there was a lead-in that came down from the aerial about a fourth of the distance from one end. We had to bring that into the building and hook it up to the radio. It took us quite a while to find material at the station to use for poles and get them set up. We finally got it set up, and we'd brought out lunch with us, but the maintenance man said, "Come on in and have dinner with us." So we had dinner, and we had a little bit more work left to do on the radio. Then we had to show the maintenance man and his wife how to operate it.

The wind had come up. It wasn't blowing too bad, but shortly after we started back to Ely we'd just gone a little ways before it started drifting snow. I had a four-wheel-drive Jeep pickup with chains to go all the way around if I needed them—and I eventually had to put them on. I probably should have turned around and gone back to the maintenance station, but I kind of hated to, and I kept on going. Finally, through the blowing snow I could see the back end of a transport pulled off the road. I stopped and went up, and there were two guys in it. They'd been there ever since the heavy wind started blowing, and boy, they were sure glad to see us! I asked one of them if there was anybody ahead, and he said that a car had passed him a while back.

We told him, "You stay here in the truck, and we'll go ahead and see if we can find them." Oh, it got tough. We bucked snow and bucked snow, and I *just about* turned around and went back, but I thought, "No, I'll go a little bit further," and finally we could see this car. It was snowed in. It was a husband, his wife and two small children, the youngest about three years old. They were from California and they were dressed in light clothes. Of course, the motor had blown full of snow; they couldn't run the motor, and they were just about froze to death.

We got them in the pickup, and this fellow from the game department had a *real* heavy down jacket on. He says, "I'll get in the back." So he got in the back, and we came on down, and we got the two truck drivers and put them in the back, too. I had a good down jacket, also, so I gave that to one of them, but the other poor guy didn't have anything. He just about froze to death coming back. We finally made it back to the maintenance station, and some other people were there when we got there. Eventually, I think there was fifteen people that night that were snowbound there. They made different beds and had blankets for the women and children. But the rest of us . . . well, they had a good heater in the living room, and they really stoked it up, and the rest of us slept on the floor. [laughter]

The next morning you couldn't go anyplace. The maintenance man told us, "The only thing I can do is crank up the plow. I'll start out, and everybody get behind me and stay right behind me. I'll take you out as far as I have to go to get out of the snow." Of course, when we got down to the transport, he had to plow it out. Then we went on up and got this couple's car plowed out and started back to Ely, with him still plowing in front of us. He had to plow to within sight of Ely before we got out of the snow. Connor Summit was entirely blown in.

That winter everything would be perfectly nice before noon, and the roads would be completely blown in that afternoon. A fellow told me that when you have conditions like that, snow comes down in snowflakes, but then when the wind blows, and the snow moves along the ground, it gradually turns to ice. When it turns to ice it's in little balls and it won't pack; and the least little wind, here it goes! It just rolls along the ground. It was discouraging–we'd plow the road, and about a hundred yards behind, it'd be rolling in just as fast as you plowed it out.

Every man on the Forest was involved in one way or another in assisting the public that winter. For this effort the Forest Service gave us special money, and it seemed like it was unlimited. Our road foreman and his crew were really kept busy, and we had to hire more men to build up the crew. Additional cat skinners were also transferred in from other Forests to help. Sometimes when we needed equipment we didn't have, Kennecott helped us. In fact, anything we wanted from Kennecott–I don't care whether it was a truck or a dozer or dozer blade or whatever it was–they never hesitated a minute to give it to us.

One thing that I have since wondered about: this was a local and state emergency, so why didn't somebody set up an overall organization? This was never done. The Forest Service functioned separately, and the highway department functioned separately, and I guess the county functioned separately. They could have saved a lot of money, and they could have done a far better job if there had been a coordinated effort.

The Humboldt and the Nevada Forests are combined

The old Nevada Forest was very small–not so much in acreage, but in personnel. There were only three rangers in the Ely area, and then the one ranger at Las Vegas. To run a Forest with a supervisor and a full office staff was fairly costly, and as time went on, wages went up and the cost of office people went up. This situation was common throughout the entire Forest Service, and the government began combining Forests to reduce costs. The decision was finally made in 1957 to combine the Nevada and the Humboldt and move the Nevada Forest supervisor, who at that time was Louie Dremolski, from Ely to Elko to administer the whole thing under the designation of Humboldt National Forest.

They were also doing away with a lot of the smaller ranger districts at that time. The Baker Ranger District was a very small one, but it had a house and a warehouse and complete headquarters. It didn't have electricity, however, so it had to have its own power plant, and it was costly to run that district, in comparison to others. They decided that since it was within easy working range of Ely, they would combine my district and the Baker district into one big district with me in charge, and then give me an assistant ranger . . . which worked out good; it worked out exceptionally good, both for me and for the Forest Service.

On the Ely district I had a small division–the Ward Mountain division–just south and a little bit west of Ely, which they split off to the White Pine district. That took one sheep allotment and one cattle allotment and one campground away from me, which didn't reduce the workload very much, because it was a very small area. What *did* happen was that I finally got a secretary. As the Ely district ranger, my headquarters were right there in the supervisor's office. (The White Pine district ranger's summer headquarters had been at the Ellison ranger station. In the wintertime he would also move in and share an office with me in the supervisor's office.) When they moved the supervisor out to Elko they left his secretary in Ely, which was really a big help. She did all my typing and filing and reading of the mail; and she was a real good girl, too. In fact I had two of them, and both were real good girls.

Although I now had clerical help, my workload had increased. I don't think that my days in the field diminished too much, but there was more working in the evenings and Saturdays and Sundays than before. However, we did have an excellent radio system, and there was a lot of district work that could be handled over the radio. I packed a radio with me wherever I went–whether with a pack outfit or pickup, I had it with me–and we had a good master set in the supervisor's office.

After they combined the two Forests they left me a sedan, so if I had any quick running around to do–like over to Baker or Spring Valley, visiting permittees and whatnot–I had this sedan. It cut down the district's operating costs, because you could operate it far cheaper than you could the pickup, and it was real nice to have that sedan. When I went out into the field, I still rode, but that district was an awful lot like the Wildhorse district in that it was very accessible by vehicle. There were only a few places that I couldn't reach in half a day's ride from the end of a road: on some of the southern part of the district I took a pack outfit, and it was also easier to take a packhorse when I rode the high country. With the packhorse I could hold to the high country and camp wherever evening overtook me.

We thwart a national park proposal The first attempt to create a national park in the Wheeler area was started about 1954 or 1955 by the editor of the *Ely Daily Times*. He wrote editorials about the possibility of a park and what it would bring to Ely, and published a lot of pictures of Wheeler Peak and the natural arch and other features that would lend themselves to acceptance by the Department of Interior and the National Park Service. But he didn't have very much luck; he didn't generate much public opinion. At that time the Forest Service didn't pay too much attention to it, and the idea died a natural death.

Then in 1959, the editor started another campaign, and this time, as I remember, he brought the idea up in the Chamber of Commerce and got quite a little support from local people. His main talking point was that the economy of Ely and White Pine County was totally dependent upon Kennecott Copper, and everybody knew that sooner or later Kennecott Copper was going to go out. He thought that by creating a Great Basin National Park the county would have another good source of income from tourism. Well, this time quite a lot of interest among the local people built up, and the editor was able to get our congressional delegation in Washington interested. Alan Bible and Howard Cannon were in the Senate, and Walter Baring was in the House of Representatives. He got them interested, and he called a meeting to discuss the proposal, and he had them come out to Ely. I attended that meeting, and the Forest supervisor attended it, too. The editor presented his wish–and that of the Chamber of Commerce and the community of Ely–for the creation of a national park.

The three legislators had never seen the area of the proposed park. By this time the regional forester was getting quite interested in the issue, too, so he sent a Forest Service plane out to fly Cannon and Bible and Baring over the area. We gave them a general look at Wheeler Peak, and we flew down close enough so they got a good view of the arch and of the supposed glacier that existed on the northeast side of Mount Wheeler. When we got back we told them that we didn't feel there was anything that wasn't getting adequate protection under the Forest Service, which was true; that we didn't feel that any feature in the area was adequate to be reserved as a national park; and also that if the area was made a park, we didn't believe there would be enough people want to see the park to even begin to support it. We also brought out livestock people and others who were going to be affected if it was made a park. There was one small timber sale within the proposed park area, and there was a lot of deer hunting in the fall, and there was at times considerable prospecting and mining going on. All these activities would be seriously affected if it became a national park.

The legislators agreed that there was nothing about the Wheeler Peak area that should qualify it for national park status; all three of them were of similar mind on that when they were talking to us. When they held their public meeting the next day, they didn't come out and tell the group exactly what they told us, however. They pretty much left it standing that when they got back to Washington they would recommend that the Park Service take a look and see if Wheeler Peak qualified for national park status. Up to that time, as far as I know, nobody from the National Park Service had even looked the place over. I'm not saying they didn't; but if they had, we didn't know about it.

Well, the Forest Service wasn't necessarily worried, but we thought maybe further action should be taken on our part to head off this park idea. I was scheduled to be transferred to the Toiyabe Forest around the early part of October 1959, but the regional forester came out and told me, "You forget about your transfer to Reno. We got a job that I want you to do before you leave." I was to go around to all the ranchers, especially those that grazed livestock on the possible national park area, and get statements from them as to how they felt and how it would affect them if it became a national park. I was also to contact any other people that I felt might want to make a statement. I spent about six weeks going around to miners and livestock people and sportsmen gathering these statements, and I gathered quite a bunch of them. They were all uniformly against it, of course. I wouldn't take one that was for it! [laughter] That wasn't what I was out there for–the editor already had statements from everybody who was in favor of it.

John Kinnear, the manager of Kennecott Copper, had always given the Forest Service full support on just about everything. Any help or assistance we wanted, we always got it from Kennecott. I knew John real well, and when I was practically through, I thought, "Well, I'm going to talk to John and see how he feels." I went up and talked to him, and he said, "Well, as far as Kennecott is concerned–as far as the business is concerned–it don't make any difference. But I am in favor of the park, because it isn't going to be too long until Kennecott's going to shut down, and the economy of White Pine County is very dependent upon Kennecott Copper." (You sure noticed it whenever they interrupted operations–everything just came to a screeching halt.) He says, "For that reason, I would support the park, because the county and the cities need another good source of income." Kinnear's personal support for the national park idea was not made public.

I finally finished up my survey about the twentieth of November, and we moved over to Reno and unloaded our furniture on Thanksgiving Day. That effort by the editor and certain businessmen in Ely eventually died down, but the idea was resurrected in the 1980s.

"I'd never refused a transfer" In my own mind I had always insisted that I would remain a ranger to the end of my career. But after I had been at Ely for twelve years, a long time for a ranger to stay on a district, I was notified that I was being transferred to the Toiyabe Forest to be on the supervisor's staff.... I'd never refused a transfer, and I finally decided that even though I was

reluctant to leave, I would go ahead and take it. One thing that encouraged me to go to Reno was that our youngest son was just through high school, and he was ready for college. There was a good university at Reno, and if I transferred he could live at home and go the university, which would help with the expense.

I had thoroughly enjoyed being a ranger on the Ely district. I enjoyed working with the permittees and also with the local people, who were interested in what the Forest Service was trying to do. They had greater knowledge of what took place on the Forest than people on other districts, and they knew and understood the terrain so that if you mentioned a certain area out on the Forest, they knew what you were talking about. It made it easy to work with them, and I think by and large we got greater support from the public on the Ely district than any other district I was ever on.

11

STAFF OFFICER ON THE TOIYABE FOREST, 1959-1965

THE WORK LOAD on the Toiyabe National Forest had been expanding for a number of years, and apparently it finally reached a point where the regional office felt the Toiyabe should have a full Forest staff. With my transfer I was promoted to staff officer in charge of range, wildlife and watershed, and we had an assistant supervisor, a timber officer, and a recreation officer. There was also a staff officer in charge of engineering, and your fiscal officer.

Productive supervision   Every Monday morning Ivan Sack, the supervisor, would hold a staff meeting. If he had anything to present, he gave it to us; or if one of us had a problem—especially a problem that might involve more than one administrative function—we discussed it. Probably all Forests do that now, but it had not been a common practice on any Forest I'd been on before. It was an excellent way of keeping all the staff fairly well informed about what the other staff officers were doing. Lots of times you could foresee an administrative conflict, and by having these staff meetings we could head that conflict off.

Along with that, on a regular schedule each member of the staff was acting supervisor for a week. All correspondence other than personnel correspondence that went out of the supervisor's office for that week would cross your desk, and you had to sign it as acting Forest supervisor. (Ivan handled all personnel cases or problems.) This, too, was an excellent idea, because it kept you informed about what the other Forest staff were doing, and you gained a good understanding of the work on the entire Forest. When it was my turn it meant that I wasn't just handling range; in a way I was involved in *all* the activity on the Forest, at least for that week.

Ivan Sack was a good man to work for. One thing that I liked about him, at least as far as my job went, was that you were pretty-well turned loose; I could do just about what I wanted to. And if you had problems you could talk to him, or if you were doing something that he questioned, he'd come and talk to you. I'll tell you about Ivan: Ivan was a range man. Sometimes when things were kind of quiet he would come in and say, "Arch, I want to show you something," and we would get in the car and drive up in the Sierra a ways to look at different plants or flowers. (One time late in a winter he wanted to show me a primrose that had bloomed through the snow at the edge of a snowbank.) Ivan thought I was a lot better botanist than I really was, because every time he'd find a plant that he'd try to stump me on, I'd nearly always know it; but I wasn't near as good as he thought. Ivan *was* a good botanist, and he was experienced. Years ago he had worked on the old range surveys for a number of summers.

We had a lot of enjoyable times out there in the mountains. A good way to get to know a man is to get him out like that where pressures are really off of him and off you and you can look at a plant and talk man-to-man without anything influencing your conversation.

The Dog Valley fire: watershed rehab   Dog Valley is a tributary of the Truckee River, and periodically floods come out of it. The flooding starts either with cloudbursts or wet mantle conditions, which occur when you have a lot of snow (particularly in the late winter) and then get a heavy rain on top of it–all that snow goes at once and does a lot of damage. The extent of the erosion in Dog Valley could be traced to historic uses of the forest. Originally, Dog Valley was well timbered with excellent virgin timber, but beginning in the 1860s most of it was cut out for mine timbers for Virginia City, and a lot of what was left was cut and burned to make charcoal for the

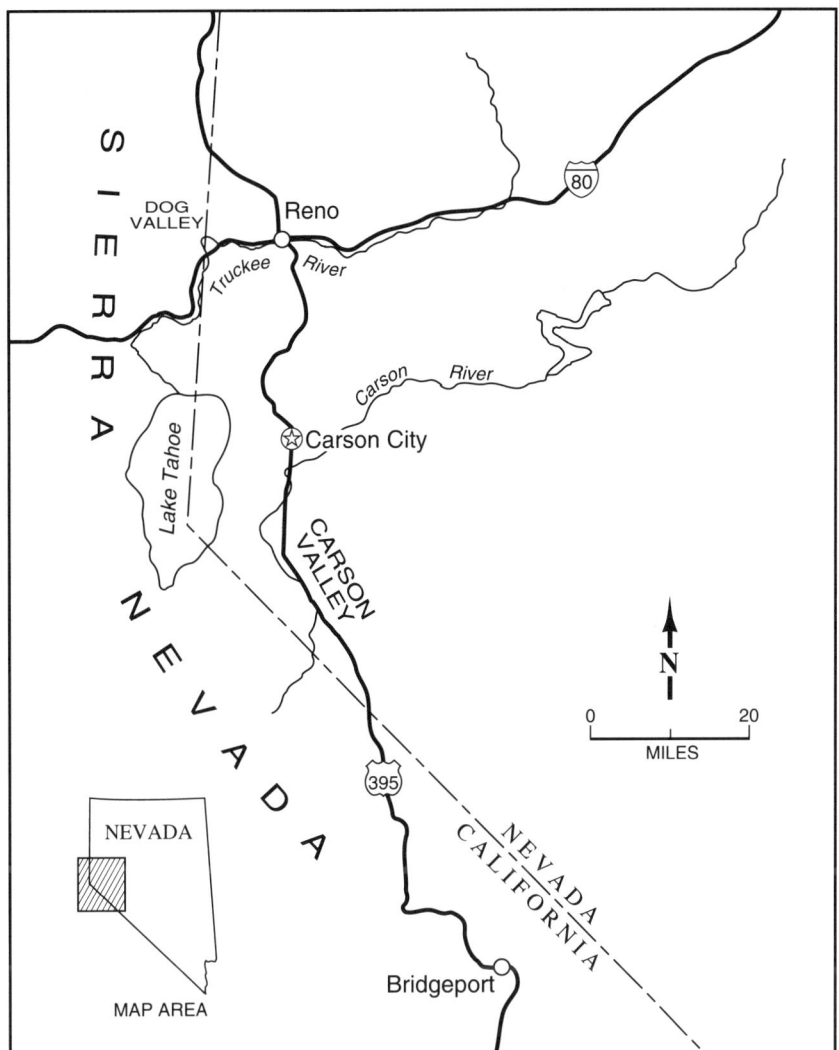

Map 5. The Toiyabe National Forest's headquarters were in Reno, where Archie Murchie served as a staff officer for the five years leading up to his retirement in 1965. Toiyabe lands are distributed throughout much of central and western Nevada; most of the Forest is not represented on the above map.

mines and mills and for domestic use.* After the timber was removed, livestock–both cattle and sheep–were turned loose on the area and they pretty well ruined it. By the time the Forest Service acquired the land–of course, all of it isn't Forest land yet; there is still some private land in there–it was in bad shape. Over the years the Forest put in a lot of erosion-control features, stream structures and so forth to try to slow the flow down; and some of the structures were successful and some of them weren't.

In 1960 the Donner Ridge fire, which started on the Tahoe National Forest, burned over onto the Toiyabe, primarily in the Dog Valley area. When the Donner Ridge fire got in to Dog Valley, it burned over practically all of the watershed except the North Fork, aggravating the erosion problem. I was put in charge of the rehabilitation effort. Dog Valley had been considered a critical watershed before the fire; and after the fire it was considered an *extremely* critical watershed, so we were able to get from the regional office all the money we needed to do a good job, and we got it before the fire was even out.

We decided we'd have to do some kind of mechanical erosion control on the real steep parts of Dog Valley, and trenching was the logical choice. After it was agreed to trench certain parts of it I was flown over the area with a set of aerial photos, and I marked on them where we wanted to trench and where we wouldn't. Luckily, there were men and equipment available to begin immediately. At the time the fire broke out, apparently there was a strike in the logging industry over in northern California and southern Oregon, so there were a lot of idle D8 Cats and their drivers available to help fight the fire. Most of them and their transport trucks were still on the site when we got our watershed money, and we were able to get five or six Cats going at one time building trenches.

The purpose of the trenches was to reduce erosion by catching and storing water from a cloudburst or a wet mantle flood. A trench would hold the water in place and let it slowly seep into the ground and disperse, permitting very little runoff. These trenches were built on

---

*The 1859 discovery of the Comstock Lode brought thousands of men and women rushing into Nevada. Virginia City grew up on the slopes of Mt. Davidson, over the shafts and galleries of mines that were sunk into this rich lode of silver and gold.

perfectly level grades that were laid out with a transit, and you used big D8 Cats to cut them into the mountainside and shove the dirt out on a berm. When the D8s got through, the depth from the top of the berm to the bottom of the trench was usually from three to five feet, depending on where the trench was located. After one of the long trenches had been built they'd send a smaller Cat in to build little check dams maybe every hundred to hundred and fifty feet. The tops of these check dams were always lower than the berm, so that if one of the sections of trench between two check dams filled, the water could overflow into another section and not flow over the edge of the trench to cause erosion.

The areas that I located on the aerial photos were all gone over with these trenches. In some places you'd have fifteen or twenty of them stair-stepping down the slope one after another. The summer after we built them a cloudburst hit Dog Valley, but we had very little erosion. Following it, I flew over the area and could look down and see all these trenches filled with water; and it was unbelievable the amount of water that those trenches stored! It gradually seeped into the ground without causing erosion of any kind.

Lots of times heat from a fire is just enough to open Jeffrey pine cones without destroying the fertility of their seed. At the time the fire went through Dog Valley the Jeffrey cones hadn't opened yet, but the seed inside them was ripe, was mature, and we could expect significant natural reseeding if the seed could be protected. We were getting a pretty heavy drop from the trees that were left standing, but squirrels and ground squirrels and chipmunks were moving into the area in large numbers to feed on the fallen seed. So I wrote to the Fish and Wildlife Service to see if they could furnish 1080 grain to kill these rodents. At that time I don't believe 1080 was completely banned, but they were reluctant to use it. (I think we'd already stopped using it as a coyote poison.) But they told us that they would prepare 1080 wheat that we could drop by airplane over the area, and that it would take care of the rodents. The grain was colored a kind of robin's-egg blue–supposedly a color that birds wouldn't pick up. Apparently they didn't, because we never found a dead bird of any kind. They cautioned us to drop about a pound and a half to the acre, maybe less than that, with the seed grain kernels scattered about six feet apart, no closer.

We closed the entire area to the public, and posted a big notice that poison grain was being put out. To drop it we brought in a Region Five plane that had been used on range reseeding and had a hopper that would scatter the seed over a wide area. I went out and checked after they dropped the 1080 grain, and they had hit the six-feet dispersion pretty close. This 1080 grain disappeared awful fast–you'd be surprised how fast it disappeared. You could walk over the area two days afterwards and hardly find a blue seed on the ground.

Initially, the 1080 helped quite a bit. But like everything else, when you poison off an area like that you create a population vacuum, and in just a little while you've got rodents moving in from the outside. Ten days afterwards we might have had as many rodents on the area as we had before, so I think we wasted our money in scattering 1080 grain. We killed a lot of resident rodents but we didn't save all that many fallen pine seeds. Of course, before rodents could get to them a lot of the seeds that fell were covered by wind blowing the ash around, and the next spring we got a pretty good stand of native seedlings ... but I still kind of question whether the poisoning was worth it or not.

As part of my rehab plan I wanted to plant cereal rye along with bromegrass and intermediate and pubescent wheatgrass. Cereal rye produces a tremendous amount of litter from dead stalks and so forth, especially in its first year. After a fire, erosion by rains or spring runoff is a big problem, and I knew that the litter laid down by rye would greatly reduce such erosion. However, not everybody was convinced; and when my rehab plan went in to the regional office for approval, timber management really went up in the air–they didn't want any cereal rye planted where they were going to plant trees! (Range management was so-so–they didn't care much whether I did or I didn't. Watershed people, of course, were all for it.)

Well, the supervisor wasn't too much for it, so he called me in and asked, "Just why do you want to plant cereal rye? What will it accomplish?" I told him that to start with, it is very hardy and it is a biannual. It will germinate with the first rain in the fall and it will make good bunches of grass. (In fact, some places they used to use it for grazing in the fall and then let it go to grain the next year.) Then in the spring, as soon as it warms up and you've still got lots of moisture in the ground, it will go on and finish growing and be all headed out and ripened by the last of June. Then you've got all these dead stalks that eventually fall down on the ground and protect the soil from erosion. Well, the supervisor accepted my recommendation. And finally timber management, range and watershed came out to Reno and we had a big meeting and went round and round and finally they said, "OK!"

Our watershed rehab in Dog Valley turned out to be quite successful, and it has probably done more to prevent erosion in the valley than anything that had been done before. But success was possible only because of the fire: only after the valley burned was enough money appropriated for us to go in and do the job that had to be done. (Now, I am not speaking of the North Fork, mind you; there was quite a lot of erosion coming down the North Fork, but we weren't able to spend any of this rehab money on it because it wasn't in the burned area.)

Meadow gullies The most prominent permittee on the east slope of the Sierra was Fred Dressler. He was quite an important man in the community—he had been president of the state cattle association, and I believe he was president of the national cattle association. Traveling around the state and throughout the nation, Fred ran into a lot of stockmen: some grazed on federal lands, and, of course, some didn't. He probably listened to a lot of problems that these various stockmen had, and being president of the associations it was natural that they came to him for solutions. So he was well versed in the problems ranchers in other regions had in their relations with the Forest Service. At times in our livestock meetings he would bring some of these problems up, and some of the things that he was criticizing the Forest Service for really didn't apply to the Toiyabe area. It was just criticism of the Forest Service in general.

There were two main problems on most of the allotments on the eastern slope of the Sierra between Bridgeport and Dog Valley: one was cattle distribution, and the other was the gullying of meadows. On most allotments in the Sierra they had large meadows, and a big part of the permittees' use came off of them. There were very few meadows that didn't have a gully coming down the middle, and while grazing may not have been the sole cause of these gullies, it was a big factor. But according to Fred Dressler, the Forest Service was a bigger factor because it tried to exert too much control over the rancher in the use of federal lands. He mentioned time and time again that if the Forest Service would just leave the ranchers alone and let them run their cattle the way they wanted to run them, we wouldn't have a lot of these range and erosion problems. He actually contended that if the Forest Service would let the stockmen alone, we wouldn't have those gullies! Well, I don't follow his reasoning. I guess he was saying that the Forest Service was as much a contributor to gullies as the cattle were.

A meadow is like a saucer. Underneath, it has a more or less impermeable clay base, and that's where your water table lays. If you have a high water table, you have nice, luxuriant sedges and other types of grasses. As the gully cuts down, it lowers the water table in the meadow. The soil above the water table isn't too deep, but as you come in towards the center, the soil gets deeper. Well, as the water table lowers, these meadow edges become dry, and your meadow vegetation dies out and leaves more or less bare ground. Then dandelion, horehound and other weeds come in and take over this bare ground. Following them, sagebrush and rabbitbrush come in and take over. As this gully cuts down, the water table of the whole meadow continues to lower, because the water table can't be higher than the depth of the gully. Eventually, if this keeps up, the time will come when the water table will fall so low that none of the meadow grasses can survive, and then you'll wind up with nothing more than a sagebrush-rabbitbrush flat, with probably some perennial grasses scattered through it. This is a thing that all land managers are trying to avoid, and one I'm sure the ranchers themselves want to avoid–decrease in the size and productivity of meadows, and probably the eventual total loss of the meadows.

The cattle really like that luxuriant meadow grass, and they'll hold on a meadow if you allow them until they've grazed it to the point that there's no more forage left. Then they'll move out onto the slopes and higher up in the canyons away from the meadows. But unless somebody forces them out, they're going to stay there until the vegetation's gone. The only way to get them off that meadow is to have a rider go in there and drive them out, and shove them up the canyons and shove them up on the slopes. But lots of times they just won't stay up on the slopes.

One time up in Paiute Meadows a rancher named Koenig was doing his own riding. The ranger, Steve Rushton, and I were staying in a nice Forest Service cabin up there, and we were riding some of the high country. Koenig came up, and he cleaned all his cattle off of the meadow and shoved them up various canyons to the west of it, but he hadn't much more than left before these cattle started coming back down. Some of these canyons had lakes up to the head of them, and fishermen who had gone up there would start down the canyon talking and yelling, and a bunch of cattle would get in front of them; they'd drive them right back down on the meadow. Koenig was having an awful time. He told us, "I'd have to stay up here all the time to keep those cattle off of that meadow. I just can't do it." It's true probably with most of the meadows along this Sierra front: they are the ice cream ranges as far as cattle are concerned, and that's where they'll stay unless somebody shoves them off.

Of course, fencing would keep the meadows from being grazed, but the Forest Service couldn't fence them because cattle have to come down for water. . . . Actually, most of the Sierra is fairly well watered with springs and streams, and so water really shouldn't be a problem, even up away from the meadows, but I don't know of any meadow that was ever fenced. I doubt that the permittees would agree to fencing. And suppose we had fenced a meadow: the cattle would come down, and they would drift 'round and 'round the outside of the fence, and in a little while you'd create a terrific erosion problem outside the fence. And it wouldn't be good for the cattle, because they'd be losing weight, and . . . . That's what they'd do—they'd probably pile up against this fence. And that's probably one of the reasons why the Forest Service and the ranchers never fenced the meadows.

Total removal of the cattle isn't going to solve anything, either. The cattle contribute to the problem, and total removal of them would lessen the rate of erosion, but it would not fill those gullies up and bring that water table up to the surface again. To be honest, I don't think the Forest Service has a solution. As long as I was involved with range management on the Toiyabe, I could never think of any possible solution to the gully problems in the meadows.

Reseeding range land   When I came to the Toiyabe from the Ely district and took the staff position, it was entirely different from being a ranger. I was kind of in an advisory capacity to the rangers on the Forest. One of the things I did was consult on reseeding proposals. If we looked an area over (measured our soil depth and had some idea about what the precipitation was), and it was suitable, and we figured we might be able to get money for it, then I'd work the project up. I'd send the proposal in to the regional office for financing, and most of the time they'd send somebody out from range management to look the area over. If they agreed, then there was a good chance that we'd get financing for the spray project or the reseeding project. If we got the money, I worked up the bids and circulated them and accepted the one that I figured would do the best job.

After the job started, I was out on the project quite often to check compliance with the contract. On a reseeding job, I think the contractor was allowed one attached sagebrush per square rod. (When you reseed, you go in with a double-disk plow, and you plow the ground. If there's more than one rooted sagebrush per square rod, then he has to go back and plow a second time.) Most of the time we contracted with an old guy from Fallon who did a lot of reseeding for the BLM before we ever got him. You never had to worry about him–if he didn't plow up to standard, he'd go back himself. Then, too, when he drilled the seed in, he complied with depth and he didn't miss any areas.

On spray jobs it was more or less the same way: I worked up the project. We didn't have to put spray jobs out on bid, though, because the region had a plane under contract that would do spray jobs throughout the region, and you just had to wait your turn. They started out first by using a converted torpedo bomber that had flown off aircraft carriers during the war. They were very good, but they flew pretty fast, and a big plane like that flying that low . . . they cracked up a few of them. They used them also to drop borate (or fire retardant) on fires. Finally they switched to helicopters, which are excellent for that kind of work.

We didn't have any trouble with spray jobs. We used an herbicide called 2,4-D which did an excellent job of killing sagebrush. However, the use of 2,4-D is now outlawed. That's too bad, because there's a lot of reseedings that has sagebrush coming in that should be taken out. Eventually, as time goes on, the sagebrush will pretty well crowd out the reseeding. One spray job every five to eight years to get the sagebrush out, and that reseeding will be good for a long time. But you can't use 2,4-D now, so I guess there's nothing else they can do but let the sagebrush come in and take over.

Trespass: fugitives, gardeners, and non-shady characters

I was also trespass officer on the Toiyabe Forest, and that responsibility covered a number of things. Trespass didn't just mean grazing on an allotment without a permit. North of Bridgeport there's a guard station they call the Wheeler Guard Station. One time some kids from a reform school in California escaped, and how they got over that far, I don't know–it was in the wintertime during a storm. But they holed up in the guard station, and there was some rations stored there, and they lived there for quite a little while. The guard station was stocked with heavy crockery dishes–probably surplus army crockery–and before they left they took most of the plates and cups and threw them against the fireplace and broke them all up. And then there was other damage.

After the winter storm was over, the ranger was going up past Wheeler and he saw fresh man tracks coming out through the gate of the guard station and no tracks going in, so he knew that whoever was in there had stayed there during the storm. He went in and checked and found all this crockery broken up. He followed the tracks and he ran onto the boys, but he didn't stop and talk to them. He went on into Bridgeport and notified the sheriff's department, and they went out and arrested the boys. Then he notified me.

I met the ranger at Wellington, and we went on up to the guard station and looked everything over, assessed the damages. Then both of us went on into Bridgeport and talked to the boys. We couldn't get an awful lot out of them and the authorities had already been notified that they'd escaped, so the thing was just dropped.

We had another interesting trespassing case up on the Sweetwaters where this old fellow and his wife had a trailer house. They'd moved up to a spring on Forest Service land and had settled there for the summer, and they had diligently spaded up quite a piece of ground and put in a real nice garden. They had built a fence around their garden out of branches that wouldn't keep the rabbits out, and I don't think it would keep too many deer out either–but they were camped right close to the garden, and the deer and rabbits apparently didn't bother them too much. The garden was going to be part of their winter supply of grub . . . but they were in trespass on Forest land. The ranger didn't find them until their garden was pretty well along–the corn was getting pretty close to putting out ears. He reported it, and I went out and investigated it,

Staff Officer on the Toiyabe

and by rights I should have kicked them off then and there, because they were in trespass. But they didn't have much, so the way I settled it was I told them, "OK, you can stay here until your garden is harvested, and then you're going to completely clean up the area." Then they'd have to move out and not come back. They agreed they would do it.

Later on in the fall we went up there, and they'd cleaned that place up just as slick as a whistle. There wasn't a can; there wasn't a piece of paper; there wasn't a branch around the garden. All there was was just this little bare piece of ground that indicated that they'd been there.

We had a lot more grazing trespass over to Ely than we had on the Toiyabe . . . or to put it another way, we had more grazing trespass over to Ely than we had trespass *reported* by the rangers on the Toiyabe. There might have been quite a bit going on unreported, because they didn't count the sheep, and very few of the cattle that went on the allotments were counted. To me, counting was routine, but . . . well, over to Ely I counted every band of sheep. Now *cattle* that went on, I didn't, but much of that Ely country was open enough that you could make a pretty good range count of cattle if you wanted to ride enough. I did a lot of range counting, and once in a great while I'd catch a man with a few head too many. The rangers on the Toiyabe weren't doing that–they took the number that the permittees said went on, and that was it.

When you caught ranchers in range trespass, there were really two ways of handling it: one, of course, was you trespassed them and charged them for the forage that the livestock had consumed (and in some places you could charge punitive damages like they did in Montana). The other way: if they failed to respond to a trespass letter or personal notification, the livestock could be rounded up and sold at auction. One example of just how that worked involved John Casey.

Casey had ranches all over the country–he had ranches in California; he ranched in Judith Basin, up in Montana; he had a ranch up in the head of Reese River; he had a ranch just below Peavine on the Tonopah district . . . he had ranches scattered all over. I wouldn't say Casey was a shady character, but he was noted for trespassing, and about 1961 Shag Taynton, the ranger on the Tonopah district, found his cattle in trespass and notified him. Although Casey was a permittee, his cattle were where he shouldn't have had them, so he was told to remove them . . . but he did not. Shag contacted me, and when I got out there I told him that our next step would be to impound the cattle–around fifty or sixty head, as I remember. I told him the procedures that he'd have to go through, including hiring enough riders to move the cattle gently without running them or taking a chance of injuring them. (Some had young calves.)

Shag had a friend who had been doing some trail and range improvement work for him, and this friend had a small ranch which was an ideal place to impound the cattle (it was close to where the cattle were trespassing, and he had water, hay, and a good, substantial corral), so they went ahead and rounded up the cattle. As soon as they had them impounded in the corral, I notified Casey and gave him the opportunity to recover his cattle by paying the trespass fee, plus the cost of the roundup, plus the cost of feed for the cattle while they were impounded, plus the wages of men to guard the corral twenty-four hours a day to see that nobody came along and turned them loose. Well, he refused to do it. So then I notified him that the stock were to be sold at public auction, and told him the day and the time. What we didn't know was that the First National Bank in Reno had a mortgage on these cattle. So, as soon as the bank got wind of their impounding and impending sale they sent a man down to the supervisor's office to find out what was going on. I told him exactly what had happened, and he said, "You mean, those cattle are going to be sold at auction?"

I said, "Yes."

"Well," he said, "I'll talk to John Casey and get him to pay the trespass fees and all these other costs."

"That time has passed," I said.

I'm sure, because of his prominence, that both John Casey and the fellow from the bank thought nobody would bid on the cattle, and then they'd offer fifty dollars or a bid like that and recover them. Well, Shag got another rancher to bid on the cattle. (And I'm not sure but he would have taken the cattle if he could have gotten them at a low price.) Casey bid a nominal amount; I don't remember what it was. Then this little rancher upped the bid pretty good, and that just scared the daylights out of Casey and his banker. [laughter] They knew what bill was against them for the trespass fee and the cost of the roundup and all the other stuff, and so Casey's next bid was over that. Of course, then they were sold back to him. The banker wrote the check and paid it off, and the regional office deducted the trespass costs and refunded the rest to the bank.

Impounding and auctioning cattle as punishment for trespass was a last resort, but during the comparatively short time that I was on the Toiyabe National Forest, we almost had another case of it. The Reese River district had a reseeding, and this fellow from Austin had a bunch of horses that got into it. I don't know how they got into the reseeding, as it was fenced, but they got in there and were in trespass. Harvey Gissel was ranger there, and he notified the rancher that his horses were in trespass and to remove them. When he failed to remove the horses, Harvey trespassed him. He still would not remove his horses or pay the trespass, so we notified him that we were going to round the horses up. Of course, it would have been hard for us—we would have had to build a corral, and we'd have had to haul hay in; but it could have been done. But he apparently had heard about the Casey deal on the Tonopah district, and he was in right away and paid the trespass fees! [laughter]

John Casey continued to be a problem. Like with some other ranchers, you had to watch him like a hawk or he'd get his cattle on their allotments too soon. On his Nevada operations Casey never put up a spear of hay—if he had to feed hay he bought it, but he'd sooner let a cow die than buy hay for it. He ran his cattle out year 'round, and if a cow wasn't tough enough to winter out, Casey didn't want it. Come spring, he would try to get his cattle on the Forest as soon as permitted; sometimes, earlier.

Casey would also try to put too many cattle on an allotment. A year or two after Shag Taynton had impounded Casey's cattle, Harvey Gissel became the ranger on the Tonopah district. One evening at dusk, while Harvey I were staying at a stopover cabin at the mouth of Peavine up out of Smoky Valley, we saw a dust plume coming up the valley. Harvey says, "I'll bet you that's Casey bringing his cattle on."

So I said, "Well, let's go down and count them." We went down, and sure enough it was Casey! His cattle were all strung out, easy to count, so we counted them, and he was something like fifty head over. Casey wasn't too happy, so Harvey asked me what to do. I said, "Make him cut out what he's over. He has a ranch on down the valley a little ways, and he can take them back to the ranch."

Well, he did; he cut them out. He had five or six Indians with him, and they must have brought the cattle quite a ways, because the cattle were tired. But he cut the extra fifty head out and took them on down to his ranch. (He held the rest on water in Peavine Creek overnight, and the next morning he put them on the Forest.) But Harvey and I knew that the ranch down the valley wouldn't hold those fifty head but a very short time, and they'd all graze on Peavine and Peavine allotment. All Casey had to do was open the gate, and in twenty-four hours or less those cattle would be up on the Forest. I *know* those cattle were soon in trespass again.

"Getting to the age . . . " I was getting to the age where they didn't want me out fighting fires, so sometimes on a fire they'd send me out as safety officer. They sent me down once to a fire in Clark Canyon on the west side of Charleston Mountain near Las Vegas. Even on a small fire like that it was surprising what the job of safety officer entailed–anything from smoke inhalation to upset stomachs to bruises and cuts, but we only had one serious accident. We thought this fire fighter had broken his ankle, but it turned out that he hadn't; it was just a serious sprain. We had to load him into a pickup and haul him into Las Vegas. But to keep a bunch of fire fighters going and healthy keeps you pretty busy.

The safety officer acted more or less as a medic, but I didn't get any special training . . . no more than any other training that I'd had before on fires. I'd been safety officer and physician and doctor, whatever you want to call it, on so many fires that there was no problem there. [laughter] And I'd taken a Red Cross first aid course plus a refresher course, so there really was no problem.

Safety officer was kind of an interesting job. We had a lot of vehicle accidents, and the Forest Service was responsible for some of them. Some, somebody else was responsible for, but the safety officer had to investigate all accidents. I remember one case where on a Monday morning the road survey crew was going to a job up out of Bridgeport. In Antelope Valley before you get to Walker there was a little store and coffee shop off to the side of the road. It was on the left-hand side going south, and they decided to whip in there and have a cup of coffee before they went on to work. Well, whoever was driving hadn't been watching what he was doing, and when he whipped over to go in he turned right in front of this Mexican family from someplace in California. There was a husband, his wife and five or six kids in an old beat-up car. They had a collision. It didn't hurt the Forest Service vehicle much, but it did a lot of damage to the Mexican's car. As the report came in, the Mexican was

Staff Officer on the Toiyabe

in the wrong and the Forest Service people were in the right. So I went out to investigate it.

When I went out I found that the Forest Service van was entirely in the wrong, because they'd pulled right in front of this Mexican. Where the collision took place, there was no reason, even though it was near a curve, that they couldn't have seen the car coming, but they pulled right in front of it. Well, on any investigation, I don't care whether it was a trespass or whether it was a vehicle or what, I always felt that you were assigned a certain job. You're going to do it to the best of your ability, and you make a thorough examination and an honest, thorough report. So I reported that the Forest Service was entirely at fault. The road crew didn't like it and the staff engineer, of course, didn't like it, but I turned it over to the Forest supervisor.

The outcome of it was they left the car there at this little place, and they hauled the Mexican family clear into Reno, where I believe they had some friends or some relatives that they could stay with. Then the guys from the Forest Service went out and got the Mexican's car and brought it into Reno, took it down to our shop and repaired it completely. They put it back in exactly the same shape as it was before, maybe even a little better. The Mexican said that was fine. He didn't bring any charges against us, and that's the way it was settled. Now, I wonder if that was legal?

I enjoyed my session on the Toiyabe under Ivan Sack and Eddie Maw, but in 1965 I retired from the United States Forest Service. I could have stayed on longer, but I'd put in enough years already and I had many things I wanted to do that I couldn't when I was in the Forest Service–like building a house. The winter of 1938-1939 at Challis, Idaho, the only house we could find was an old log house. The chinking had fallen out, and when we moved in you could see the ground outside between some of the logs. I promised Jane then that some day I would build her a house. I still had that to do. I was still bodily able to do it, in good health, and I wanted to do some other things for myself before I got so old that I couldn't. When I retired they calculated that if they added up the sick leave that I never got and the annual leave that I lost, I had put in around thirty-four years in the Forest Service. I figured that was long enough.

12

Unexpected Adventures: Encounters with Animals

WHEN I WAS A RANGER, a lot of times you were out alone–strictly alone. You didn't have anybody with you, and you had to depend on your own resources to handle your problems. It was not uncommon for the problems to involve animals.

Mule in torrent, with pack In the spring of 1938 we had a late fire season. The road into Loon Creek had already opened before we had to put men up on the peaks, so I had them working trail. Since they were getting pretty low on the grub they'd brought when they came out on trail work, they'd all placed an order for grub to take up on the lookouts with them when fire season began. The supply truck came in, and I set out to pack their supplies out to them on six or seven mules, all of them loaded fairly heavy.

It was early enough that the streams were still running pretty high water, and I was leading the string up along a trail five or six feet back from the edge of the cut-bank that dropped down to the stream. There had been quite a lot of erosion cutting into this bank at one place and there was a drop-off of probably four feet directly into the water from the trail. My saddle horse crossed this eroded place OK, but when the lead mule came to it the bank caved in. It just flipped her over and dumped her into the water. The pigtail that the packhorse behind her was tied to broke, which was the purpose of the pigtail. When the mule went in, she fell on her side, and with the weight on her and the fact that she couldn't get her head down, she couldn't get up. I still had dallies around my horn with her lead rope, and when she'd lunge to try to get up, I'd pull to see if I could help pull her to her feet.

Well, the bank finally gave way under my saddle horse, and the minute I felt it go, I knew I was going in too. I threw the halter rope, and I jerked my feet out of the stirrups. When my horse went in, he fell on his side, but I was free of him. He was a bigger animal than the mule, and of course he wasn't loaded, so he could get his head out of the water and still struggle and get up. He got to his feet, went on up the stream a little ways and got out on the bank. Well, that left me in the water, and I was wearing batwing chaps. As long as I was facing upstream I was all right, because the batwing was back; but when I turned, the water caught these batwings and away I went, under water. I finally gained some footing and got my head out. The mule was just below me, and as I went by her I grabbed at the pack and got stopped, and she was trying to get up. I knew that she was about to drown, so I pulled out my pocketknife and cut the lash ropes; I cut sling ropes; I cut cinches; I cut everything just as fast I could. Instead of trying to unsnap the chest strap, I just cut it, too, and peeled the pack off. She struggled, and I got ahold of her halter, and I finally got her up.

She was so weak then–she'd taken on so much water–that she just stood there spread-legged. The water ran out of her nose, and she stood there for quite a little while. Finally she got enough air into her, enough strength back, that I could lead her upstream a little ways and get her out. Then I got to wondering about my pack and all this fellow's grub that I had packed on the mule. I worked on down the stream, and there was an eddy. The first two things that I found were the pack boxes. (I had a lot of the grub packed in pack boxes.) I got them pulled out, and there was a sack of sugar as a top pack. I found everything except the saddle blanket. I even found the pad.

Everything was soaked, of course. There was a carton of these old wood matches that came about eight boxes to a carton. They got all wet, and the glue that held the boxes together came undone, and you could open them and the matches would just fall out. They were wet, and the minute you touched a head, the head'd fall off. So I just opened the boxes and put them on a rock in the sun. When they finally dried, a lot of the heads of the matches were fused together, and the man whose

Archie Murchie starting out on a pack trip in 1936 on the Stanley Basin district, Challis National Forest.

provisions these were had to be awful careful when breaking them apart that he didn't ignite them. But this wet sack of sugar turned as solid as a rock. He took it up to the lookout anyway, and when he wanted some sugar he'd just chop off a chunk and make syrup out of it, and he used it as syrup rather than sugar. It worked just as well.

The labels came off of all the cans, but he got so good at listening to a can.... One time I was up to the lookout and he asked me what I wanted—corn or peas or what—and he went and got a can and shook it, and then he shook another one. I says, "You can't tell what's in there by shaking it."

He says, "Yes, I can." And he'd open it up, and it'd be corn or whatever he said it would be.

Running naked in the night

Late in the fall of 1939 or 1940, when we were building four-foot catwalks around the lookouts on the Loon Creek district, I set out to pack some lumber from Dutch Charlie's sawmill to the Fly Peak lookout. I went through all the stock, and there was almost enough mules that had shoes on, but I needed one more animal. (Late in the fall when the mules started losing shoes, we wouldn't replace them because we'd have to pull all of them anyway when we put them on pasture.) We had a Forest Service-owned sorrel horse that was still shod, so I took him. He was what we called a reefer. He was balky in a pack string, and every so often he'd just brace his feet and that was it: something had to give; he either broke his halter or broke the halter rope or the pigtail on the saddle, or something had to give.

I packed up that morning, and it was pretty late when I got started. The sorrel was trouble all day—I don't know how many times I went back down the string to deal with him. I put more knots in his halter rope than you could shake a stick at. Getting on fairly late in the evening I wanted to make it up to where one of our patrolmen had installed bars that could be put across the trail so his horses couldn't pull out on him when he turned them loose above it. My string included some of my mules and some of the packer's mules, and whenever you mixed up a pack string, you had problems—my mules would stay with my saddle horse, but the other mules wouldn't. So I wanted to get above these bars, so I could put them up and keep my stock.

It got dark before I made it to the bars, so I stopped and I went back down my string, and this sorrel horse was gone. I tied up my outfit and hiked back down the trail, and sure enough there he was standing. I says, "Well, this is it. I'm not going to go any further." We were in a very narrow canyon with a stream in the bottom. There was no place to camp or even unpack, but I could see a little flat just above the stream a ways, so I pulled up there. There was a windfall, and I unpacked and laid my pack saddles out on this windfall and threw the blankets on top so they'd dry, then took the stock and shoved them on up the trail quite a ways. I hadn't come to the bars yet, but I thought, "Well, they'll stay. There's fairly good feed, and they'll stay."

When I got back down to the flat I was so tired that I didn't even build a fire–I just rolled out my bed and went to sleep. Well, I had my saddle horse belled, and it must have been around midnight that the sound of his bell coming down the trail awakened me, and I knew the pack string was right behind him. I was sleeping in the nude, but I was in such a hurry I just jerked my boots on and tore down to the trail. It was lucky I did because I *just* got there ahead of them. If I hadn't, I'd have had to hike clear back to the station to get them the next morning.

I got them turned around, and it was quite a little ways, but I thought, "I'm going to shove them clear up till I come to those bars." And I did. I put up the bars, and it was cold. When I was going up I didn't feel so cold, but coming back down I was cold. As I walked I'd look up above the trail for this little flat where I had my bedroll, and I couldn't find it. Dark . . . oh, it was dark! No moon. I'd go down a little bit, look up, and I couldn't find it. I thought, "What in the world am I going to do?" If I had just laid a pole across the trail or built a fire that I could have smelled, I'd have been all right. But I had done neither.

Finally I was getting so cold that I figured, "Maybe I should take off afoot for the station." But on the way down I would have to ford Loon Creek, which would come up above my waist, and as cold as I'd be, I knew I'd never make it past that. So I thought, "Well, I'll go down a ways further, keep looking for that bench," but I was pretty sure I was below where I should have been. Once more, I knelt down and looked up the slope against the stars, and I thought I could make out a break, make out a bench. So I climbed up, brush tearing my legs, and sure enough it was a bench. I walked down the bench a little ways and came to the end of a windfall; walked down the windfall, and there was my saddles. And glad, I was.

I crawled into bed, and it took me about two hours to get warmed up. The next morning I went down to the stream to get water for breakfast, and there was ice on the water, so I knew that I'd never have made it back to the station. I've often wondered what the regional forester would have said if the Forest supervisor notified him that they'd found one of their rangers stark naked, with just his boots on curled up underneath a tree stone dead! [laughter] But that didn't change my sleeping habits any. I have continued to sleep in the nude.

Huckleberry bears   Up there on the Kootenai in 1929, wherever a fire had gone through it would always come up to huckleberries.

The bears were just crazy for them, and they had paths all through the huckleberry patches where they'd feed on the huckleberries. (*Boy, they were good! I liked them as well as the bears did. Some of them would be as big around as the end of your finger.*)

Bill Brown was my partner, and one day he and I were trying to work our way up to the top of a ridge. We came to one of these big huckleberry patches, and he took one bear trail and I took the other, because we figured they'd lead us through this patch. (If you didn't follow a bear trail it was kind of hard fighting your way through the huckleberries–they'd get as high as your head or better.) I was going along up not too steep a slope, and I heard a noise up ahead. I figured, "Well, Bill's bear trail has joined mine." I looked up the trail, and here came this bear just as hard as she could come, and I thought she was going to get me! I thought she was after me. And scared . . . I was scared! [laughter] I tried to get off the trail, but the huckleberries were so thick that I just barely got off it. She whistled past me with about a three-fourths-grown cub right behind her. I just stood there. I didn't know what to do; I just froze.

What had happened was that Bill's bear trail had joined the bear trail I was on, but below him on my trail was this female black bear eating huckleberries, and he spooked her, and of course, down that trail she came. I thought sure she had me. [laughter] I was born and raised in North Dakota, and up until then I had never seen a live bear.

"Instead of a Hereford, it was a grizzly!"   I kind of wanted my own pack mule, but I didn't have a mare to breed. Shorty Conyers had been a lookout of mine when I was on the Rapid River district, and he had a young mare about six years old. She'd never had a colt, but I bought her, figuring I'd breed her to a jack and get a mule out of her. She was broke to pack, but she wasn't

broke to ride, so I broke her to ride. She was a good horse, but one of these round, kind of pot-bellied individuals, and peg-legged; she was kind of hard to ride.

This day in 1941 I was riding the mare up through Cash Creek Basin into the head of Woodtick Canyon. There was a trail from Woodtick Summit that went around on a level and dropped into the West Fork of the Camas, and that's where I was heading. Riding along, I happened to look way down in the willow bottom, and I saw this Hereford cow. There weren't supposed to be cattle of any kind in Woodtick, so I thought, "Whose critter is that?" It could either be one of Wilson's cows from Meyers Cove, or it could be one of Sam Lovell's, a rancher at the mouth of Loon Creek. I decided I'd better go down and find out whose it was and notify him, because if that cow ever started on down Woodtick, they'd never get it. It was pretty steep, so I tied up my outfit and left the trail and had to lead my saddle horse down quite a ways before getting on her.

I didn't want to approach the cow direct, because she probably would just take off before I could check her brand. Down in the willows there was a point that ran out and broke off real sharp into the creek, and I figured if I got out on this point I could look down and either get a brand or an earmark or something off of her. So I circled around and tied my horse up and worked out onto this point. I looked over, and instead of a Hereford, it was a grizzly! Now, if you're talking about an animal that's unpredictable, a bear is the *most* unpredictable. You can *never* tell what a bear is going to do, and sometimes you run onto them and surprise them. This grizzly either heard me or smelled me—their eyesight isn't very good—and he had his head up scenting the air.

There weren't supposed to be any grizzly bears on the Challis, but there he was. Well, I wasn't just going to ride off and leave, because I got to thinking that the grizzly would just have to go over the pass and he'd be right down into Wilson's cattle on the West Fork of the Camas. So I decided, "Well, I'll run him out." I figured the best way to run him was down the canyon, so I circled around and came down through the willows, which weren't very high. You could see the top of his back above the willows, and he was digging up and eating willow roots. (Why he'd eat those, I don't know. I sampled some of them once and they're just as bitter as gall.) Since I could see the bear's back down there, I thought sure my horse had seen it. We were getting closer and closer, and the horse was looking down the canyon, and why she hadn't seen the bear I don't know.

I was just about ready to yell to frighten the bear when my scent apparently carried down the canyon; the bear caught it, and he whirled around and stood up and just about scared my poor horse to death. She just about unloaded me, and back she went! She didn't stop for the creek or willows or anything–she went through everything as hard as she could run. [laughter] I was doing my best to stay on her, and finally I got her stopped maybe a quarter of a mile off. I turned her around and looked back, and I guess we scared the bear just about as bad as the bear scared her. [laughter] Anyway, the bear was gone.

I went back up and picked up my pack string, went on over and dropped into the head of the West Fork of the Camas. There were a lot of little meadows and flat areas along there, and as I was riding I looked over and here lay this dead cow. She wasn't bloated or anything, and I thought, "What in the world did she die from?" So I got off my horse and walked over. There was quite a bit of blood that had run out of her nose, but I couldn't see a mark on her anyplace. I could see that she hadn't been dead very long, and I wondered if rigor mortis had set in, so I took her head by one of the horns and lifted it up, and I could hear the bones grate in her neck–she had a broken neck. Then I started looking around, and here were these grizzly tracks all around. He must have killed that cow and gone over the pass and down into those willows where I saw him. Apparently, from what I could tell, he just caught her alongside the neck with his paw and broke her neck. He'd eaten part of her udder, and that was the only mark on her. Of course, most bears that make a kill like that–whether it's sheep or whatever–take maybe a little flesh off them at the time of the kill, but very little. They will leave them for a week or longer until they really get ripe, and then they'll come back and feed on them. They like that rotten meat. Maggots . . . they don't care; they eat maggots and all.

Every fall we had to make out a wildlife report, so that year I reported one grizzly on the Loon Creek district. My district report was combined with the Forest report and it went in to the regional office. They wrote back and told us to delete the "grizzly bear on the Challis" report because there were no grizzly bears on the Challis. So I wrote back and told them all I knew about this grizzly bear, and they finally accepted it.

She-bear and cub   There was a small area on the Wildhorse district that was pretty rough and heavily timbered, and there was no grazing on it of any kind. It was just left to deer or bear or whatever other animals wanted to use it. I had never been in this area, and I was curious as to what was there, so one day in 1946 I decided to take a ride over into this country. After covering most of it I was on my way back, coming up through lodgepole pine timber—it was a mature stand, just like riding up through a park. I looked up ahead, and there was something black and something yellow lying beside a tree. I rode on up, and finally when I got quite close I realized it was a she-bear and a yellowish-brown cub, and they were curled up there asleep. By that time I'd seen lots of bear, but this cub was such an odd color that I wanted a better look at it, so I rode on up a little further.

The cub woke up first, and when he saw the horse coming he let out a squall and took off up the hill. The mother bear woke up and jumped and followed him for just a short distance, and then she stopped and just stood there broadside of me watching me. I rode around her and on up. Her cub had stopped and was standing beside a tree, just peeking around the edge of it, and every once in a while he'd let out a squall. I got pretty close; I got a good look at him, and then he took off again and went further on up. I looked back down the slope, and even with all the squalling that the cub was doing, this she-bear was still standing there. She was the biggest black bear I think I've ever seen. Oh, she was fat!

I rode back down towards her, and she slowly turned and started coming up around me. I was riding a horse that was a good cutting horse, so I just cut him over and he headed her off, and I started working her just like you'd work a cow. There was a ledge with points that ran out on it, with a drop-off that was not very far—probably seven or eight feet. I thought, "I wonder what she'll do if I work her out on one of those points? Will she try to go off of it; or will she get mad and turn around and fight?" So I slowly worked her out onto this point. At no time was a hair up on her back and at no time did she ever utter a growl or any kind of noise. I worked her out on this point, and although at first my horse had been afraid of her, by this time he wasn't afraid at all. I was behind her not over thirty feet at the very most, and she walked over to the edge and looked over, and she knew she couldn't go off.

She turned around and stood facing me. My horse was facing her, and I thought, "If she decides to charge, this isn't a very good position to be in. I'd better turn around the other way." So I turned my horse around and just sat there and watched her, and pretty soon she started walking real slow. She passed within twelve feet of me, went right on past me just like that, and she looked up at me and went on up the hill to her cub and went on.

Now, she was a wild bear, I assume. But it boggles my mind to think that a wild bear would act that way. I suppose there's a remote possibility that at some time she'd been captured and had been kept, and that she had no fear of horses or of humans. But that's highly unlikely.

"You could get hurt"

One time the Lincoln district ranger and I ran right into a bear that was feeding on a dead, rotten sheep. We were on top of a ridge, and what few trees were there were limbed down to the ground. It was a real windy day. We came around this tree, and it was so windy that the bear hadn't heard us; and the wind was blowing from the bear to us, so he hadn't smelled us. When we came around that tree we surprised him, and we were within twenty feet of him. If we'd made the wrong move then, I'm sure he'd have charged–if we'd gone further towards him, he'd have charged; if we had turned around and started walking away from him, I think he would have charged. But we didn't move . . . we just stood there. He had his nose down to the ground, and he was ready to come, but finally we got our bluff in. You could see indecision set in: he looked to one side and then looked to the other side; then he slowly turned around, watching us until he got completely turned, and all of a sudden there was just four clods of dirt flying. He was gone.

If you get too close, a bear will fight, try to defend himself. Not so much that he's angry at you, but he's defending himself. I knew a man named Louie Koch, who was once ranger of the Challis district on the Challis Forest. Many years ago he was cruising timber in Wyoming someplace, in an old burn. There was real low reproduction–probably two or three foot reproduction that had come in–and he was walking on this old windfall, and it started turning on him and he fell. It was a fairly steep slope, and he went end-over-end down the slope, and he rolled right on top of a bear that was curled up asleep. Louie had part of his scalp tore off and a big gash across the back of his neck and a bunch of big scars on his back where the bear had clawed him. The bear wasn't mad at Louie; he was just protecting himself, and as soon as he could get away he was gone. You could run into a situation like that where you could get hurt, but it wouldn't be because the bear was attacking you.

A bear defeats a trap Once on the Wildhorse district, I was riding up the north fork of Lost River, looking for a band of sheep. I'd ridden all morning and couldn't find them, and I'd come back to my camp. While I was eating lunch, I could hear sheep bleating on down the canyon, so as soon as I got through I got on my horse and rode on down. It was the time of day that sheep should be down in the willows shaded up, but these sheep were all in a tight bunch up on a little bench above the creek, and I thought, "Why in the world would the herder bunch his sheep like that in the middle of a hot day?" (It's hard on sheep when you bunch them like that in the middle of the day.) I looked around and didn't see the herder, and I called, and no answer. So I rode on up on the bench, and the minute I got to the sheep I could see that they were scared like everything–just terrified. I started riding around, and I counted twelve dead sheep, killed by a bear. One of them, this bear had hit so hard that it was thrown up in the air and turned upside down and landed on its back in a clump of sagebrush.

I knew the herder couldn't be too far away, and as I rode around I could see horse tracks going up this ridge, so I started riding up the ridge. I'd just gone a little ways, and he was coming down, and I told him what happened. We went on down, and he had a hard time breaking the band up, but he finally got them separated and back down in the willows where it was cool.

Back at the station that evening, I reported the kills. There was a biological survey predatory animal control man that had a little ranch up Copper Basin, and I told the supervisor I would go up and tell him what happened. So I did, and he said, "Well, if I can come by the station tomorrow morning, would you go up with me and show me where it happened?" We went up and he built a V out of loose windfalls, logs around six, eight, nine inches in diameter. Then we drug in all the sheep corpses and piled them way back into the V, and he put a log across the entrance of it about eighteen inches off the ground and he set this bear trap just inside it, so when the bear stepped over the log, he'd step right into the trap. Then, as I remember, he laid two other logs on top of the log extending off the ground to the front, so the bear'd kind of be funneled into this V and step over at the right place. He didn't dare to secure the trap hard and fast, because if caught, the bear would just fight until he ripped his foot out. So he tied the trap onto a toggle log about six or seven inches in diameter and maybe five feet long, so the bear could drag it.

The animal control man says, "He won't come back until those sheep start getting pretty ripe." This was on a Saturday, and the next Saturday we went up, and the bear still hadn't come back. The sheep were getting pretty ripe; you could smell them pretty good. So the next Saturday he came by, and he says, "Arch, let's go up and check our bear trap." So we did, and he'd caught him all right. But the blood that we found was black–it was all dried up–so it had to have been a few days before we got there.

You could easily follow where the bear had dragged the toggle log. He'd traveled at least a quarter of a mile with that trap on his foot, and once in a while after he'd gone a ways, you'd see where he'd hit the ground with the trap. Finally he came to a windfall log, probably two and a half feet or so in diameter. It was an old windfall; the wood wasn't really rotten, but it was getting soft. We could see where he had stopped and beat the trap against that log, and he beat it so hard and so long that it bent the springs on the trap, and finally the jaws opened enough that he got his foot out. Now, his foot must have been badly mangled. Apparently, he was caught by just the front half of his foot, but I'm surprised that he hadn't torn his foot entirely off, as bad as he'd beat that log. He got his foot out, and we could see that he was losing a little bit of blood–surprisingly, not very much for as bad as his paw must have been mangled.

We followed the tracks of the bear a ways, and he kept going, but I'm sure he lost the front part of his foot. Years ago, when they did considerable bear trapping, there was a lot of bear missing half a foot. If the jaws closed on the whole foot, the trapper got the bear, but if they caught just the front part . . . even with the teeth that a bear trap has, and the jaws, bears can actually rip a foot out. It mangles the foot pretty bad, but they free themselves.

Cougar kill at Hospital Cove Cougars are by far my favorite animal. They're a very intelligent animal and they've always interested me. That they can sneak up on a deer, stalk a deer in wide open country, and kill that deer . . . it is just incredible.

The winter of 1937-1938 was an awful tough snow winter down on the Middle Fork of the Salmon, and that's where most of the deer off the Challis and Payette Forests wintered. The range was very badly overgrazed, and we were getting heavy deer loss. The state Fish and Game people were worried, so they asked if we would go down and make a study and investigate the loss. We went down, and there was an awful heavy loss; many hundreds, or maybe over a thousand head had died that winter. A young fellow by the name of Hank Halverson, who later became one of my lookouts on the Loon Creek district, had a couple of black and tan cougar hounds. One day I was going out to look at deer and he says, "Arch, I think I'll take my hounds and go with you." So we took off on snowshoes, and on into the afternoon we went up a ridge and were getting fairly high, probably at the upper limit of where the deer would range. The snow was pretty deep, and Hank says, "Well, I think I'll leave you and cut back down."

I said, "OK," and I went just a very short distance when here was this cougar trail, more than one cougar, that crossed in front of me. It had snowed a little the night before, and the tracks were snowed in. I was still in hailing distance of Hank, and I called to him and said there was a cougar trail in front of me.

He says, "What direction is it heading?"

"It looks like it's heading down slope," I said.

So he said, "I'll keep going, and if I hit it the dogs will tell me whether it's a hot or a cold trail." He always led his dogs; he never turned them loose until he got right close to a cougar. After a little while I heard the hounds cut loose, and Hank called out, "It's hot; I'm going to take it." Well, I went up a little bit further and there was no more deer signs, so I thought, "I'll just cut back and hit Hank's trail and check with him." So I did.

The dogs were still on leash, and they hadn't started to bellow yet, and we kept going along and dropped down into a place called Hospital Cove. The cougars had stopped there. We found out later that they were an old mother and two about three-fourths-grown kittens. In fact, the tom kitten was just about as big as the mother was; the female was a little bit smaller.

At night, some deer had been lying down there, and some standing up just above them. The carcass of a big four-point buck was lying in the snow, facing to the left, and then there was a dead yearling buck right close to him, facing the other direction. Other deer had been there; you could see their bed grounds around. You could read everything in the snow, and it was interesting. The mother cougar had stopped (she'd probably *seen* the deer, because I don't think cougars' noses are all that good) and she'd bellied down with the female kitten on her left and the male kitten on her right. She started creeping up, and she crept to within fifty feet of the deer. Where she made her charge the snow must have been two and a half feet deep at least; but she drove her hind feet practically to the ground and she must have leaped fifteen feet that first leap, and then the other bounds were good long bounds. But she was right on top of that buck; he never even got up! She nailed him right where he laid.

The yearling deer that was standing nearby was able to make a couple of jumps before the male cougar took him. The cougar leapt on to his back and socked his dew claw so hard into the deer's neck that it turned completely backwards. (When we killed the male cougar later, he had a big chunk of deer hide still lodged underneath that claw. He had just about ripped it out!) The female yearling cougar just run along for the fun; she didn't make a kill.

The cougar family had taken a feed off both carcasses, and they'd consumed a lot of meat. Hank turned his dogs loose, and they bellowed and away they went. They ran up to the kill and then took the trail out from the kill. The cougars had bedded down right close, and all of them were so full of meat that they couldn't run very far–the female yearling didn't run hardly three hundred yards before she treed! Most of the time when you got two or three cats on the run, if one of them trees, the hounds stay at this tree. Hank's dogs were well trained, and when they treed the yearling, one dog stayed with it, and the other dog took off on the other two tracks. It ran them probably another short quarter, and then the tom yearling treed. When we got to them, one dog had the young female treed and the other dog had the male treed, but the mother had kept on going. So we just killed both young cougars and left them and took off after the mother. Perhaps she'd been run before–this is kind of ledgy country in places, and when she treed the first time, she treed right close to a ledge; and when she saw us coming, she leaped out of the tree onto the ledge and away she went. We took after her again and she didn't go too far, because she was full of meat. She treed on a good-sized limb next to a ledge again. Hank was carrying a small-caliber rifle, a .25-20, and he would have to get close to make a kill. He said, "Well, Archie, we'll fool her. You stay here and I'm going to circle around and get on top of that ledge. When I signal, you start moving forward real slow."

The cougar was watching me; she didn't know Hank was moving in behind her. She was watching me, and I kept getting closer and closer, and I could see Hank, and if she had turned around she could have seen him too. Finally, she must have heard him or smelled him, and turned around and squatted on the limb and I thought, "Well, she's going to leap out." Even though Hank was there, she was going to go out. So I started yelling and walking fast towards her, and she turned toward me and Hank shot her. She was a big female, and I was curious as to just how much meat she had eaten–her stomach was swelled way out. When we got her down, I opened her up, and for the short time she'd had meat inside of her stomach . . . . Stink? Oh, that stunk! There must have been at least twenty pounds of meat in her. And she'd bitten through the leg bones and shattered them, and she had parts of bones that were up to five inches long in her stomach . . . of course, the bones were shattered, but some were just as sharp as a needle. How she got them down without rupturing her gullet, I don't know.

(I've always been interested in cougar, and whenever I'd run onto a cougar scat while riding or hiking, I'd always stop and examine it. I've seen where they had passed bones several inches long, and their stomach acids had so removed all the minerals from the bone that it was just like a piece of hard chalk. The ends of the bone were just as blunt as the end of your finger, and the bone was soft; you could take it and break it in your hands. It was pure white and it just looked like a real piece of hard chalk.)

As time passed, my sentiments about killing predatory animals changed. Although hunting cougar for sport had something to do with Hank and me shooting the three up near Hospital Cove, the principal reason we killed them was that they killed deer and we were trying to protect the herds for deer hunters. But looking back, maybe we were doing the wrong thing, because that range was in awful bad shape from overgrazing by deer. It wasn't horses, cattle or anything else that did it; it was just strictly deer. If I had known then what I know now, we should have just let the cougar do their jobs. But Hank was going to kill them anyway, and I went along with it.

How many feeds in a deer? When I was over there to Ely, every spring we conducted a deer count in three different areas, and a fellow by the name of Les Robinette from the Fish and Wildlife Service always came out and rode with us. He was very interested in cougar; in fact, he'd made a study of cougar. But there was one thing that he didn't know: when a cougar killed a deer, just how many feeds did he take out of it? In other words, how long would the deer last him? There was a lot of discussion among us on that.

One fall I had a deer tag, and I was going to go deer hunting in a canyon just above the guard station. It was a dry canyon, except that down towards the lower part of the basin there was a small spring, and sometimes it'd have water in it. Shortly before the season opened I decided to go up and find out, because if there wasn't water, it was no use hunting up there. One morning about daylight, I hiked up. Below the spring was a series of little cliffs that the water would run down, and I moved cautiously up through that area knowing that if there were deer they would be fairly close by the spring early in the morning. I was going up *very easy*, a step at a time, and looking, and finally I came up high enough that I could see the spring, and off to the side from the spring was a fair patch of spruce trees. I was looking ahead, and out of the corner of my eye I saw something move in the spruce. I turned real slow and looked, and here was a cougar. It had made a kill that night and had taken a feed off of it and was in the process of burying it. They always bury their kill. The cougar was turned away from me and the wind was blowing down the canyon, but he was coming around, and he came around, not quite facing me, and all of a sudden he threw his head up. There was only my head and shoulders that he could see, but just like that he was gone.

Water and deer were there, so I decided to hunt the area. Jane and Johnny wanted to go with me. The day before the season opened they drove out from town and walked up to the spring, and I had my saddle horse and a packhorse and packed our camp in. I was going to tell them about this cougar kill, but I thought, "Well, if I do, they'll be scared and all shook up," so I decided I wouldn't tell them. We camped probably two hundred and fifty feet from the kill.

One of my horses was a gray mustang, and the other one was a farm-raised horse, and during the night this mustang started fussing. He started pawing the ground and carrying on, and I knew what was bothering him, but I didn't get up–I was confident that the cougar wouldn't bother my horses. Oh, they like colts–they'll kill a colt in preference to a deer–but to attack a full-grown horse a cougar would have to be pretty hungry, and I've only known one case of a mature horse being killed. A big old tom killed a fair-sized horse, a three-year-old: he grabbed its neck just in front of its withers, and he crunched its vertebrae.

Even with us camped there, the cougar had come back, and he'd taken a feed off of the carcass and covered it back up. We went out early and got our deer, and I thought, "When we break camp, I'm going to take them over and show them this cougar kill." So I did, and I uncovered it a little bit so they could see what it looked like, and then I covered it back up. My scent was all around it, and their scent was all around it. Then they took off, and Jane said that she and Johnny walked pretty fast on down the trail! [laughter]

I was staying at the guard station. The next morning I thought, "Now I got a chance to see just how many feeds that cougar is going to take off of that carcass." The deer was a medium-sized buck, probably weighed about a hundred and forty pounds, live. But this was a big male cougar. The next morning I hiked up, and it'd come back and taken another feed. That'd be three feeds.

After he kills it, the first feed that a cougar takes off of a deer . . . he opens up the stomach and eats the lungs and the heart and the liver and entrails, and he buries the rest. The first feed he hadn't eaten an awful lot, but the second time he really consumed a lot of flesh. The fourth morning, he had pretty-well consumed the carcass, but there was still a little bit of meat left on it, and he had buried it, so I knew he was coming back. The next morning I went back out and the cougar had taken the rest of it. He had piled the bones in a nice, neat, little pile, and didn't bury them. During my life in the Forest Service, while riding I would occasionally run onto a little pile of bones, and think, "Why in the world would a person pile up deer bones like that?" But I guess a cougar does it; at least this one did. And it was just a nice, neat, little pile of bones.

According to hunters that have hunted them, after a cougar takes its last feed on a kill it'll lay until its stomach is practically empty, and then it'll travel for probably two days before it makes another kill. They have trailed them where they've gone right past a band of sheep, and they've gone right past deer, and never looked at them—until they get good and hungry. Then they make their next kill.

A killing frenzy Bear and cougar as a rule aren't much of a problem with sheep. Coyotes are the big problem. Of all the time I've spent in Forests out on my various districts—and I've been on districts that had a lot of bear—that incident on the Wildhorse was the only time that I ever saw where a bear bothered sheep. But the trouble with both bear and cougar is that once they start killing sheep, I guess they find out that it's an easy way to get their food, and they'll continue to do it. Once they start, you just about have to get rid of them, or they're going to keep repeating.

The only time that I ever saw a cougar kill a sheep was also on the Wildhorse district. There was a canyon they called Muldoon, and a fellow by the name of Thatcher Kimball ran two bands of sheep up there. They were coming out, and I knew the herder of one band real well, and he'd asked me if I'd be up there at such-and-such a time to count his sheep for him before he took off, so he was sure that he wasn't out any. He'd bedded his sheep down on a little, flat ridge between Muldoon and Lake Creek. I pulled into his camp that evening, no more than three hundred yards from the sheep bed, down next to the creek. He slept in his sheep wagon, and I'd brought my tepee in. That night a cougar and her three-fourths grown kittens–I won't call them kittens; they were probably long yearlings–destroyed a little over two hundred head of sheep. Now, they just went crazy. A lot of sheep they'd just taken one swipe at, and some of them they ripped the stomach open enough that they'd died, and others were wounded so bad that the herder had to kill them. We didn't examine all of them, but most of the damage that we could see was done with their paws rather than their jaws. They had no intention of eating any of them. They went off and left them; they were long gone.

The herder and I couldn't figure out how we had failed to hear it . . . the sheep must have made an awful noise with this going on. And why didn't the dogs start barking? The dogs didn't even stir up a ruckus–they were probably afraid of the cougar, but you'd think they'd bark–and the sheep didn't disperse. They are *herd* animals, and when they're attacked by a cougar or a bear, they just come together. All the cougars had to do was take their time and walk around the outside and just kill to their heart's content. Boy, there was sure a sad herder the next morning! I talked to a game biologist after this, and he thought that the mother cougar was probably teaching her young ones to kill. And I have been told by cougar hunters that if a cat's in heat, she'll go crazy and do pretty much the same thing. She'll even kill a deer and not take a feed off of it. There's just something in their makeup that causes them to do that.

Porcupine: pretty tasty, salt hungry and in the sourdough

My first experience with porcupine was in 1929 on the Kootenai Forest–that was the first time I'd ever seen a porcupine. There was quite a few of them on the Forest, but people in that area had a feeling that you should never kill a porcupine unless you wanted to eat it. As I understood it, the main reason was that porcupines don't hibernate: they're out all winter, and if a man was caught out and he was hungry, he could always kill a porcupine. All he had to do was just take a stick and hit it over the head, and he got it. Back then there was a lot of trappers and miners and people that lived in that back-country and at times could be caught without sufficient food, and it was a very easy source of food.

While we were on trail location it was next to impossible to bring along any kind of fresh meat, so we killed a lot of porcupine and ate them. (And it was illegal, but we did kill quite a few grouse, because when you're working you can sure get meat hungry. There was no hunting of grouse permitted back in that country; they just lived and died. So the effect that we had on the grouse population I think was very insignificant.) Porcupine were pretty tasty. The first one we killed was old, and it was strong! We couldn't eat it, because we didn't know how to prepare it. The next one we killed was a younger one, and we made a mild salt solution and soaked it overnight in salt water. There was very little pine taste to it then, and it was good. You roll it in flour and fry it up; it was just as good as chicken.

Porkies naturally taste piney because in the winter they practically live on the cambium layer of pine, fir, or spruce trees. Down to Region Four they were killing many trees. They get up in a tree, and all they have to do is just go around once and girdle it, and the top of the tree is dead. They chip the bark off with their teeth and eat the cambium underneath, and the cambium has an awful lot of nourishment. The Indians used to take the cambium of fir trees and grind it up and dry it and make flour out of it, and I've eaten the cambium of fir trees. You take the bark off, and then the cambium is real thick on a fir–sometimes an eighth of an inch thick or more. You take a piece of it and lay it up next to your fire one side, then turn it over, and kind of toast it, and those are good eating.

I have seen areas where porcupine liked to congregate. In the wintertime you'd have maybe a dozen of them in a small area where they'd killed just about every tree. I've seen the worst such porcupine kill on the Toiyabe Forest, and there was a lot of timber kill on the Ely district there at Ely. But timber wasn't as valuable as livestock, and it didn't make all that much difference, so we didn't have any big porcupine poisoning programs. They did poison some: on the Sierra front of the Toiyabe in 1959 they would take a little block of wood about four inches long and two inches wide, and they'd bore some holes in it and then mix strychnine with salt, and pack it in these holes. They'd hang these blocks in the trees, and the porcupines would get salt hungry, and they'd eat the block and all. They killed a lot of them that way.

The porcupines were always salt hungry. You don't want to leave a saddle or bridle or a pair of chaps out at night, or the porcupines will eat them. [laughter] In 1929 when Bill Brown, my partner, and I were locating trails in the Kootenai we stopped in at an old abandoned miner's cabin. The miner had made a table out of one-inch planks. Apparently, when he'd eat, he'd sit at the table at the same place all the time. He must have been a hard worker and perspired a lot, because where his arms had rested on the table, the porcupines had eaten out big holes . . . eaten the wood and all! This centerpiece where he had his plate just stuck out. [laughter]

I had a packer by the name of Dick West who was pulling a pack string for me, and one day we arranged to meet at an old trapper's cabin. It was built out of logs and had a dirt floor, and it was just long enough for a bunk across one end of it. The Forest Service had taken it over as a stopover cabin. We had a good horse pasture, and we had a telephone there and a five-man fire outfit. I was pulling a pack string, and Dick was pulling a pack string, and mine was about half empty. So I had arranged that he'd take my empty stock back to the station when he went back. He pulled in ahead of me, and it was dark when I pulled in. I unpacked my stuff and turned the animals loose, and before I went to bed, I decided I'd better mix up my sourdough. I mixed up enough for two of us, and I had the jar pretty full. There was a little shelf above the table, and right next to it was a metal sheepherder's stove to cook on. I set the sourdough jar on this shelf, with the lid off so a lot of air could get to it and work good.

Sometime during the night I woke up. When you're sleeping out like that and you wake up in the middle of the night, you know something woke you, and you lay perfectly still until you find out what it is. A porcupine was right by my head, and it went down along my side, and when it got down to my feet, I started yelling at Dick, "There's a porky in the cabin!" He woke up. I got up with a flashlight, flashed it over the top of the stove, and there he was on top of the wood pile. There was two pins out just above the bunk, and Dick had laid his .30-30 rifle on top of the pins. So Dick got up with the rifle, and he just leaned over the top of the stove and shot. He demolished the porcupine, all right. We both went back to bed, and the next morning I got up and was going to make the hotcakes, and here was porcupine quills and porcupine hair everywhere, including in the sourdough. But I didn't want to throw away good sourdough, so I just scraped it off and got rid of all the quills and hair and made hotcakes, and they tasted pretty good!

Reading tracks in the snow   When I was stationed in the tie camp up in the Uinta Mountains in northern Utah, there was a fair number of Canadian lynx that lived in that area. (From what I've read, that's as far south as the Canadian lynx territory goes.) There was a fellow up there who would hew ties, and he'd also trap. That winter he caught five lynx—some of them were big—and he had them tacked up on the wall of his cabin. He'd open them up and spread the hides flat, which kind of surprised me, because I thought he'd case them like you would a bobcat or a coyote or animals like that. But he'd open up the stomach and then spread them out flat.

Anyway, this day I was coming back to the cabin and I came across lynx tracks. You can always tell a lynx track, because even in the soft ground or a wet snow you can never see a pad mark. They have so much hair on the bottoms of their feet, it looks just like you had shoved a wad of cotton into the snow or into the dirt. The lynx was going the same direction I was, so I thought I would just follow him. Going along, all of a sudden he'd squatted in the snow—you could see where his belly had just flattened out in the snow—and then he'd leaped. I looked ahead and saw where a porcupine had been cutting across. This lynx apparently had killed porcupine before and knew that a porcupine don't have any quills on its stomach. He had run up alongside this one, and you could see in the snow where he'd run his paw underneath the porcupine's belly, ripped it, and flipped the porcupine over on its back. Where the porky got back up on its feet again, there was a little bit of blood, but not an awful lot. Just a little ways ahead there was a windfall, and in that deep snow country there'd always be a big hollow underneath a windfall sitting there at an angle. That porcupine had run underneath there. Now, how the lynx got that porcupine out, I don't know, but all the snow around the windfall was just padded down with lynx tracks, and the underside of that log was all full of porcupine quills. Off to the side was the hide of this porcupine laying flat on the ground; the head and tail were still attached to it. You might as well say the lynx had consumed all of that porcupine.

I thought, "Well, he sure got some quills in him someplace." So I followed him, and not once did I see where he stopped to try to bite quills out of a paw. He was walking right along; he wasn't limping. It'd be very interesting to know how he was able to kill that porcupine without getting quills in him. Up in that country–this is pretty high country–about all that was up there was porcupine and rabbits, so that must have been what the lynxes were living on mostly.

Photographing a moose calf

One time on the old Wyoming Forest I was on a timber survey with Clark Miles, the assistant supervisor. I hadn't been in part of the area at the head of the Green River, and I didn't know whether there was enough timber in there to make it worthwhile to survey or not. One weekend Clark came out, and we decided to take a hike over this area–they called it the Roaring Fork.

It was early enough in the season that the moose were just having their calves. We were walking along through a willow bottom–part of it was kind of swampy, with water standing in places–and we came onto a moose calf that probably had been born the day before. It was *real* young, and it was lying there stretched out. Clark said, "I want to take a picture of it." So he took a picture of it, and then he says, "Arch, can you get it to stand up? I want to take a picture of it standing up." So I tried to lift it up, but it flopped back down. Finally it got over its instinct to just lie there, and it stood up and Clark took some pictures of it. It was a cute little thing.

The mother moose had taken off when we had approached through the willow bottom, and she was up on the side of the hill, probably worrying about her baby. We had Tom Matthews's dog with us (a Labrador-Setter cross) and he was smelling all around it. When we resumed our hike this darn calf decided to follow us! The mother moose started making a noise up there on the hill, and the dog took off . . . he was going to chase her. She ran off a ways, but when she saw the calf following us, she turned on the dog. Clark and I figured we had better get away from the calf . . . Clark was a lot older than I was–I could take off running, but he couldn't. We were trying to get away from this darn calf, and we could not! He stayed right with us. [laughter] Clark was scared, and I was kind of scared, too. He says, "What are we going to do?"

"Well, Clark," I says, "I'll grab ahold of the calf and hold him down on the ground, and you take off and get some distance. Then I'll let loose of him and take off as hard as I can run." So I held him down and Clark ran, but the calf started bawling and that alarmed the mother still more, and pretty soon here she come, the dog in front of her, running as hard as he could run. The dog saw me, and he dropped down right behind me, and I was holding this darn calf! [laughter] I jumped up, and I didn't care where Clark was or anything; I just took off as hard as I could run–the dog right behind me. If that cow hadn't found her calf, she'd have worked somebody over. I don't think she'd have caught me, but she'd sure have caught Clark! [laughter] But she ran onto her calf first, and when she saw it she stopped, and Clark and I and the dog got away.

You take a bull moose, you don't want to fool with him any time, either . . . at least not the ones they had up there on the Green River. When we were cruising timber out in a place called Porcupine, there was one big, old bull moose that sometimes would decide to stop right in the middle of the trail. And boy, if you came along with a pack string, you went *around* him. He wouldn't move. You could throw sticks at him, and all he'd do was just put his head down; and you didn't want to get too close, because I'm sure he would have charged.

Apache, my mustang colt All the time I was on the Challis, I had to buy my own horses, but when I came to the Ely district the government bought them. I was going to bring my own horses down, and the supervisor said, "No, the Forest Service furnishes the horses." So I sold my horses and used the government horses. Later on I caught a mustang and broke him, and I used him. He turned out to be a dandy horse. This is how it happened:

I was going up a canyon one spring, and I ran onto this band of sheep; they were on water, and it was Dan Clark's sheep. Of course, he hadn't notified me that he was coming on, [laughter] and I hadn't counted them. So I told the herder that I would like to count his sheep.

Well, counting off of water in the middle of the day isn't really the best way, but I tried it anyway. We moved the sheep up a little ways to where the canyon narrowed down, where we were going to count them through. The sheep were scattered out, and the herder was on one flank, and I was on the other flank, kind of shoving them along. I looked down into the canyon bottom, and I saw something black lying there. I'd already kind of gone past it, and I thought, "Well, there's one of his markers." So I dropped down into this fairly high brush, and a little black colt jumped up. From the look of his mother (I took her to be a mustang), she hadn't been dead long enough to bloat yet, so I guess that he was probably two or maybe three days old. She'd died giving birth to him, but you could see his little tracks in the dirt where he'd known enough to . . . he had enough mustang in him to go down and get water. But he didn't have any teeth, and the cheatgrass–he'd grab ahold of it and just work it around in his mouth, and he'd finally get some of it off.

The little guy stood there, and he was going to fight me to protect his dead mother. This fellow was so small that I just gathered him up and took him back and put him in the trailer. When I came back after counting the sheep I put my horse in the trailer, but I was afraid the horse would step on him, so I tied the colt's feet together and laid him on the seat in the pickup next to me, and I brought him in.

He never had a drop of mother's milk in his life. The only thing that he really liked was rolled oats: he'd really eat rolled oats! I'd buy alfalfa hay, but he couldn't eat the stalks, so I'd break the leaves off or let him break them off, and he ate these alfalfa leaves. All he had was water and oatmeal and alfalfa leaves for the first six months, and he hardly grew at all. He had a big potbelly on him, and I thought, "He'll *never* make a horse." But finally when he got his teeth so that he could start eating, he started straightening out. Oh, he was a good horse! He never got very big–fully grown, he weighed probably around nine hundred pounds, and if he was in good shape, he might go a thousand. But *tough* . . . oh, he was tough! I named him Apache.

I'd look at Apache and think, "You sure don't look like a mustang." In fact a lot of people, after Apache grew up, took him for a quarter horse, and you could have passed him off as one. The year after I found him his mother's body had bloated and maggots and whatnot had eaten her, but the coyotes hadn't bothered her. I went up there and found the skin and bones, and I rolled her back end over as much as I could, and sure enough, there was a brand. I later learned that the mare had come from a ranch just south of Ely that had been purchased by a woman from California. She raised quarter horses, and she turned these quarter horses out on range outside of the ranch. Sometimes they'd get up on the Forest, and when they did she'd send a rider out to bring them back down. This old mare had been in the bunch, and they figured she was too old to come in heat; but apparently out in that wild grass (I don't know if that is what did it or what) she had come in heat, and a mustang stud had picked her up and taken her into his band. I thought maybe I ought to take Apache down and give him to the ranch, but my kids just went into orbit: they wouldn't stand for it! So I figured, "Well, I got enough money into him that he's worth a lot more than he was worth when I found him." So I kept him. Maybe that's dishonest; I don't know.

Apache turned into an exceptionally good horse, although he piled me one time up on top of Mount Moriah–it really wasn't his fault. A permittee named George Elridge told me that he was going to meet me there at eight o'clock, so I went out and saddled Apache and had him tied up, but Elridge didn't come. Apache really got antsy, and finally about nine o'clock the guy came. Well, he was riding about a half-broke horse, and he was a fairly old man. When we were ready to go, he was having a hard time getting on his horse, so I held his horse down while he got on. The horse, of course, jumped around quite a bit after he got on, and that disturbed Apache some more. We had a little pasture there, so I thought, "Well, I'll take him in the pasture, and if he piles me, he can't get away from me." Sure enough, I hadn't even gotten my foot into the off stirrup yet when he bogged his head. I turned him into the fence, and he didn't buck too far, but before he hit the fence he turned real quick and off I went! [laughter]

Apache just ran off a little ways. We got him, and George Elridge says, "Let me hold him." I wanted to get on, but George kept jerking him around, which was no good, because he was a spirited little animal. So I took him out in the pasture and walked him around and brought him back and got on him, and he was all right.

I didn't take Apache to the Toiyabe Forest with me when I transferred. I found out afterwards that I could have, and I regret that I didn't. But I sold him to the ranger on the White Pine district.

13

Ridding the Range of Undesirables

ALTHOUGH I DIDN'T feel too strongly about it, I had some reservations concerning the need for predatory animal control as far back as 1932 or 1933. Of course, rangers didn't get directly involved in this; back then all such work on the national forests was done by the biological survey people.*

Coyotes get the blame and the poison   Montana had a state grazing association, and I think the state put a certain amount of money into predatory animal (principally coyote) control through the association. In that country there was a lot of downed timber, a lot of snags. On the national forest we had a form that livestock owners had to fill out in the fall listing their losses by snags, by predatory animals, and by poison plants. It is true that ranchers were losing a certain number of animals to coyote kill on the Forest, but far more animals were dying from snagging than from either coyotes or poison plants. Nonetheless, they were showing their biggest losses as due to

---

* Biological survey men were under the authority of the Division of Biological Survey of the Department of Agriculture. The division was transferred to the Department of the Interior in 1939, and was merged with the Bureau of Fisheries to create the Fish and Wildlife Service in 1940.

coyote kills so they would get more money from the state for predatory animal control! (Money was not given directly to the ranchers, but the state paid the predatory animal control man.)

Ranchers recognized that coyotes were not a serious threat to the sheep population on the national forest, but at the same time they wanted as many of them killed as possible, because off the Forest they were a threat to newborn lambs and calves. I had ranchers come right out and tell me that. I'd go over the grazing report with them in the fall, and I'd see these tremendous losses to coyotes and maybe only ten or fifteen to snagging, and sometimes none to poisonous plants, and I'd ask about that. I'd say, "You know you didn't lose that many sheep by coyotes."

"Well," he says, "we got to show it by coyotes, because if we don't they'll cut our control money off."

The biological survey men did some trapping, but most of their work involved the poisoning of coyotes and bobcats with strychnine. They also used a method they called denning. They would find a coyote den in the spring when they had their young, and they'd either poison them with gas or . . . . Most of the coyote dens would have more than one opening, and they could just build a fire, draw the smoke down into the den and kill the pups and adults too.

When I was on the Rapid River district on the Challis, the fellow there was primarily a trapper, and he mixed up his own scent. (Just about every trapper had his own secret formula for scent.) He'd just take a little stick and dip it into the scent and stick it in the ground right close to his trap, and the coyotes could smell it for quite a few hundred feet, I guess. That'd bring them into the trap, and he'd catch them. The coyote was the main predator problem on each of the Forests I was on.

I never liked strychnine poisoning because it wasn't always fatal, and you don't know how many eagles, hawks, magpies, and other meat-eating birds died from it. I've seen coyotes that had not quite eaten enough to kill them, and boy, those were some of the awfullest-looking animals you ever laid eyes on! They'd be so stove up they could hardly walk, and the hair would be sloughing off of them. There'd be big bare patches of skin on them, and they'd just be a bag of bones. Since they weren't able to hunt, they'd eventually just die of starvation. I always carried a gun with me, and any time I run onto a coyote on strychnine, I'd shoot it.

They used strychnine for years and years, and finally they came out with this 1080 poison, which was very deadly on coyotes and other members of the dog family. It also killed a lot of bobcats. I've been told that an eagle or a hawk had to consume an awful lot of meat to die from 1080, but I have found bald eagle and golden eagle and hawk that I'm sure died from it. I've also been told that a person could consume a great amount of 1080 poison and it would not kill them. That I don't know, and I wouldn't advise anybody eating any to prove it. [laughter]

Cyanide guns   After I was transferred to Reno I was in charge of wildlife; that was part of my job, but I was never in favor of cyanide guns and I was against 1080 poison. Back then, if a cow died for some reason or another, they'd put a bunch of these game-getter cyanide guns around the carcass to kill scavenging coyotes. For some reason cattle would often come into a spring and die around springs; I don't know why, but they would. One time the ranger and I went out to this spring that he was thinking of maybe developing. A critter had died right close to the spring, and we were walking around, and I looked down and here was this cyanide gun, this game-getter. It just scared the daylights out of me, because suppose someone with a family came in there with little kids; and this kid walked out and saw a little puff of cotton, and reached down and pulled on it. That would have been it. Not only children, but adults wouldn't know what it was. They might see this wad of cotton lying on the ground; it'd be out of place, and they'd wonder what that cotton was doing there. If they picked it up, it'd be all over.

Whenever they put 1080 stations or cyanide gun stations on the Forest, we rangers had to approve their location. The Fish and Wildlife predatory animal control man had to come in and show exactly where they were going to place each one. Then before the livestock came on to the Forest for a grazing season, all these 1080 stations and cyanide guns had to be gathered up. (That wasn't always done, but they were supposed to be gathered.) But before I left Ely, I prohibited them on the Forest; they continued to put them on the BLM, but I wouldn't authorize any on the Forest. Then when I came in to Reno on the Toiyabe Forest in 1959, one of the first things I did was to do away with the cyanide guns. I think it was the next year before I got rid of the 1080 stations. When I put a stop to it on the Toiyabe, the predatory animal control man went right to the Forest supervisor, and the supervisor called me in and wanted to know why I objected to these measures. I told him exactly the reason I objected, and he supported me. The Fish and Wildlife office was in Portland, and I guess *they* even contacted the regional office and wanted to know what was going on. Although the Toiyabe was probably one of the first Forests in the region that put a stop to it, I'm sure it wasn't too long afterwards that all Forests did.

Larkspur non-eradication   About the only work that we ever did on control of poisonous plants was the control of tall larkspur. In places tall larkspur will grow as high as five feet, but even in a green state it has a brittle stem that is easy to break off. There was quite a little larkspur control done out on the Wildhorse district with CCC boys, and I know there were districts in other Forests where they were doing the same thing from around 1938 through 1942. The method was that the ranger would report a bad larkspur area, and they'd actually go in with shovels and dig it up. Then they'd wait until the larkspur dried and they'd burn it so they could be sure they destroyed the seeds, too. That was effective, but if they left so much as one plant . . . . They got the big ones, sure, but you're not going to get all of them. Afterwards, you could turn livestock in there for a few years and not have any problems because the surviving larkspur weren't big enough to give enough feed to bother a cow.

Eventually, the larkspur would all come back again, so then we tried fencing off those areas. It seemed like the higher up the canyon you went, the more larkspur there was, so we'd fence off a canyon so the stock couldn't get up there . . . which sounded all right, but we were just creating an ideal situation for the production of larkspur, and pretty soon we had larkspur showing down below the fence. So we either had to forget about it or else build another fence across. [laughter] And each time we built one, we were fencing off good range.

On the Antelope allotment of the Wildhorse district they had an awful lot of larkspur and they lost a lot of cattle. They'd read or found out from somebody that if they mixed up a certain kind of salt it would keep the cattle from dying of larkspur poisoning. (I wish I could remember what all went into the salt.) There was one time I believe I saw this work. I was riding with a herder and we'd run onto a bunch of thirty or forty cattle in an aspen stand, and there was some *real* heavy larkspur. The herder says, "I don't dare to move them . . . ." We could see they'd been there for quite a while. He went down and got a couple of bags of this salt and brought it up and scattered it around where every one of the cattle could get to it. And he didn't lose any cattle in that area.

On the Toiyabe there were some fairly wet swampy meadows up in the Sierra. Where you have wet, swampy ground, you have water hemlock, and that's very deadly. The tops don't bother cattle too much, but the plant isn't well rooted, and they'll pull it up and it'll float on top of the water. The tuber is about an inch and a quarter in diameter and up to four or five inches long, and for some reason cattle like these tubers. These tubers are very deadly, and I've known of some eradication of water hemlock.

Sagebrush removal   We did a lot of eradication of sagebrush, either by spraying or by plowing. If we had a fair stand of grass and heavy sagebrush that was competing, then we'd spray, because it'd take the sagebrush out and leave the grass. But where the grass was gone and it was a suitable site where we were sure we could grow crested wheat, then we'd go in and plow it. We'd actually do soil preparation just the same as you would on an agricultural field, then plant crested wheat, and we got some *excellent* stands.

In the fall of 1947, not too far from the old mining town of Hamilton, the Forest Service did some reseeding on sagebrush ground. Much was done with what they called a Dixie harrow, which was made up of eight to ten well casings, each about ten feet long, with eighteen-inch teeth welded to their walls. These spiked casings were fastened to a railroad rail with a swivel so they could turn, and the whole device was dragged behind a tractor. That'd knock the sagebrush down. When we used that, we had a gasoline-powered seeder–I think they call them cyclone seeders–that we mounted on the back of a tractor, and we scattered seed ahead of the Dixie harrow. Some of the seed we drilled, and where we drilled, we got an excellent stand . . . in fact, in some places we had too good a stand–the crested wheat came up too heavy and the plants never got very large. I go hunting just about every fall over in this area, and places where we drilled back there in 1947 you can still follow the drill rows, just one right after the other. Those plants have survived.

"They poisoned an awful lot of rabbits"

While I was ranger at Wildhorse we had a few real heavy snow years while we were at a peak in the rabbit population. There was a lot of ranchers below Arco–that's the next town down from Mackay, right on the edge of the Arco Desert–that was having a terrible problem with the rabbits moving in off the desert and feeding on their haystacks. In that country all the hay was stacked–it wasn't baled; it was stacked with what they called a beaver slide. (They slid the hay up this beaver slide and dumped it over and piled it into a stack.) The rabbits got so bad that they started eating around the base of the haystacks. They would stand on their hind legs, and they'd eat everything up to two and a half or three feet. There were actually cases where a stack would begin to resemble a mushroom, and the stack'd even tip over from loss of its base.

You could go into any drugstore then and buy strychnine. So the ranchers would take alfalfa leaves and damp them a little and spray just a little bit of salt on them, and then they'd put them in a sack and put a little bit of strychnine in and shake it till all the leaves had strychnine on them. They always had their haystacks fenced, and they'd sprinkle that poisoned alfalfa around the fence, so the rabbits would hit it before they could get to the hay. Boy, they poisoned an awful lot of rabbits! They'd haul their bodies away in their hayracks. But the rabbits would dig holes under the fences, and it didn't make any difference how often they plugged the holes, the rabbits would still get through.

Barton Flat was more like a basin, really. It was well watered, subirrigated, and there were a lot of great big meadows. All the big ranchers around Mackay were up on Barton Flat. They were bothered with rabbits pretty bad up there, too. There was a rancher there who owned well over a thousand head of cattle, and one rancher, he'd done the same thing with the strychnine and alfalfa. In the morning before he fed his stock, he'd gather up all the dead rabbits and haul them over to a draw and just throw them into it. Then he'd go ahead and feed his stock. He'd been doing that all through this winter; as long as the rabbits were bad, he'd been doing it. The next summer the carcasses rotted away, but the paunches–stomachs–stayed. They looked like great big, oval-shaped eggs ... but brown, like they were covered with real thin rawhide. Some of them were broken open, but a lot of them were not.

The first couple of those that I saw, I couldn't figure out what in the world they were! It was rabbit stomachs.

The rancher had dumped all the dead rabbits over there in the draw, and that fall, when he brought his stock in from the Forest and turned them back out on the meadows, he started losing cattle. He couldn't figure out what was wrong, and he finally got a vet to come down. They opened one of the dead cattle, and the vet tested its stomach contents. He said, "They've been poisoned." The rancher thought about these rabbits. That was the only thing that he had ever used strychnine on around his place, and he went out, and he found that his cows had been eating the alfalfa-and-strychnine-filled rabbit stomachs.

The only time that I ever actually used strychnine or strychnine-treated hay was on a reseeding we had over in Reese River Valley a number of years ago. The first year after the reseeding, the plants are quite small. Rabbit populations run in cycles, and this was an up-cycle in rabbits in Reese River Valley. They were just mowing that grass off for hundreds of feet into the reseeding. We didn't have too much snow that winter, and they had pretty well cleaned out that reseeding. They not only ate the grass, but they dug the plants up; they ate the roots. I believe we worked through the state that time, and it was the game warden at Austin that prepared alfalfa leaves treated with strychnine, which we scattered along the fence. The rabbits wouldn't stay in the reseeding during the day–they'd brush up down in the sagebrush, then at night they'd come in. We killed a lot of them; we killed enough of them that it solved the problem.

Tame horses gone wild   Nobody in the Forest Service ever considered so-called mustangs to be real mustangs–they were just tame horses that had gone wild. In other words, none of them have any ancestry tied to horses that the Spanish brought over. In the southern parts of the United States, maybe down in Texas, there still may be a few wild horses that have a faint trace of that Spanish blood, but all the horses that we have in Nevada are just tame horses that have gone wild. And a lot of this wasn't purely accidental; there were a lot of ranchers in the early days that turned their horses out with intent, because when they were captured, they made a lot tougher animal–had better feet and more wind. Better feet for the reason that there's no comparison between the hardness of the feet of a horse that has been running out in the mountains and a horse that's been running in an irrigated meadow. And that stays with them. I've ridden mustang and shod mustang, and a mustang's always got harder hooves. Sometimes they're so hard you can hardly rasp them.

Periodically, to keep the wild horses from becoming inbred, ranchers would go out and shoot the mustang studs and then turn good-grade stallions out on the range. And in the early days, the Army in eastern Montana was going out and shooting mustang studs and turning in Percheron studs or draft horses. In the early days on farms there through North Dakota, a lot of their farm horses were mustang, and if they weren't mustang, they had mustang blood in them. My dad owned a mustang mare that he bred to Percheron, Belgian and Clydesdale studs, and she produced some awful good horses. She lived until she was up into her twenties. We drove her to school one day, and she died right there in the barn.

That cross between a mustang mare and a domestic stud made some awful good light workhorses. Even in Nevada, up until people started getting all upset about the plight of the mustang, a lot of ranchers depended on wild horses for saddle horses. They rounded up a lot of them, and they broke them for their own use; and the better ones, they sold as saddle horses. I knew a rancher by the name of George Elridge there in Ely–he lived over in Spring Valley–and he had a grazing permit on the BLM just north of the National Forest on Moriah. Periodically he claimed those horses. They were unbranded, of course, but he didn't say they were mustangs, and he claimed them as his. He'd go up there and have a roundup and take the best of them down and make saddle horses out of them.

Capturing mustangs I guess probably as long as there were settlers in the Ely area and they had horses, there were men who made their living rounding up mustang and breaking them and selling them as saddle horses and workhorses. I don't remember any mustangers rounding up on the Forest, but there were three mustangers who were rounding up horses on the BLM. One wasn't really a mustanger, he was a rancher. He ran quite a few horses on the outside, and they became mustangs–he claimed them, but they weren't branded. His mustang corral was just off the Forest, and I'm sure some horses that he caught in his corral came off the Forest.

Mustangers that I knew wouldn't go out and chase horses and catch them; they'd trap them, either at a spring or in a pass. One mustang corral at Ely was built right in a pass, with a gate at either end. They'd leave the gates open so the horses would get used to going over from one side to the other–they wouldn't even know the corral was there. Then one day mustangers would sneak up there and, depending on what side of the ridge the horses were, they'd close one gate and leave the other one open. Then they'd get down below with their saddle horses and spook the mustang, and when they'd run into the corral there'd be somebody right there to close the gate, and they'd have them!

George Elridge trapped mustang at a spring, and he pretty much did it the same way. He built a corral there that they had to enter to get to the spring. The gate was always open and they'd get used to coming into the spring and watering and going back out–no interference. Elridge had a little cabin off to one side, and he strung a wire from it along through the trees to the gate, and he had a trip and a spring on the gate. On the night he was ready to trap mustang, he would wait and watch from that cabin, and when they went into the corral he'd pull the wire and the gate'd slam shut, and he'd have them. The gate was made out of two-inch metal pipe, and in the middle of the gate one of the pipes was bent way down. He told me he had caught one bunch and the old stud took off, and when he came to the gate, he tried to go through it. He didn't get through, but he bent that pipe down a good six inches or better.

Bad mustangers Back when they used to capture mustangs, there were good mustangers and there were bad mustangers. Some of them went out and rounded the horses up in a corral like George Elridge did, and either broke them out right up there on the mountain, or they brought them in and broke them out down at the ranch. But like everything else, there were bad mustangers. Some of the things that they did were very cruel to the mustangs, mainly in methods used to prevent them from escaping as they drove them from wherever they caught them in to the ranch. These are things that ranchers and old mustangers told me have happened: One way to prevent escape was to tie their heads back to their tails, so they couldn't run. Another way was to tie a front foot up, so they could only run on three legs. Those methods weren't too good, but the worst cruelty, probably, was when they wired horses' nostrils shut so they had a hard time breathing through their noses. That way they wouldn't run, so they couldn't escape.

That was probably the cruelest thing that *I've* heard of, but no doubt there was a lot of other things they did.

Another thing I was told they did involved a muscle in the shoulder; you had to know right where it was. You could cut that, and that'd make that leg completely useless. They'd just have to drag their front leg. It *would* heal up, but it would keep them from escaping until it did. And, of course, they hobbled them, too, and drove them in with a pair of hobbles on. And sometimes they'd do what they called cross-hobble–they'd hobble a hind and a front leg together, crossways. If he was careful, a horse could walk while cross-hobbled, but he couldn't run. It was probably this kind of cruelty that caused Wild Horse Annie to get upset about the taking of mustangs.*

There was never any mustanging on National Forest land that I observed when I was in the Forest Service. Now, George Elridge . . . his mustangs would sometimes get on the Forest, and I know he and his boys have been on the Forest and driven them off, but there was no mustang corrals that were being used on the Forest. However, I've been told that during World War I there was a lot of mustanging being done on the Wildhorse district. One fellow that was actually involved in the roundups said that they drove them from Lost River over to Hailey, where they loaded them in stock cars. He told me that as far as he knew, those were shipped to France . . . probably to eat. But there was never any that I personally know of that took place on a National Forest. The only removal of wild horses was by the Forest Service having a closure and disposing of them by hunting and shooting them . . . and that would be done only when their numbers had grown to the extent that they were doing severe damage to the range.

Hunting mustangs   The mustang is an interesting animal: he's far more alert than deer or elk or just about any other wild animal, and he can see the farthest, I believe, of any four-legged animal that's out in the mountains. He also has good hearing and an excellent sense of smell, and if you're upwind from one he can smell you for a long way. And they are fast learners–you don't have to shoot up a mustang

---

*Velma Johnston, better known as Wild Horse Annie, was born near Reno in 1912. Beginning in 1952, Mrs. Johnston led a three-decade-long public crusade to protect the mustang that resulted in local, state and federal legislation being passed for that purpose.

herd more than once until you're going to have an awful time getting in to them again.

I preferred to hunt alone, but I once got together with the Forest supervisor and two other rangers, and we went over into Murphy Wash and set up camp and hunted mustang together. The first day we had pretty good luck, but after that we couldn't even get within gunshot distance of them. Hunting alone, and having your rifle with you when you were riding, you could see a band and stop and sneak up on them and get off a shot. Once in a great while you could catch them in a patch of timber or a patch of mahogany, and sit down and get set and decide to shoot them as fast as they came out of the mahogany. Then before they could get away from you, you could get two or three. But I don't think that I've ever shot more than three at one time.

The best way to locate mustang is to go around to all the springs and find where they are watering. You can tell by the tracks and their droppings just how long it's been since they've been there. Mustangs often water at night; if they've been there that night, they can't be too far off, because they only go out to where the first good feed is and they stop. So you get on your horse and you follow their tracks until maybe you came to a ridge. You stop, get off your horse, and sneak up on top of the ridge and look over; and if they're there, then you do your best to try to stalk them like you would a deer. Sometimes you can get close enough, and sometimes you can't. But I never shot after night, and I never shot them around a water hole.

If you shot the stud out of a mustang band, and the mares couldn't tell where you were concealed, they would sometimes just mill around for a little bit. When you shot again they'd take off, but if you waited–and sometimes you'd have to wait for half an hour or maybe an hour–lots of times the mares would come back looking for the stud, and then you'd get maybe one or two more. When you shot a stud out, as soon as the mares got near another herd, if the other stud didn't come right out and round them up, the mares would go in and join his band anyway. They were never very long without being around a stud in their society, and that was the reason you sometimes would run onto a band that contained twenty or more mares. Normally in that country a mustang band would never get that big; it would contain anywheres from six or seven to maybe twelve or fourteen breedable mares. (There'd be yearlings and colts with them, too.) Of breedable mares, fourteen would be a good-sized band, and it would take a good stud to hang onto even that many.

The stud watched over the band to a certain extent, but there was always an old mare in the bunch that probably had as much to do with controlling the herd as the stud did. When you jumped a bunch the mare would take off. She'd be the lead animal, and she was the one that determined where they were going to go. The old stud would stay behind to see that everybody kept up, and if a mare lagged a little bit, he'd give her a good, big bite on the rump. So he kept his herd together, and then after they got out a ways, sometimes he'd swing around and take the lead and go. But the old mare in a band probably did as much handling of it as the stud did.

When the young studs in a herd got around two years old they'd start getting interested in the mares, and the old stud would run them off. They'd form little stud bands—from two up to six or seven young studs in these bands. Of course, when they got older they'd start challenging a stud with a band of mares. Finally as he got older and weaker and they got stronger, they'd whip him out, and one of those young studs would take over the band.

We'd always try to eliminate the young stud bands if we could. What good does it do to shoot the stud out of a mustang band, and then turn right around and have a young stud come in in the next few days and take over, see? If you eliminate the studs you go a long ways towards controlling population. We'd also reduce the mare population; sometimes we'd cut a herd down to maybe only four or five mares. Of course, some of the mares are pregnant, and they're pregnant with stud colts. I never did shoot colts, but I probably should have. They just grew up to be mature horses, and you had a problem all over again. But I just couldn't get myself to shoot colts.

Towards the last of the mustang hunting, one of the big complaints was about all this meat left out there just going to waste. To me, very little of that meat was wasted—very little of it. If you shot a horse, you'd be surprised how soon animals and birds like eagles, buzzards, hawks, coyotes, bobcats, skunks, weasels, badgers . . . you name it—how soon they could clean up a carcass. People are always complaining that there isn't enough food anymore for the eagles and the hawks to live on. Well, we were just doing them an excellent service, so I never worried too much.

"An island surrounded by mustangs" Even though mustang over-population on the old Nevada Forest was a recurrent problem we *never* tried to eliminate the herds entirely, because we felt the mustang had a place on the Forest. I like to see mustangs same as anybody else, and I'd be the last one that ever tried to kill out a herd entirely. We just tried to reduce the population to a number that the range could handle. Sometimes we'd kill out a whole band in a certain area, but that didn't make any difference, because the Forest was just like an island surrounded by mustangs, and others would drift in from the outside. Up until 1955 there in Siegel Basin we didn't have a problem, but then mustang drifted onto the Forest from BLM country to the north and found it to their liking, and so we had another mustang band. As the pressure against the mustang builds up due to lack of feed and water outside the Forest, they gradually start moving in.

If the Forest Service and the BLM had continued the policy that we had back in the 1950s, you'd still have plenty of mustang, but you wouldn't have all the problems that there are now. They have reached a point now where I don't think there's any solution other than an extremely tough winter, but if that happened I'm sure some people would start crying to the government to feed the mustang, and the federal government would probably expend millions of dollars flying hay out, and you'd still have your mustang problem.

14

Managing the Range

EACH CLASS OF LIVESTOCK grazes the range differently, and there can be a lot of difference between the way sheep and cattle use it. For example, sheep are with a herd, and the herder can put them wherever he wants to, whether it's down on a stream bottom or up on top of a ridge, or any place in between. With a good herder you can assume that he'll graze the range to the best benefit of the sheep and to the best benefit of the range–the federal land. He wants to come back next year, and he wants to put the most weight on his lambs that he can put on them. But you have all different kinds of herders: some of them are good herders, and some of them are poor; some of them you can talk to; and some "no speak the English," and the only way you can get to them is through the camp tender, who most of the time is either Mexican or Basque and can speak the same language as the herders. There used to be a lot of herders that even the permittee couldn't talk to.

Tepeeing and salting out If sheep are herded right they are fairly easy on canyon bottoms, because the only time a good sheepherder would bring them into a canyon bottom would be to water them. And if there's cover like willows or aspen in the bottom, the sheep may shade up there. They like to shade up during the hot part of the day: they just lay down and rest and chew their cud. Then along in the afternoon about two-thirty or three o'clock they take off and graze some more. As a rule there is little trailing of sheep back and forth along a canyon bottom–they'll come in and water, and then they'll go back out again.

The range that's probably hit the hardest by sheep is on the ridge tops and the passes. A band of sheep will move from one range to another through passes, and this trailing through the passes (where the soil is naturally shallow anyway) can cause a lot of damage. Lots of times the heads of gullies start right there in the passes and go clear down to the streams or the canyon bottoms. And the tops of ridges often get hit pretty hard, because that's where herders bed and salt their sheep. Used to be they would return to the same bed ground for maybe a week or two at a time, and after they'd been doing this year after year, that bed ground was just as bare as the top of a table–sagebrush and everything else was killed on it. Long after this practice had been stopped you could still look out from a high ridge and see large, yellow, circular patches, which were old bed grounds–hadn't been used for many years–where cheatgrass had come in after the other foliage had been killed by trampling and by the salt that had been put out on the grounds. Sheep could do lasting damage that is visible to this day.

A herder who bedded his sheep down on the same ground every night would have a main camp where he'd keep his food, his bed and his sheep salt. When his sheep left the bed ground in the morning and the herder got them lined out where he wanted them to graze that day, then he'd go back to this main camp and cook himself a good breakfast and make coffee. (They always had a coffee pot with them.) And if he had clothes to wash, he'd wash clothes; or if he had to make some bread, he'd dig a bean hole and bake a loaf of sourdough. Sheepherders made a lot of sourdough bread; they'd cook it in a bean hole in a dutch oven, and the crust'd be about a quarter of an inch thick. Oh, it was good! Some of those sheepherders could sure make bread, but I had a fellow that worked for me–Shorty Conyers–who could do more with sourdough than any man I ever knew. He made an upside-down cake with sourdough that was out of this world!

(I can't remember who taught me how to make sourdough, but I've made it ever since I was at Lincoln up in Montana. It was common within the Forest Service then, and most of the lookouts had sourdough.

When I was a ranger I'd always pack some when I went on a pack trip, and even when I snowshoed over the summit I would take sourdough with me. It was not a Forest Service thing–it was just something we adopted that was good. You ate a bunch of sourdough hotcakes for breakfast, and boy, they stayed with you all day. And you could roll a couple of them up and make good sandwiches for noon. It was something that you could pack easy, and it didn't take long to mix your flour and a little bit of milk with it at night; and then in the morning just before you cooked it you had to put in a little bit of soda and sugar, and a little bit of cooking oil, and you were ready to go. And it's misnamed: if you ever let your sourdough go sour, then you were through . . . you had to keep it sweet. But it did have kind of a sour smell to it.)

We used to work with herders to try to get them to tepee out with their sheep away from their main camp–in other words, to pitch their tepee where they wanted to bed their sheep, and each night to move their tepee to a different place, and bed their sheep at a different spot. In addition to the damage caused by trampling, they always salted their sheep when they brought them into a bed ground . . . which wasn't too bad for only one night, but they wouldn't clean up all the salt; there'd always be some that got into the dirt and around the rocks. When it rained, the salt would dissolve into the earth, and that killed a lot of foliage, especially sagebrush. We first got herders to tepee out when I was on the Wildhorse in the 1940s. Both on the Loon Creek and the Rapid River districts they were still pretty much going into those common bed grounds that they had used for a number of years.

If we had persuaded a sheepherder to tepee out, about five or six o'clock he'd have his evening meal, and then he'd load what salt he wanted to use for that night on his packhorse and go on back where his tepee was and knock down the tepee and roll up it and his bed. Then he would move the tepee to where he wanted the sheep to bed that night. He'd always try to put it where it was in view of the sheep, because they got so used to it that when it got time to head for the bed ground they'd head for the tepee. When they'd all come in, he'd salt them. He'd try to salt them on rocks–it kept the sheep from eating a lot of dirt, and it saved a lot of salt, too, that they didn't waste. He'd salt the sheep, and then he'd bed down for the night with them. And the next morning–when the sheep came off the bed ground–he'd head them in the direction he wanted them to graze.

To get the sheepherders to tepee out, we talked to the permittees and convinced them that by coming in to the same bed ground each night they sometimes had to trail their sheep quite a long ways—useless trailing that takes the weight off. And they had the range so fed off around the bed grounds that they'd have to trail their sheep out quite a ways in the morning before they could find anything to eat. We took a lot of the permittees out and showed them the damage that was being done, and they in turn talked to their herders or talked to the camp tender. (And we worked with the camp tenders, especially the ones that could speak English.) It wasn't hard to convince a lot of the herders . . . and this tepeeing out was no big deal with them, really. They always had the tepee on the bed ground, anyway, so whether they had it on the old bed ground or a new one didn't make that much difference, see. But it was a big improvement; it saved an awful lot of trailing and sure helped the range.

In Nevada (well, let's just say in terrain like the kind that you find over around Ely) the foraging of weeds and forbs was potentially more damaging to the environment than the grazing of grass. If you consider all species of weeds and all species of grasses, a grass will stand heavier grazing than your forbs will. Some forbs—I'm thinking particularly of *Senecio* (butter weed)—you can graze very heavy for a couple of years straight, and you will just about kill it off. Most forbs bloom early in the season, and the sheep consume the flowers when they graze the plant off. The plants don't reseed. Grasses—unless they are annuals—produce seed on into the fall, and often by the time grass is grazed its seed is ripe, and the sheep or cattle will tramp the seed into the ground and next year you'll get new grass coming. In that way, sheep are harder on the range than cattle, because perennial grasses will stand some awful heavy use for three or four years straight and still be able to maintain themselves, but your forbs can't stand that heavy a grazing.

The only way vegetation of any kind (other than cheatgrass) could come in to a denuded bed ground was through what is called encroachment. Cheatgrass seeds came in to the bed grounds on the wool of the sheep. The main thing with cheatgrass is that it's an annual that germinates real early in the spring. It comes in thicker than the hair on a dog's back, and it just crowds out any other plant that reproduces by seed. It also has the advantage that if the seed's in the ground and you get any kind of rain in the fall, it'll germinate then and make a nice luxuriant growth and be just ready to grow full bore when it warms up in the spring. So once you get cheatgrass established, it's pretty hard to eliminate–spraying won't even do it, because the seed can lay dormant in the ground for years and years and still germinate. It's a noxious weed that's very hard to eradicate.

Close herding ruins the range When you speak of overgrazing on an allotment, it don't mean that the whole allotment's overgrazed. It means that there are certain areas, certain watersheds . . . or, like with cattle for example, it's certain stream bottoms. With sheep it may be ridge tops or the heads of watersheds where trailing might have been more damaging to the range than the actual consumption of forage. They call it overgrazing, which you think of as consuming too much of the forage, but a lot of it with sheep was trailing damage.

You take a band of sheep that's close herded with dry conditions like you have in the middle of the summer or early fall–just once over a range and they can do a lot of damage. They'll trample vegetation out and they'll break the soil up. You see, these mountain soils, especially your steeper soils, are covered with what they call erosion pavement. It's an accumulation of rocks–small rocks, big rocks–that actually pave the steep slope so that when it rains the rocks break up the force of the rain, and the rain runs off. If you disturb that, break that up, then when the rains hit you don't have this pavement to take the brunt of their force. The rain starts stirring the dirt up and the dirt runs off along with the water. That's the beginning of erosion. It goes from that to little rivulets, into gullies, and bigger and bigger gullies; but the first breakdown of soil, especially on steep slopes, often follows the disturbance of this erosion pavement. Cattle cause the same damage, but it's more prevalent with sheep, because lots of times they're close herding them. Another thing that's bad about sheep is that sometimes herders will hold them out from water too long, and when they get thirsty they really move out–then they don't graze; they just travel. The best herders will graze their sheep into water and graze them back out and open herd them.

We didn't really have any control over how sheep were herded under a grazing permit on Forest land. We encouraged them all the time to open herd their sheep as much as they could, and all your old herders would do that. Of course, the good ranchers were all for open herding too, because it put a lot more fat on their lambs than close herding would. The ewes would give more milk, and after the lambs started grazing they'd get better forage. But your newer herders from Spain, the Basque herders and a lot of the Mexican herders, apparently if they had been involved with sheep, they close herded in the countries that they came from. And if they'd never been involved with sheep, a lot of them were afraid they were going to lose the sheep if they let them spread out too far, so they had a tendency to close herd.

Close herding of sheep was common where there was a lot of timber, and practically all the sheep range on the Lincoln district was in timber. There, they had to close herd because . . . . Actually when they came off the Forest in the fall, a lot of those herders had lost more sheep to snagging and to straying–in other words, getting lost–than they lost to predators. This was because in heavy timber, when the sheep leave the bed ground, most of the time the herder won't see all of them until he rounds them up and puts them back onto the bed ground that evening. That was one reason, too, why herders used the same bed ground over and over again–if you missed some when you put your dogs around your sheep and brought them in, and it got on towards dark, the sheep that were missed would return to the known bed ground.

Transient bands   I don't want to leave the impression that between sheep and cattle one did more damage than the other, because in certain places under certain conditions, cattle can do a lot of damage; and in certain places under certain conditions, sheep can do a lot of damage. But back in the early days before the time of the Forest Service, when they had these big transient bands of sheep, the sheep did do a *lot* of damage. They did a lot more damage than cattle did. Sheep were herded in such great numbers, and the herders didn't have any interest in the range . . . it was a one-time-through thing. They would say, "Next year we'll go someplace else." Just as long as they could put some fat on their lambs and be gone, they didn't care how heavy they grazed the land and they didn't care how bad they trailed it, because they wouldn't come back again. The cattle ranches, on the other hand, were permanent. They had a home base and they had to depend on the same range year after year, and they had to take care of it, even though there was no Forest Service or BLM. A lot of those old ranchers did a darn good job of taking care of the range. You can't say that of the early sheepmen, because they may have started in California and wound up clear over in Wyoming someplace. Then they'd start their circuit back again.

About the time of the creation of the Forest Service, the period of the big transient sheep bands was practically over. There might have been a few yet coming in on what they called the public domain land, which had practically no controls on it–nobody paid any fees. (It wasn't until the BLM came along that the government took over practically all of the old public domain that wasn't Forest Service land.)

A counting game There was no rule or regulation that said it was mandatory that a ranger make accurate counts of the livestock that were grazed on the Forest. Assuming the permittees are honest, they will tell you how many head of sheep or cattle they have put on. I made it a habit to count anyway.

You take a rancher–especially a rancher who has more than one band of sheep–and in the fall when his sheep come off the Forest he makes up what they call winter bands, which are strictly ewes with some yearling ewes to fill in for replacement. He makes them up, and they run those bigger bands all during the winter. Then when spring comes, and the ewes start lambing, he breaks them into his lamb bands. These are the bands that go on the Forest.

Now, in the fall when ranchers make up their winter bands they always keep more sheep than are permitted on the Forest, for the reason that there are always some winter losses, and ewes will die lambing, and there's coyote kills and disease and one thing or another. Sometimes when they make up their lamb bands in the spring they wind up with a few less than they have a permit for; but other times they'll wind up with a hundred head or so *more*. Well, most of them know when they are going to be over, but sometimes they don't know exactly; they wouldn't know until I counted their sheep how many head they might be over. As soon as they found out, they'd come to me and say, "Well, I'm going to be over my permitted number by so many head. Would you figure out the annual months that I have coming, and then adjust my season accordingly?" So I would. They'd run a larger number of sheep for a shorter season, which worked for them and also worked for the Forest Service. Sometimes it could work out where they'd take that adjustment in the early part of the season, which we would very much prefer to do, because that allowed the range to get a better start than if they came on a little bit earlier.

It was the fellows who wouldn't tell us they were over, and who would come on with larger numbers figuring they could get by with it, that were really the trespassers. Or it was the ones who at the end of the season didn't take their sheep off. Now, a ranger is spread pretty thin, especially when he's all alone in some of those bigger districts and he can't inspect every allotment at the time that the sheep go off. So some of those permittees would sometimes stay on a week or ten days longer than they had paid for. Of course, if you caught them on the Forest, they were trespassed.

I learned how to count sheep on the Wildhorse, Loon Creek and Rapid River districts, where the owners sometimes asked me to help them count their sheep, both going on and when they came off to see if they were short. I learned that there is an art to it. The best time to count a band was first thing in the morning when they were just breaking the bed ground. They were the calmest and easiest to handle then. You'd have your band more or less in a bunch on the bed ground–some of them standing up, a bunch of them lying down. Finally a few head would start wandering off from the band. The herder had his dogs, and he would get on one side and you got on the other side, and you would let the sheep pass between you. If any tried to stray away from the band, he'd put the dog around them and keep them bunched, and the sheep would just gradually file between the two of you.

There's kind of an art in counting, too. Since the sheep would come in twos and threes and ones and fives, you couldn't just count one, two, three, four–you had to be able to add a two and a five and a four and a three, and so on. But you got so you could do it without making any mistakes. You don't want to count them going into the sun . . . and, like in the middle of the day sometimes, you don't want to count them going into blowing dust, because when they can't see well they get confused and panicky. And it's nice if you can count them uphill a little–that slows them down a bit.

You can make real accurate counts. I'll bet I've counted a band of sheep, and I don't think I was off over two or three head, if they come through right. If there was a thousand said to be in the band, I would pick up eleven stones. (Ten stones would make a thousand, but I would always pick up one extra in case the permittee was over.) [laughter] Each

Wagons such as these were commonly used by sheepherders throughout the West. It was here that the herder had his main camp.

There is an art to counting sheep.

time I reached a hundred I passed a stone to the other hand. You never count over a hundred, see. And then when you're through, if you've got one stone left over in your hand you've counted a thousand sheep and whatever over that last stone you counted. You want to be careful that you don't drop any of your stones, so when I'd get through I'd always open both hands, and I'd count what stones I had in one hand and what stones I had in the other; and they'd better tally to the number I picked up to start with or I was in trouble. That's happened–I've had to run a band through and count them over again, because I accidentally dropped a stone.

I liked to count sheep going on to the Forest if I could. The permittee would also like me to count them going on so he would know whether he'd left any strays when trailing onto the Forest, which sometimes they had. And a lot of permittees would want me to count when they came off to see if they'd left any sheep back on the allotment. Many times I've counted a band of sheep and had them short as much as two hundred head. If there was a few short they'd say, "Well, that's just coyote loss," or some other loss. But if they were, say, fifty head short, then they'd know that they'd left some behind.

As for cattle, the Wildhorse district was ideal for counting cattle. They had to feed everything in the wintertime; they had to feed every cow that they owned there on the meadows, and there you could get a 100 percent count. If a rancher was over his permit you knew exactly how many head he was over. (When they were on the Forest they had to pay for anything that was six months and older; if a calf was under six months when he went on the Forest, they didn't pay for it.) A lot of the ranchers were maybe five or six head over, and they'd say, "Well, we'll keep a few beef back on the ranch for meat during the summer," and so it was no problem. But some of the bigger ranchers were sometimes sixty, eighty, a hundred head over.

When you get down to Nevada country where most of the livestock, both sheep and cattle, are run on the BLM during the winter, counting cattle is more difficult. In late spring they run on the BLM on alluvial fans adjoining the Forest boundary. Then when it's time to go on their allotment, they just go up and open the gates and the cattle know where they're going. They have been there before and they just slowly drift on–they don't come on in a coherent group so that they can be counted. It may take a week or longer for all the cattle to drift on, and you just about *have* to take the rancher's word as to how many head he's turned on.

Some years this range on the BLM that they're supposed to be running their surplus cattle on gets awful short and awful dry. And if a rancher sees that his cattle are dropping off, I believe many of them would be tempted to go up and open the gate and let them go on the Forest, just purely because of the economics of ranching. You don't want the stock to drop off, and if there's good feed on the Forest a rancher is probably going to take it. You could pretty well depend on what a rancher tells you as to what number of cattle he had turned on . . . but, then again, that varied with certain ranchers.

Salt in the saddles You take up in Idaho there on the Challis, nearly every allotment had a full-time rider, which meant that they could put their cattle about any place they wanted to put them.

But on most of the cattle allotments on the Nevada National Forest, they just turned the cattle loose. There were no herders, and they pretty much went where they wanted to go, which meant they did some damage to the range. On the Ely district and the old Baker district there were very few meadows, and the canyon bottoms were generally quite steep. The borderline of the streams in the canyon bottoms was pretty narrow sometimes, and just the trailing of cattle up and down the streams–looking for a place to drink or a place to shade up–was very detrimental to the stream banks.

The only way you could control cattle on the Ely allotments was by salting. Cattle get salt hungry. If you placed your salt away from water, they'd get salt hungry and feed on out and lick salt for a while, and then feed on back to water. Up to a point this worked reasonably well in drawing your cattle out on steeper slopes and brushier terrain. Unfortunately, one thing that was done on the Ely district–and I originally encouraged this as much as anybody else did–was to place salt in the saddles of ridges where cattle traveled from one drainage to another. They'd come through a saddle and they'd lick salt, and then maybe they'd drop off into the drainage on the other side. In a way this worked, but it caused a lot of erosion. First it caused a bare place in the saddle where the salt was located, and then for thousands of feet out around the salt ground everything that was edible was pretty well consumed by cattle coming in and going out. It also led to a lot of stock trails coming up into a pass, and if we had a cloudburst or a good rainstorm, water started running down these stock trails. Pretty soon you had a gully, and the gully kept getting deeper and getting bigger until it reached the creek bottom. You had these gullies originating right in the pass coming down either side, and I doubt very much if from the time I left the Ely district they have healed up or improved any. That didn't happen only in that Ely country: I believe it was true of just about any cattle range where they salted in passes.

In the 1940s and early 1950s we actually encouraged permittees to salt in passes. We stopped that when we saw what was happening, and we began encouraging ranchers to stop putting salt in the passes, but we still tried to move the stock out from water by using salt. From then on, grazing permits showed the units, their allotment, how we wanted it grazed, and as much as we could we showed the recommended salt grounds.

Sacrifice areas I *know* that through my Forest Service career erosion greatly accelerated. Surely it did on the Ely district in certain areas, and although I was only on the Toiyabe for six years, I'm sure that there has been an acceleration of erosion taking place in some areas there. For a long period of time, in our Forest Service policies, we could accept certain sacrifice areas; we could accept a certain amount of erosion. If we had a canyon that had a gully starting from the top of it, we would accept that erosion if we could still graze a band of sheep or a hundred head of cattle on that drainage. When the regional office would come out on inspections and they would see a gully starting up near the head of the canyon, they didn't like it, but they never said that we had to get rid of the livestock. But I believe that whenever you have erosion taking place–or if you have past erosion that hasn't healed, that is still present–there's no recovery from it: there's no way for these old erosion scars to heal and still permit livestock. But there are many allotments on the Forest where active erosion is taking place, and I assume that the Forest Service is accepting it because they're still permitting livestock to graze on these allotments.

If you have a water trough and you have sheep and cattle coming in to it, you're going to have a sacrifice area whether you call it that or not. The only way you're going to heal that sacrifice area is to quit using the water trough. And the same is true to a certain extent about salt grounds–they too can become sacrifice areas. Now, a herder or an owner can very easily eliminate (or at least decrease the size of) sacrifice areas around salt grounds by changing the salt grounds a couple or three times a season. Years ago, when I was ranger on the Wildhorse district on that Copper Basin allotment, that's what we practiced. We worked with the herder and encouraged him to move the salt. It was good for the permittees, too, because it improved the ease of getting good distribution, and we didn't have too many sacrifice areas around salt grounds.

Other sacrifice areas included allotments where there were narrow canyons with streams coming down the bottoms. There should have been nice pristine vegetation along the streams, but there wasn't because sheep and cattle had to come to them for water. If they come to water, they're going to destroy a lot of the vegetation along a stream by overgrazing and trampling. And often that vegetation growing along a stream was the only reason it hadn't started cutting back into its banks—the vegetation stabilized the stream banks. When you remove that vegetation, then you're going to have bank-cutting; it don't make any difference whether it's a stream three feet wide or one fifteen feet wide.

Improving the range   We fenced a lot of passes to keep cattle from going over; and taking the salt out of the passes kept them from straying over, too. Actually, although cattle were left to run free on their own without a herder, they did not routinely drift off onto somebody else's allotment. When they get used to a home range, they'll pretty well stay there, and they won't go up onto ridges where they could stray over into another allotment. About the only time they would drift off was when salt was placed in a pass on the line between basins; then a few cattle might break over.

Distance from water is the principal factor controlling where cattle will go to graze. It doesn't bother sheep as much: they can be shoved out onto the range and go a lot further from water. Also, most springs up in high country on sheep range have developed sheep troughs. We did that to a certain extent with cattle, too, but a lot more for sheep. In the 1930s a good job for CCC boys was range improvement; they fit in there just perfectly. They were husky kids, they liked manual labor, and it was a comparatively simple task. They built a lot of good fences that still exist to this day, and they built a lot of water troughs.

There was a lot of damage done to the range, however, just by the poor location of water troughs. Apparently the ranger (or whoever was approving the location of some of these water troughs) on the Ely district didn't know what he was doing, because some troughs created more problems than they solved. There were troughs up in steep country where sheep should never have been trailing in and out to begin with; and they put some troughs down in fairly narrow basins where they had to trail down steep slopes to get to them. There was a lot of damage around some of those old troughs, from sheep coming in and going out, and I got rid of a number of them when I was ranger there.

A permittee would say, "Why don't you fix that trough that needs fixing up?"

And I would say, "Well, that trough has gone by the board. We're not going to use it any more. Period."

We used to get a lot more range maintenance and construction money than they do now. Back then, Forest Service personnel did all the range improvement construction and all of the maintenance. We rangers generally had a combination trail crew and range improvement crew that we'd work all summer on fences and water development and trails, but sometime in the 1960s, perhaps due to lack of funds, it was decided that it wasn't the business of the Forest Service to be constructing or maintaining range improvements, and they turned the maintenance of these improvements over to the permittees. Well, some permittees, I guess, did a fair job; but many of them didn't even turn a wheel. A lot of those good improvements, both fences and water developments, have gone to wrack and ruin. I go hunting every fall over around the old mining town of Hamilton, Nevada. This is all cow range, and there were a lot of water troughs that I remember being in good shape, where they watered a lot of cattle; there's nothing left of them now. Where the Forest Service spent good money building water developments, now there are empty troughs just sitting there, the pipes all frozen and cracked and the troughs unable to hold water.

When the Forest Service traditionally took responsibility for range improvement, a certain part of all grazing fees went into a fund that was allocated back out to the Forests and to their districts for range improvement. Maybe they've changed the structure of grazing fees, because they apparently don't do that now, and I kind of wonder what has happened to that money. I don't say we got *plenty* of range improvement funds, but we got a goodly amount, and just about every year we built maybe another water trough or some more drift fence. We were always getting a certain amount of improvements made.

In one area of the Ely district, on the head of Berry Creek, we had some aspen stands that sheep had badly overgrazed. Most aspen stands don't have that much vegetation under them, anyway, and they'd badly overgrazed these. So one fall I took my recreation guard from Berry Creek guard station and we went up there with hand cyclone seeders and planted cereal rye all through the aspens. We got a fair stand; we didn't get a heavy stand, but we did get enough dead rye on the ground to prevent a lot of erosion. (Of course, we took the sheep off.)

Cereal rye is useful in watershed and range rehabilitation for several reasons. It is a biannual, and under good conditions will grow very luxuriant and to a height of four or five feet. Being a biannual it ripens early in the summer and reseeds itself. However, although it will reseed itself, the next crop is a lot lighter than what you planted, and in about three years it is gone. It does not sustain itself. That is the beauty of it: it depends entirely upon man and can not take over an area it is planted in.

For watershed rehabilitation, the best thing about cereal rye is that it matures early, and after the last of June all you have standing are these dead stalks, three feet high or more. Sometimes they are really thick, but you can go right in there among them and plant trees and make good plantings. During the summer the dead rye creates a micro-climate entirely different than any you would find out on a bare surface where you would normally plant trees. In the spring when the small trees are planted there is plenty of moisture in the soil for both the cereal rye and the trees. In June when the rye matures there is still plenty of moisture in the ground for the trees, and after the rye dies the litter from the leaves and stalks that fall to ground protects the soil from evaporation. Those stalks still standing reduce some of the air movement over and around the little trees, preventing further loss of moisture from trees and soil. Thus protected, small planted trees will survive under conditions that would kill them if they were planted out in an open area. If you go into Dog Valley, when you reach the first summit and look down and see all those trees, they were planted among dead rye.

Erosion of the watershed

Watershed problems on the Toiyabe were more extreme than on any other Forest I was ever on. A person must understand just how the Toiyabe Forest is made up: In the western part we have the Sierra front, which has more precipitation than elsewhere on the Forest, and a lot of it is timbered, but your central Nevada districts–the Austin, Reese River, and

Tonopah districts—you'd have to call these more or less semi-desert areas. Precipitation is very low, and much of the terrain is covered with sagebrush. In your lower foothill country you have pinyon-juniper, and you don't get into timber of any kind until you get up to higher elevations.

Although precipitation, both winter and summer, is comparatively low, central Nevada is subject to cloudbursts. (Hot thermals of air rising up off the valley bottoms maybe have something to do with this.) One year while I was on the Toiyabe, from the middle of July to the last of August there was hardly a week that we didn't have cloudburst damage some place on the Forest, especially in central Nevada. This kind of water erosion had been going on for hundreds and thousands of years, and by the time the Forest was created there were gullies coming out of some of those canyons that you could drop a car in . . . and maybe some of them deep enough that you could darned near drop a house in. Each time you get a cloudburst these gullies are cut down a little deeper, and as you go up them you've got finger gullies or channels coming into them. I doubt that there's anything the Forest Service can do that will ever solve that problem.

Controlling erosion by trenching Years ago the Forest Service had a bad erosion problem in a large area in Davis County north of Ogden that had been under private ownership and had been heavily overgrazed. Some terrible floods had come off of this eroded area and brought down huge boulders with them. The whole community and the state and everyone knew that something had to be done. Well, when the CCC program started, they had the manpower and they also had the equipment to do something. Bus Croft was the assistant regional forester in Region Four, and as far as I know, he was the originator of contour trenching for the control of erosion. He was a firm believer that you had to start at the source—the cause of runoff—and after they controlled the Davis Creek erosion with their contour trenching, they went in and seeded it with different kinds of range grass. (I last saw the Davis County rehabilitation area probably thirty years ago, and you couldn't tell it was the same country that these floods had originated from, because it was *real* heavily grassed over, extremely heavy. A lot of the contour trenches were hardly visible anymore.)

In 1938 I went on a field trip to see the Davis County project—they were still doing some work on it yet—and I learned how it was done.

Then over to Ely, we also did some contour trenching. So when it came time to deal with the Dog Valley watershed after the 1960 fire, I was familiar with how the trenches were built, and their purpose. And this contour trenching, as it was proven in the Davis County addition, was so successful that the Watershed Division in the regional office had come out with a lot of literature on the construction of contour trenches.

Prior to the trenching work in Davis County, I don't know of anything that was done on national forests to prevent erosion from water runoff. There was a lot of work that had been done in stream channels by building check dams and trying to slow the water down in the channel, but nothing up on slopes. (At times they'd tried planting different kinds of vegetation to slow the water down, but nothing else.) And I don't think anything has been developed since that was better than those trenches. There's very little of that being done now, but, of course, those trenches cost money.

The trenches served their purpose; the only trouble was that they required maintenance–they were intended to last up until the time that the slope had completely stabilized itself and the vegetation on it could protect the slope. Some trenching work had been done up in Snow Basin out of Carson City at the head of Ash Canyon a few years before the Dog Valley fire, but they used a small Cat and they didn't build their trenches anywheres near big enough. They may be serving some purpose yet, but proper maintenance has never been done to keep them up, and a lot of the trenches have broke, and there's gullies going down. Vic Goodwin was ranger here when that work was done, and he talked to the Forest Service a number of times to try to get them to go up and do some maintenance so the terracing would function the way it was supposed to. But as far as I know nothing was ever done.

To do it right they should have gone in there with a Cat and rebuilt all the trenches, because the back side of them had sloughed in, the trenches had filled up, and the berms had been broken off. Of course, there was trespass livestock that were getting in there periodically, and people had been taking Jeeps up there and running over the terraces, which broke the edges down. For all general purposes they'd been abandoned and nothing had been done. Shortly after I retired the ranger did get a little money and they went up and did some hand work, but since then I don't know.... Of course, I have been away from the Forest Service for a long time, and maybe a lot of work has been done.

In Ash Canyon these trenches should be permanent features because of the type of soil and the terrain. I don't think you will ever get much ground cover on the area. We have been lucky we haven't had a good wet mantle flood or a heavy cloudburst hit Snow Basin for a long time, but one of these days it is going to happen, and they are going to get a lot of water come down through Carson City.

Snow surveys The snow surveys we did were actually for the Soil Conservation Service. The Forest Service did not have too much interest in snow pack data, except that it gave us some knowledge of what groundwater we were going to have and what vegetative growth we might expect, especially at the higher elevations. If you had real heavy snow at high elevations, it meant that your grazing season was going to be late. If you then held your sheep on lower elevations where the range had already reached vegetative readiness, you might graze your lower and intermediate ranges heavier than you really should. If you had lighter snow, it meant that your higher range would reach vegetative readiness a lot earlier, but you may not have as good a range, because you didn't have that subsoil moisture that you would normally have. So it kind of worked both ways. The water content of the snow pack was useful information, and we maintained a continual record of it.

Dr. Church's snow course system was a very effective way of measuring the water content of a snow pack, and as far as I know the method is still being used today.* I doubt that there's been any change in the equipment or the system. The beauty of it was that it was simple (you didn't have to be a college graduate to use the equipment), and the tubes were light and easily packed so you could take as many with you as you wanted. A lot of times it was just one man measuring snow courses. I've measured a lot of snow courses alone, but over to Ely, Kennecott always furnished a man to help me if we did the two extra measurements for them, and they paid his wages. With two men, I don't know how many tubes you could pack. I guess you could probably pack enough tubes to measure twenty-four feet of snow, they're so light.

The SCS laid out most of the snow courses. (There were a few of them that the Forest Service laid out.) You wanted to have an area

---

* See pages 187-188.

where the snow didn't drift, if you could possibly find one. In timber a good place was an open aspen stand, because in the wintertime a leafless aspen stand interfered very little with the snowfall. You'd run your course so no sample would be taken close to the trunk of an aspen tree, because the sun beating in on an aspen trunk will melt a certain amount of snow, and a lot of the snow that melts evaporates. And you wanted the course to be as level as you possibly could get it. There were a lot of snow courses that run . . . well, take the one up in Copper Basin: it was just straight across a meadow; there were no trees or anything around it, but it was in an area that got very little snow drift. Sometimes you get some drift: if you're measuring fifty-two inches, and you jump up to sixty-eight inches, you make a note on your sheet that this measurement was in a drift, and maybe you'd throw that one measurement out and compute the average from the remaining nine samples. (There were always ten stations, as I remember.)

Ordinarily, you didn't want an extremely high elevation for your course, and you didn't want a very low one. The only ones that I've been on where they had both high and low were at Duck Creek, and I think it was primarily for Kennecott that they had the low ones. When they put the courses in, I think Kennecott contacted them and said that as dependent as they were on water from Duck Creek, they'd like a low reading, too, to get some idea of what the early runoff was going to be. They had wells, too, but their mill at McGill was highly dependent on water out of Duck Creek; and what flow came out of there, they were very interested in.

Not only was the water used in the mill, the Kennecott company town of McGill also depended on it. That mill used a tremendous amount of water and, conceivably, a drought could bring ore milling and refining to a halt. Kennecott had a big pipe coming out of Duck Creek, something like forty-eight inches in diameter. Years ago it was a wood pipe, and it had bands around it. [laughter] I don't know how much pressure . . . I was told once two hundred pounds, but I can't conceive of a wood pipe, even with all the bands, standing two hundred pounds of pressure per square inch. They finally got rid of it and put a metal one in. I know when they got a break in the metal pipe, that water shot up there for a good ways! I imagine that's all shut down now. [laughter]

The snow course concept was developed by Dr. J. E. Church at the University of Nevada, and involves collecting snow in an alloy tube to measure water content.

Mechanical erosion control was undertaken after the Dog Valley fire in 1960. Contour trenches such as these were built and were maintained until a slope had been completely stabilized by the growth of new vegetation.

Reseeding   The Forest Service had been doing reseeding experiments for a long time, and after I went to Ely in 1947 we put in some experimental plantings. We had a six-foot grain drill, and we would take a five-acre plot, and plant maybe six rows of crested wheat, six rows of intermediate wheat, six rows of pubescent wheat, six rows of smooth brome and so on at different sites to see what grass really did the best at that particular elevation, with that particular soil type. One such planting had been put in down by Tonopah before I came to the Toiyabe Forest. I went out there with Shag Taynton, the ranger, and we looked it all over, and on the lower part of the planting there was a bare place where nothing was growing. I just figured they hadn't planted anything there, but the next year–a good three or four years after the original planting was made–we went out there and everything was really standing up, looking nice. Down next to the fence where this bare place was, I asked Shag, "Now, what in the world was planted down there?" Something real bright green was growing. He had a map of the planting and he said, "Let me look at it." He said, "That's Indian ricegrass." It had taken all that period of time for it to germinate and come up.

Now, Indian ricegrass is a perennial, and it grows all throughout Nevada. It is a native grass and was a grass that the Indians used a lot for food, and some years it is just loaded with grain. They're little black seeds, harder than can be, shiny, and they have a very hard seed coat. There must either be a growth retardant built into the seed or else they have such a hard seed coat that it may take three years for the acids in the soil to break the shell down enough so moisture can get in and germinate the seed. It had taken that long for these Indian ricegrass seeds to come up.

Native grass was not commonly used in Nevada for reseeding, but there must have been some places the Forest Service had been using Indian ricegrass, because at one time I got a letter from the experiment station over in Utah wanting to know if I knew of any area on BLM or Forest Service land that Indian ricegrass could be combined. Well, going out on Highway 50 just past Frenchman Flat, there was a big area to the north of the highway where the grass wasn't too tall, but that year it was really loaded with seed. So I wrote and told them that if they cut grain on it they could probably harvest all the Indian ricegrass seed that they wanted. In the meantime somebody else must have written in and told them of a closer area, because they never came out. They were probably going to gather that seed to use someplace for reseeding, but of your truly native grasses, I can't think of one that I actually *know* was used for reseeding.

Probably the reason the Forest Service didn't use natural grasses was they had come up with some other grasses that were tougher and had qualities they desired that the native grasses really didn't have. Most of your native grasses will put on a good growth in the spring and early summer, then go to seed; and that is just about it for the rest of the year. There are very few that will green up much in the fall. But as soon as you get a rain in the fall, crested wheat will green up, and if you get enough rains you'll have real good grazing all through the fall. Lots of times crested wheatgrass won't even die during the winter. If you get any kind of a snow on it, it will stay green all winter and take off again in the spring.

The main thing I gained from my 1931 seed viability study was a better understanding of plants that are primarily used for grazing by livestock—especially grasses. I knew which ones were tough and produced seed that could carry over, so that even if you had a poor seed year, you could figure that next year or the next year or the next year you would still get a good stand of grass. And I saw that other forage plants had seeds with very short lives, so if you didn't get seedlings from a seed drop within the next two years or so, those seeds weren't any good. It gave me some idea of how tough and how rugged various plants, especially grasses, were out on the open range.

One application of that knowledge was that in the Forest Service we used to practice (and maybe they still do) deferred and rotation grazing. They would graze an area this year, and next year they wouldn't graze it. They'd let the grass or other plants produce seed and reseed themselves. Then the next year they would graze it lightly, and this would tramp the seeds into the ground; and the next year they would graze it late, and this would give those seedlings a chance to establish themselves and regenerate the range. With some grasses like western wheatgrass you knew that if you didn't get a good take of seedlings this year for one reason or another, you didn't have to worry too much because that seed would live for a long time.

On the Toiyabe we started clearing sagebrush and putting in reseedings about 1960, and then for three or four years afterward we put in reseedings just about every year. I remember one reseeding on the Reese River right close to the ranger station: we plowed the ground, got rid of the sagebrush, did a good job of drilling, but we had a very poor winter (very little snow) and a dry spring, and we got probably 10 percent of a take–very light. The next year we had about the same kind of a spring, and we probably got another 10 percent–still, a very light stand. The supervisor, Ivan Sack, went out with me to look it over. He was quite disappointed, and he started questioning if it was practical to reseed those sagebrush sites so close to the bottom of the valley. I told him, "Well, it doesn't look too good now, but let's not give up, because crested wheat has a long viability and it may come yet." Next winter we had a good snow blanket, good rains in the spring, and most of the rest of the seed came up–an excellent stand.

On all the reseedings then, it was left up to me to decide what to plant. Sometimes I mixed species. If we got up closer to the mountain where the soil got deeper and there was a better chance for a little bit more rain, we'd sometimes put in a little smooth brome, intermediate wheat, pubescent wheat, and grasses like that. But when you got down in the lower part close to the valley bottom, crested wheat was about the only grass that would make it.

On the Ely district in 1946, the year before I got there, they seeded with a grass less viable than crested wheat. On the west side of Ward Mountain there was a series of terraces, and these terraces were comparatively level with good, deep soil. It had been badly overgrazed by both sheep and cattle, and they closed it and went in and seeded it with smooth brome, which is a grass you'd normally find down around creek bottoms or moist meadows. (Bromegrass required considerably more water than crested wheat.) The terraces were fairly high in elevation–the first one was at about seven thousand feet, and the upper terrace was well over eight thousand feet. Both terraces received a lot of snow in the winter and considerably more rain than the valley received in the summer, and the smooth brome did real well on Ward Mountain. When I was in charge of rehab of the Dog Valley burn in 1960, we also planted a lot of smooth brome. But I wasn't doing this blind, because years ago, before there was even a Forest Service, that country was logged real heavy, and in feeding their horses the loggers brought in a lot of bromegrass hay. Of course, the seed got away from them, and there was a lot of brome growing in Dog Valley before the fire ever occurred; so I knew it would grow, and that is one of the reasons I used it.

Equity of grazing fees The last year or two when I was over to Ely, it runs in my mind that the grazing fee for cattle on Forest Service land was somewhere around a dollar and a half a cow-month. On private land in that area, they were charging five dollars a cow-month, and some ranchers were paying considerably more in grazing fees; but I'm sure that there were cattle that came off the Forest in just as good condition as some of those cattle that were being grazed on private land. I don't know precisely what grazing fees are now, but apparently there is still a great disparity, and that just really burns some people up, even in Congress. They say we should be selling grazing privileges to a rancher for the same price he has to pay for them on the outside. And in a way, they're right. Of course, a rancher running cattle on the Forest has more loss than if he's running them in a meadow down in the valley; and it costs him more to take care of his stock—to herd them, to take them out on the range, to bring them back—and there's other charges that he has that he wouldn't have if he ran them down on the meadow. But, even considering those additional expenses, they could never amount to the difference between the two fees. The federal government is subsidizing permittees with unrealistically low grazing fees.

Actually, economics may solve the problem even if the government won't act. In many cases, especially along the Sierra front, due to the increase in population and the increase in land values it is becoming more profitable for a rancher to sell his animals, go out of the ranching business, and subdivide his land for house lots than to continue to use it as a base for grazing livestock. If this happens the rancher's grazing permit on the Forest is revoked (it could not be transferred), and eventually that should reduce grazing pressure on Forest lands.

In the 1930s logging and grazing were about the only uses other than hunting or fishing being made of Forest land. These were good uses, profitable to the U.S. Government. The policy then was to manage the Forest for the greatest good to the greatest number of people in the long run, and our management of the Forest met those standards. However, due to an increase in population, better means of transportation, more leisure time, and probably many other reasons, the recreational use of Forest lands by the general public has gradually increased, until now use by loggers and ranchers makes up a very small percent of the total. As a result there has developed a severe conflict between use of the Forests by the public and use by the timber and ranching people. What the public desires, and what many people go to the Forest to see, the timber people and the ranching people are actually destroying. This has been brought to our attention recently by the problems with the old-growth timber and the spotted owl in the Forests of the Northwest. In view of all this, I believe the time is coming where there will be very few if any livestock grazed on much of our Forest land, especially on the eastern front of the Sierra.

Right or privilege? On many areas of our Forests where grazing is permitted there is evidence in the form of gullies, sheet erosion from slopes, and sand and silt in stream bottoms that significant erosion has occurred. Yet we continue to graze livestock on these ranges, and I wonder how much longer the public will allow the Forest Service to follow this policy.

A distinction needs to be made between grazing rights and grazing privileges, and that distinction has not always been clear in the minds of the cattle and sheep owners. You take a permittee whose family has had a grazing permit on a particular range from the time the Forest Service was created . . . well, you can see from their point of view this has become a *right*. But a regular Forest permit is only granted for a period of ten years, and at the end of ten years the permit is rewritten and reissued or renewed. It's strictly a privilege to graze so many head of cattle or sheep on national forest land for a certain period of time: there are no rights attached to it in any manner, shape or form. And that's it. In talking to permittees the idea that it was a right didn't come up too often, but when you got to these livestock meetings and when they talked through their lawyers, that's when they started putting pressure on the "right" business. They knew the same as I did that grazing your animals on a national forest is a privilege. They were not fooling anybody.

"We should have started cutting numbers" There have been a lot of problems with grazing on national forests over the years, and I think one of the big reasons is that years ago when the Forests were first established and the grazing allotments were set up and the livestock numbers and the seasons were set, the men that did it had very little knowledge from study or experience as to just what a range would carry. They just came up with the best guess that they could as to numbers and season and amount of use that they were going to allow on the allotments. Well, lots of times they allowed too

much; and I'd say that from that time on a lot of the allotments have been gradually going downhill.

A second mistake that they made was that those permits were issued for a ten-year period. As I remember the regulations, at the end of the ten-year period another look should be taken at the range, and if it wasn't adequately carrying the allotted number of livestock for the period that they were set up, they should bring it down to what they considered the proper carrying capacity. This wasn't done. During my entire Forest career, I can't remember a single permit that was changed at the end of a ten-year cycle. I suspect this was because if numbers were reduced, permittees would have written their congressmen, and there would have been a whole bunch of appeals, and the Forest Service for a good many years has been gun-shy of going into court on appeals. They have continued to keep reissuing these ten-year permits. And all this range improvement and water development improvement, and even fencing and working up deferred and rotation grazing plans and all that, has been for the purpose of trying to get allotments to carry an unreduced number of livestock for the season they were originally set up for.

Right at the start, if an allotment wouldn't properly carry the livestock or the use that was permitted on it, we should have started cutting numbers and season length down to where it would. We are eventually going to have to do it if we're going to have our public range improved, and when we do it, it's going to mean that some livestock people have to go out of business. It's that simple. There's no salvation—there's no cure, that I can see, unless that is done. When you got gullies coming down the middle of a meadow; when you got pristine stream bottoms that's being destroyed; where streams and rivers are cutting their banks and you got a lot of erosion taking place, if they can't take care of that and have it improve and still have livestock on, the livestock have to go.

15

Fighting Fire

I'VE ALWAYS ENJOYED fighting fire, whether it was on my district or somebody else's or another Forest or region. There was always a sense of accomplishment, and it was on-the-ground, visible accomplishment, done by a lot of hard work and a lot of planning . . . and sometimes a little help from the weather. [laughter] To go in and put out one of those roaring wildfires like we used to do with no equipment other than a shovel and a Pulaski and a saw–when you did that it was an accomplishment, and every man in the crew would be proud when the job was done. I guess it's similar to men going into battle and they fight and win the day–it's the same feeling that you get, and you begin to feel a strong camaraderie develop among the people who are fighting the fire. When we would all get together to fight it most of us were strangers, but by the time we got a fire out you knew just about everybody there. A lot of Forest Service people that you had never seen before stayed in your memory, and you would remember each other years afterwards.

The best fire detection system What we had in the 1930s was, to me, the best fire detection system that the Forest Service has ever come up with. Of course, wages were low in the Depression, and that probably made possible the system that we had then. When I was on the Loon Creek district some summers I payrolled up to thirty men. These were lookouts (and doubles and triples on the lookouts, sometimes), patrolmen and special suppression crews that we stationed at different places on the district for fast attack. When we doubled up on lookouts, one man could go any time and we'd still have one man for detection. Sometimes we'd triple them, and two men could go on fires, or one man would go on a fire and the other two would wait. (After a severe lightning storm, some lookouts would pick up two or three fires from the one peak.) Of course, some fires were in places where they were no danger–up in the crags, they're no danger. On those the lookout could go to one fire and get it under control and then go to another one. But lots of times you'd get a strike and it wouldn't lay there; it'd take off. That's when the suppression crews really came in handy, and then we had trail crews out, too, the latter part of the season. They built trail, but they were strictly on fire money and they were ready to go, too.

Starting about 1938 we developed a seen and unseen areas map. We mapped in everything that a lookout could see from his peak, and then we mapped in everything that he *couldn't* see from that lookout. Maybe some of the area that he couldn't see would be seen by another lookout, but we still had some blind areas, and sometimes they were fairly good-sized blind areas. We'd put out special crews (sometimes one man, sometimes a two-man crew) in these blind areas to patrol. After we worked up the seen and unseen areas maps, we developed a travel map so that from his peak a lookout could hit everything that he could see in fifteen minutes. A lot of lookouts had horses, and for them along a trail the fifteen minutes would extend a greater distance, but where one had to go cross country the amount of timber, the terrain, steepness of the slopes and everything else were considered in this fifteen-minute zone. Then we drew a half-hour zone; and then an hour-and-a-half zone. These were all like big concentric rings around the lookout peak. As I remember, we didn't go anything beyond a two-hour zone: we figured that if a lookout had to travel two hours to a fire, we would take care of it with other fire crews.

Smokechaser fires   A smokechaser fire is where you've just had a lightning strike, and the fire hasn't had time to spread. Most of the time it's a single tree burning, maybe with a small area burned around the base of it. Sometimes by the time a man would get to it, it may have burned a quarter of an acre; if it had gone beyond that, one man would have a hard time handling it. When the fire season would permit them to leave the lookout, the lookouts would go to a smokechaser fire. Later on, when the fire season was so bad that we didn't want a lookout to leave his peak, we'd double up and put two men on the peak or sometimes even three men, and one of them would go to it. But most smokechaser fires were one-man fires.

When you were out fighting little fires all by yourself, you got up at daylight and had breakfast, and then you'd put a couple of cans of ration in your pocket and go out on the fire. You'd stay there until dark, and then you'd come in and eat and go to bed. You'd try to have your camp reasonably close, because there's no use wasting a lot of energy walking back and forth from camp to fire. Your camp would be close, and it also had to be close to water, of course–all you had was a little quart canteen or water bag. Sometimes the fire wasn't near water, and then you had to camp some distance from it. Of course, one of the prerequisites was that you had your camp in a safe place, but one where if the fire blew up you would know it immediately. During the day you would see a fire blow up if it happened, and you were keyed up enough at night that if the fire made any amount of noise you'd be awake–I never had to worry about not waking up if one started to flare up. I'd jerk my boots off before sleeping at night, but on fires I slept with my clothes on, and lots of times I'd wake up in the middle of the night, kind of worried, and I'd look around. After the second day you should have had your fire down where there wouldn't be very much light.

You were not getting a lot of sleep, because you were up and down all the time, and being in smoke so much of the time was hard on your lungs and your throat and your eyes. Generally, when you came out of one of those fires your stomach was upset and you didn't feel too good for a few days. I don't know if it was from the smoke and ash, or drinking so much water. . . it may have been a combination of the whole thing. Sometimes your water was covered with ash, and you had to spread the ash back to get to it. Well, when you're always gathering up a little bit of ash with your drinking water, that will give you the runs. You can sure get sick from that.

All the time that I was a ranger we would send only one man out on a smokechaser fire, unless it looked like a big fire that really had a start; then we'd send two. We'd give them a certain length of time, and if that smoke didn't start dropping down, we'd start doubling up on it. We continued to send out only one man right along through most of my career. I didn't see them routinely double- or triple-man smokechaser fires until I was transferred to the Toiyabe Forest in 1960. By then, instead of hiking them in to a fire, they'd drop a lot of them in by copter and pick them up by copter, which helped a lot. And they were not going out with so little equipment and no access to water other than what they could find. (The copter could bring water in to them.) As far as their actual fire-fighting tools, about the only thing they had that we didn't have in 1929 (other than radios) was their power saws, which, of course, were a big help.

In Region One in the 1930s, if you had a smokechaser fire you trenched it and made it fire-safe. You pruned up any trees along the fire line and a lot of the trees outside of the fire line, but anything that was burning on the inside, you let burn; you didn't try to put it out. The idea was to let the fire consume all the fuel that was inside the fire line, and when the fuel was consumed, the fire was out. When you got down where you thought the fire was out, and maybe a little bit of smoke would come up, then you'd go down and dig a hole and bury that smoke. But mainly you let the fire burn . . . . That's why they left smokechasers on site for three days, to give them plenty of time to let this fire completely burn out. You had to stay on the fire twenty-four hours after the last smoke showed, so if you were on the site three days and there were still a few little smokes curling up, the guy that took your place would have to stay on twenty-four hours after the last smoke showed. But when you were through with a fire, you were pretty sure that it was out.

Well, later when I was in Region Four–and I imagine this was true on all the regions–we would get a good line around the fire, and then we'd go in and do what they called mop up. Anything that was burning that wasn't too big, you'd dig a hole and bury it, or sometimes you'd cover it. Or maybe it was a big log, and you'd chop the fire out down to clean wood; what you chopped out, you'd bury. It was a good way to put your fire out a lot quicker. But the thing I didn't like about this system (and it happened a lot of times) was that some of the burning wood that was buried . . . you thought it was out and it wasn't. And maybe after you'd left the fire, a wind would come up, and pretty soon you'd have some sparks flying around, and you'd have some more fires.

Perhaps one of the main reasons the approach to fighting small fires changed was the old bugaboo of cost. Back when you were paying a fire fighter maybe seventy-five cents an hour, you could afford to leave him on a fire for three days. But now if you left, say, three men on a fire for three days, you got a pretty big bill staring you in the face. So they have to put the fire out just as fast as they can. Of course, they can always call in a retardant drop, or they can call a copter in with a water drop, or . . . . In fact, they have so many things for fighting fire and I'm so far behind, that I wouldn't even try to guess how they fight fires now. [laughter] It's been going on twenty-six years since I retired.

In the smokechaser outfits we carried a Pulaski. I've been told that back in 1910, when they had those terrible fires, there was a ranger by the name of Edward Pulaski who invented this tool, and it worked so well that the Forest Service adopted it as a fire-fighting tool. They're still using it today, and they'll probably use it for a long time to come, because it's a very versatile device, a two-in-one tool: it has a head drawn into an axe blade on one side and a heavy hoe blade on the other. You could use the hoe end of it to dig trenches, and if you came to a root, and you didn't want to dull the axe side of it, you could chop through the root with the hoe end. A Pulaski wasn't near as good for felling trees as a double-bit axe was, but you could do it. And the hoe blades didn't last too long–you could wear them down pretty fast by digging and chopping–but you had a file and a whetstone in your smokechaser's outfit, and when either your axe blade or your hoe blade got dull, you could sharpen it . . . and you could always get new heads.

The only thing that I disliked about some of the Pulaskis–and maybe they've corrected it now–was that the eye in the head was quite small and didn't have sufficient taper to it. When you'd hang the handle (when you'd drive your handle into the eye), and you put a wedge in it to spread the handle head so it wouldn't come out, sometimes on certain Pulaskis the eye didn't have enough taper. One time I had a crew on a fire, and there was a windfall across the fire line about five feet off the ground. It was a CCC crew, and these CCC boys were having an awful time cutting it, because they were trying to cut it standing on the ground. So I told this boy, "Give me your Pulaski, and I'll get on top of the tree and cut it." He handed me his Pulaski, and I was standing there, straddled, and made a swing down between my legs; and when I swung, the head came off. It just missed the CCC boy by inches and drove itself into a log. Now, that could have been a serious accident. I've seen the heads get loose on Pulaskis, and we'd have to tighten them up, but that's the only time that I've seen a Pulaski with an apparently solid head just fly off the end of the handle.

Each smokechaser outfit had a little first aid kit for minor emergencies. It held antiseptics and some burn salve and bandages and band-aids, but that was about the size of it. We also had a large first aid kit that was better than the small ones. In the large kit there was a pamphlet that showed how to make slings, treat wounds, cuts, burns, et cetera, and how to set minor fractures. This large first aid kit was included in the ten, fifteen, and twenty-five man fire outfits.

Back in those days rangers did not get special medical training of any kind, but when I was on the Rapid River and Loon Creek Ranger Districts we held a fire training camp each spring for our lookouts, guards, patrolmen and other seasonal personnel. Some first aid was taught at these camps, and first aid literature was available, especially through the Red Cross. (Later in my career the Forest Service recommended that we take the Red Cross first aid course. I took the first one of those when I came to Ely, and then I took a refresher course later. When I was on the Toiyabe I didn't take any first aid courses.) The big first aid box that was kept in the ranger's office contained a good manual that even showed how to set a broken arm, and how to take care of burns and cuts and abrasions and things like that, so you weren't going in to it blind. If you read that book and read it good–understood it–you had pretty good knowledge of how to take care of just about everything that happened on a fire.

Maintaining telephone lines

On the Challis in the 1930s we always had plenty of trail money; we always had plenty of fire money; but for telephones and even the maintenance of a building, sometimes we were pretty short on funds. Yet our big job until the fire season started was maintaining trails *and* telephone lines; and back before we started using radios, a working telephone system was critical for reporting fires.

Before the fire season even really started, we'd put a lot of the boys up on their lookout peaks. There were trails radiating out from the peaks, and we'd leave all those for them to maintain. Then when we got into the fire season, if they didn't have their trails all maintained, they'd maintain trail in the forenoon and be on the peak in the afternoon. So we were sure that before we really got into the fire season all our trails were in first-class shape, and our lookout buildings were in good shape, and our telephone lines were in good serviceable condition. Still, during the summer we were always getting breaks in the telephone line. We could generally tell pretty close where the breaks were, and we didn't have to hunt too much to find them: if the dispatcher could talk to one lookout but he couldn't talk to this other one, we knew where the break was, and either somebody would go out from the station or one of the lookouts would take off to find it. We had little portable telephone sets, and you would walk along and throw a wire over the telephone line and call on one. If you could connect one way and not the other, you'd keep moving until you'd either find the break or you knew you were close enough to really start looking for it. Breaks were sometimes hard to find, and lots of times as you went down your telephone line you'd find trees across it that hadn't broken the wire yet. Of course, you'd cut those out.

They had what they called a tree line on which was used a split insulator—a porcelain insulator in two halves—and on line maintenance you always had a lot of these insulators that had pulled off and would have to be put back on. One half had a little tit and the other side had a hole in it, and when you'd put them together around the telephone wire they wouldn't move. You'd wrap binding wire around the insulator two or three times and twist it, leaving a tail at either end. Then you'd drive a good, heavy staple into the tree, and run one wire through the staple one way, and the other wire through the other way and bend them down. You wouldn't wrap the wires; you'd just bend them so that if a tree fell across a line and the pressure became too great on the wire, the insulator would pull out of the staple before the telephone wire broke.

Often, you could still use the telephone. (Maybe your communication wouldn't be too good, but you wouldn't be grounded too much.)

When the wind would blow, the line would slide back and forth through the insulators. You wanted to keep as much slack in your line as you could, but on the steeper slopes the line would slide back through the insulators and your slack would all run down hill. That wouldn't do. As part of our telephone maintenance, we'd have to pull slack back with a set of come-along wire stretchers. Every so often in the bad places we'd tie the line hard and fast so that all the wire wouldn't run clear to the bottom, and we'd only have to pull slack on it a short distance. There were a number of ways you could tie hard and fast to a tree: You could use a post insulator and tie to it, but sometimes you would have problems, as these insulators were on brackets and the brackets had a tendency to split, dropping the line to the ground. Or you could put on two tree insulators; on a big tree they would be about one and a half feet apart. You'd bring the wire in and cut it in two, and you'd tie one side hard and fast to one insulator, and you'd bring the other side in and tie hard and fast to the other insulator. Then you'd run a jumper from one line to the other to complete the circuit. That was one way of anchoring it; in fact, that was really the best way. But it didn't take too many trees to fall across until your wire was broken.

One thing that was different then, compared to now, was we didn't have any form of communication of any kind on a fire. (You could report a fire by telephone from a lookout, but you couldn't call out from the fire itself.) [laughter] That's why they would leave a man out only three days on a smokechaser fire, and if he didn't come in they would send somebody out after him. That practice may have saved the life of one young fellow who got ahold of some bad rations.

In the early 1930s Region One put out a fire ration that was probably a better ration than was ever put out by the Army, Navy, or anybody else. R-1 (Region One) rations were made up of canned stuff that you could have taken off a shelf in a store, and they were bulky and heavy, but you could get three good meals out of one ration. The standard ration was done up in a cloth sack. One of the items it contained was a big can of brown bread; also in the sack there was a piece of wire and a nail, and the idea was that after you removed your bread, you'd punch two holes in its empty can, put the wire in for a bail, put your coffee in, and you'd have a coffee pot . . . and also something to drink out of.

Telephone lines such as this were repaired in preparation for the fire season. Falling trees were the biggest cause of broken wires, so wires were often left with quite a bit of slack between insulators.

Then there was a pretty-good sized can of what we called bully beef, from Argentina; there was a can of beans; there was a can of peaches; there was a can of Vienna sausages; a can of pears; a candy bar; a can of corned beef hash; and four cigarettes and a little packet of matches. There was also a napkin and a bar of soap and some paper towels . . . you ate with your pocket knife. Each man carried his own ration, and in a smokechaser outfit you had two rations to last you two days.

(Region One was, and I guess it still is, *the* primary fire region in the nation. It developed its ration quite a long time ago, and it furnished R-1 rations—or sold them, I guess you'd say—to whatever region or Forest wanted them. Region Four used a lot of R-1 rations, and when I was ranger at Rapid River and Loon Creek I ate a lot of them. I believe it was in the late 1930s that the military came out with the C ration and the Forest Service started substituting the C ration for the Region One ration. Well, these R-1s were palatable; but C rations . . . you had to be darned hungry to eat one of those.) [laughter]

Because they were canned goods, you had to gather up all the R-1 rations every fall before first freeze, and you had to put them in a cellar or someplace. They couldn't be allowed to freeze, especially the beans—the beans were the worst. If a can froze, the soldered seam would bulge; and it could bulge just enough to let air in, but not enough for anything to run out so you could tell. Every once in a while some rations would freeze, and occasionally there were cases of ptomaine poisoning from this. (In fact, I got ptomaine from a can of beans once just because it had frozen.) Other than that, R-1s were exceptionally good rations, but they were heavy . . . they were heavy!

One time this young fellow went out on a smokechaser fire. He went out, and when he got to the fire he decided to have something to eat, so he opened up this can of beans and ate from it. Well, it had previously been frozen and it was spoiled, and he got ptomaine poisoning. I don't know whether somebody had told him what to do or what, but included in the R-1 rations was a bar of soap, and he started eating it. When his three days were up he hadn't come out, but the fire had gone out by itself. The ranger sent somebody in and he found him, still alive. They got him out, and the doctor said the only thing that saved him was that soap. I don't know what it did—maybe it made him throw up the poison food, or made an emulsion in his stomach to stop the absorption of the poison, or something. Anyway, he was still pretty sick when they found him.

CCC: *"They were ready to go"* A project fire was a fire that you would have to bring in a crew to suppress. When I was at Loon Creek the initial attacks on project fires were mounted by crews from our CCC camps. They'd go in as a unit in twenty-five man crews, and these boys were some of the best fire fighters that the Forest Service ever had. They were well trained, and those kids could work harder and recuperate quicker than any group of men I've ever been around. You could work them until their tongues hung out, and they'd be so tired in the evening coming back to camp that they could hardly pick up their feet. But fill their bellies and give them a night's sleep, and they were ready to go the next morning. I don't know whether it was the grub they fed them or what, but . . . well, it was the grub and the kind of life they led. Around camp, I think the lights were out at nine o'clock, and they got a good night's sleep. There were some good men among those boys. The Army sure found that out when the Second World War started, and they started picking up those CCC boys.

One of the best things to come out of the New Deal was those CCC camps. They accomplished a tremendous amount, and did many things well in addition to fighting fires. Those CCC boys could build a ranger station complete and do a real good job: the cabinet work, the plumbing, and if you were where there was electricity, the wiring. This was all done under good supervision by foremen who were experienced plumbers or carpenters or electricians or whatever. Those boys acquired enough knowledge that I believe any one of those, say, carpenter crews, after a year or two could go out and build a ranger station. Many of these were only eighteen-year-old boys, but it would be just as good as any ranger station built by any carpenters in the country. The same way with building a road; the same way with surveying out a road; putting in culverts; building drift fences; water troughs . . . you name it. The best cat skinner I ever saw was an eighteen year old CCC boy.

For some reason–I guess it was the background of these boys–they were ringy when they came. You couldn't do hardly anything with them for a while, and a few of them had to be sent back–they were just incorrigible. But after they went through the first two or three month period, they settled down and were very interested in their work. You never heard any griping. You could work them hard on a fire and you never heard any griping.

We had two CCC camps close to Challis, and a few of the winters when I had come out from Loon Creek for the winter I taught forestry courses to the boys. Of course, I didn't get in to real complicated forestry, but you'd have big groups and they'd have more questions to ask! I don't care what it was, they had a strong interest in what went on. I kind of think that's why so many of these boys became proficient in what they were doing–whether they were carpenters or plumbers or Cat operators or surveyors or whatever it might be–it was that interest and the goal that they had. Sometimes when their enlistment would run out and they'd have to go back to the town where they had enlisted, boys would just cry like babies and not want to go back. They'd come to the Forest Service and say, "Isn't there *some* kind of job, even though it's just for pennies, that you can give me so I can stay out here and work?" A lot of them went back to their point of enlistment, and turned right around and came back out again to where they had served. A lot of construction companies like Morrison and Knudson (the big construction company in Boise that was hired by the government to develop a lot of those islands out in the Pacific before and during the war) hired these boys, the cat skinners and patrol operators and carpenters. Too bad that so many of them had to go back home. One boy told me, "If I go back, I'm going to wind up just exactly the same as I was when I enlisted." You see, some of them were forced enlistments. The authorities–the police or the court, I don't know just who did it–required them to join up with the CCCs. A lot of them didn't want to come out to our country at all, but some of those in the end turned out to be some of the better boys.

The CCC boys were heavy eaters. I guess you can say that of all fire crews: they need simple grub, and lots of it. In fire camps that had it, beef was a big item. I remember back in 1929 on the Geiger Lake fire the pack string came in with a hind quarter of beef on either side of a mule, and on the Hells Canyon and Pioneer Creek fires they would drop us a hindquarter of beef. By hanging it up at night and rolling it in blankets or sleeping bags and keeping it in a shady place during the day, we could keep the beef until it was all consumed. No spoilage.

The first fire I was ever on that we didn't have our own mess was on the Ball's Canyon fire on the Toiyabe National Forest in 1949. The meals were catered from Reno. As regards the Ball's Canyon fire, I must say that was the poorest managed fire I was ever on, both in the field and in camp. When I got to the fire, and for a couple of days afterward, there was no sanitation in camp. If a man had to relieve himself, he just went out in the timber and let her go. It wasn't until dysentery broke out in camp that they brought out some Sani-Huts. They also had Kaopectate in gallon jugs. [laughter] The men lined up each morning and evening and were given a small cup of Kaopectate. It took me the best part of two weeks after I got off that fire to get over my runs. I think the food was OK, but the flies carried the bug. I much preferred camps where you had your own mess. You never had to wait for meals or lunches, and it cost the Forest Service a lot less when we had our own mess in camp.

Fire plows and Pacific Marines The tools for fighting fire varied from one region to another back in the 1930s and 1940s. Region Four and Region One had tools exactly alike, but Region Five had some that were different–of course, they had entirely different types of fires to fight. They fought a lot of brush fires, and the rakes that they used were designed just for special fuels.

Years ago they used to have something like a Forest Service experimental shop. Anybody who had an idea and thought it was worthwhile would send it in to the regional office, and they'd send it down to this shop, and they'd make it. Even if it took welding or casting or whatever else, they'd make it and try it out. Some of the devices worked and some of them didn't. They came out with a fire plow years ago that was pulled by a horse. It was a type of plow that you could dig a fire trench with, and it was reversible: you could throw your dirt one way; you could throw your dirt the other way. I've tried it. We used the same kind of plow to plow trails with, which worked good. You'd pull it with a horse that'd weigh sixteen, seventeen hundred pounds–a good-sized workhorse. Well, in certain types of ground that was free of roots and free of rock, it also worked perfectly as a fire plow. But in that Idaho country, that was all granite country, and very rocky .... A lot of the Toiyabe is also very rocky. I don't think a fire plow would ever work there; it's just too rocky. And if you ever got into mature timber and hit heavy roots you were hung up all the time, so the horse-drawn fire plows were never practical.

About all a crew had to fight fire with was two-man saws, shovels, double-bitted axes and Pulaskis. Then we had five-gallon water packs made of the same material as ordinary water bags. (Back then, I think they were made out of linen.) There was a pump that we attached to the valve or faucet on the lower part of the water bag, and it would shoot a stream of water sixteen feet or better through a nozzle that you could adjust to a spray or a jet stream. This water bag had two corks in its top; to fill it you pulled both corks off so the air could get out, and then you just took a can and poured water into your bag. That was it as far as fighting fires with water went on the Lincoln district.

It wasn't until I got on the Rapid River district on the Challis National Forest that I saw the first fire pump. Of course, from then on we had lots of fire pumps, but I didn't ever have a district where they had pumper units on pickups like they have now. The only thing we had were these Pacific Marine fire pumps–an excellent four-cylinder, water-cooled pump with a two-cycle motor. They could lift water at least five hundred feet, and you could rip the bark off a dead log with it. You could attach one of them to a hose and fight fire with it, or if you were at lower elevation you could put a Y on the pump and you would have two hoses. They had a lot of power, but the bad part of them was that they drew their water from a creek; the water also cooled the motor, and if a little bit of sand clogged the screen and enough water wasn't going through, you could burn one of them up. They ran so fast, you could burn one of them up in just a very short time.

These Pacific Marines were portable–a man could carry one wherever he wanted to take it, and pack the hose: the hose was made up in units on pack frames. They used an oil and gas mix of fuel, and you already had that made up ahead of time. It was put up in five-gallon cans, and you could strap a five-gallon can to a pack frame and pack it wherever you wanted to go. You could go up anywhere on the side of a mountain where you could find enough water, and you could set up a Pacific Marine pump.

During my Forest Service career the Pacific Marine pump was probably the best fire-fighting tool, if you want to call it that, that they ever came up with. I think that helped us more than anything else. Towards the end of my ranger career we did have Caterpillar tractors. Of course, they helped us, but when you had a fire back in the mountains up the side of some steep ridge, a Caterpillar tractor didn't do you any good because you just couldn't get it there. So, for all-around equipment, that Pacific Marine pump was probably the best thing that we ever had.

When a fire was crowning

What we did back in the 1930s when a fire was crowning was entirely different from what they do now. Now they've got fire-retardant planes, and they've got copters with water buckets; they can drop water on it–there are many things that they can use to slow the front of a fire. Back then when a fire was crowning, all we could do was flank it and try to contain it. We might stop it if . . . it depends on the weather, but late in the afternoon or early in the evening a crowning fire will come down to the ground. The humidity will rise, or sometimes the wind will drop and it'll come down to the ground. Then we could get in front of it and fight it direct. Or if we had enough manpower, we moved ahead sometimes as much as a quarter of a mile or further if it was a big fire, and put in a good fire line and then backfired. If we did this just right–if a fire was coming towards us and there was also a reverse draft of air that came up to meet the fire–when we backfired, it would suck our fire right back into the main fire. It wouldn't work if we got too far ahead, but if we were at the right distance from our main fire and it was crowning, our backfire would burn towards the main fire. That backfire was the only means that we actually had of stopping a going, crowning fire.

When we flanked a fire, we worked the sides. We would have the bottom end cut off, and we would just work the sides and hope our fire lines were advanced enough that when the wind died down or higher humidity brought the fire back to the ground we could pinch off the front of it. You try to do both flanks, but as a rule one flank is more dangerous than the other. Sometimes we'd work one flank part of the time and then work the other flank, but if we had enough men we'd have men on both flanks. We'd be working up just as close to the front of the fire as we possibly could, which helped diminish the size of a fire, because it prevented it from moving sideways. (The only way the fire could move was straight ahead, see.) This cut down a lot on the work that we'd have to do later. If I just said, "Let her go," she'd have spread in all directions. Of course, she'd have spread faster in the direction the wind was blowing, but she would also spread sideways.

Fire in Hells Canyon One of the most interesting fires I was on was the Hells Canyon fire on the Snake River in August of 1949. The Payette had four or five fires going at the same time, and they weren't making too much progress on this Hells Canyon fire. It was in a very remote area, and it was burning on the east side of the Snake River. And, boy, it was steep! The country is made up of chutes—very small drainages that start right up on the ridge and go clear down to the Snake River, and some of them are darn near straight down. The lower parts of these chutes were fairly heavily timbered, but as you came up-slope the timber became more and more sparse until when you got up about halfway you kind of ran out of timber, and it was just brush and grass. It was country that at that time wasn't grazed other than by deer, because it was too steep for even sheep to graze it. And there wasn't all that much water on it either.

The fire started quite a ways up the Snake River, and it was burning down-river on the east side. They brought their fire camp into where the fire started. First thing I knew about the fire, the Forest supervisor got a call from the regional office and they wanted me to go to McCall, Idaho, as quick as I could and take another ranger with me. The supervisor left it up to me as to which ranger I wanted to take, so I took Merlin Bishop from Baker. We got into Ogden, and got down to the airport at something like six o'clock the next morning. They were flying a bunch of radios and stuff into McCall, and at first they said they couldn't take either one of us. I kept insisting that I had to go, and finally they said, "Well, we'll work it around; maybe we can take you. We'll be overloaded but we'll take you." So I got on and they flew me into McCall, and Bish caught a plane right afterwards. But after they had me there in McCall, they didn't send me out to the fire. I couldn't figure out why they were holding me, they had been so anxious to get me up there, but they told me to go up into the smoke jumpers' loft and get some sleep.

Real early the next morning they loaded Bish and me and a ranger from the Teton–Woggenson; we called him Woggie[*]—and the assistant ranger off one of the districts on the Teton and two or three other Forest people, and they hauled us out to the main fire camp. We had something to eat, and then we had a guide that took us afoot around the upper

---

[*] Mr. Murchie is unsure of the spelling of these names, and begs forgiveness if he has not gotten them right.

edge of the fire out to a spring. There must have been around seventy-five men that they had already brought in there, and that was our first camp. I was assigned to be the division boss, but I was in effect acting as a fire boss because from that point on there was no contact whatsoever between my fire camp and the main fire camp, and in the end I was still fighting my fire after the main camp crews were all through and gone.

We fought fire there until we got the east side of it controlled, but all this time the main body of the fire was moving on down the river. We were not close to controlling it. We could have fought it that way until the snow came before we'd ever stop the thing. So when we caught up everything we could reach from that camp, they told me to leave one of my better men there, and I left Woggie Woggenson and fifty men at that camp. They told me this morning that they were dropping another camp in at Oxbow Spring about three to four miles on down the ridge. This wasn't a very big spring–it was big enough to *just* maintain a camp if you didn't wash your hands and face very many times, but there was enough water there to take care of the cooking and drinking. Once in a while we would have to fly water in, but most of the time we had enough to take care of the camp. I took about twenty-five men with me and left fifty, and hiked in to the new camp. Those were good men, loggers and mill workers from a timber company located just out of McCall. They were all older men, but they were really good workers and they'd fought fire time and time again, and I had to give them little instruction, if any. I just told them, "There's your sector," and they took off. The same day or the morning after another bunch of about fifty men hiked in. They were also loggers and millworkers.

We fought fire there from the new camp for three or four days. But down at the bottom of the canyon there was a lot more brush and timber, and this fire was moving on down the river, see, with nothing stopping it. Where it hit one of these chutes, it would start up the chute . . . and it seemed like most of the time a strong wind was coming up. Hells Canyon runs pretty much north and south, and there was a heavy wind coming out of the south down the canyon all the time: especially on after ten o'clock or so, you would always have a heavy wind. The structure was very steep, and the fire would start up one of these chutes, and pretty soon it would start climbing, and then it was just a roar and up she'd come! Before you knew it she would be right up at the top of the ridge.

On the east side of the ridge there was a drainage that paralleled Hells Canyon–it was a tributary of the Snake River. This drainage was heavily timbered, and they didn't want the fire to get going over on that far side because there were just thousands of acres of timber there. Our fire was still on the Payette National Forest, but you could look up across this drainage and there was a lookout . . . I believe it was called the Horse Heaven lookout. It was on the Nez Perce Forest, and they were looking right into the fire.

The Nez Perce Forest was in Region One and the Payette in Region Four, but they were looking right in from the Nez Perce side, and they had a good view and kept us posted. Anytime the lookout saw a spot fire down there in the canyon he'd tell us at the first puff of smoke, and we'd try and get somebody down there. One day I was sitting up on top of a point looking down the bottom and I decided that there was only one way I was going to stop that fire, and that was by putting a crew clear down on the Snake River bottom. There was a big sandbar down there. I came in that evening and called the fire dispatcher and asked him, "What's the chance of dropping supplies, grub, and everything for a twenty-five-man crew on the sandbar?" A camp set there would be safe.

The dispatcher said, "No chance. You don't want to put men down there."

I says, "We'll never stop that fire unless we do, because there's no way we can hike men up and down that long slope."

Finally he said, "Well, I'll talk to some of the pilots."

We were still flying the old trimotor Ford. That was a great plane, and the best plane on fire that was ever made. We talked to the pilot and he said, "Yes, if I get in there at the break of day when the air is heavy I can put you a camp on the sandbar. I'll have it there in the morning."

So I says, "OK." I didn't have to tell him what I wanted. I just told him to drop in a twenty-five-man camp and enough grub for a week. When we went out on the fire line that morning I got on a point and looked down and saw orange chutes on the sandbar. It was so far down that they looked like little orange dots. As I remember, there was ten chutes, ten drops.

A husky young fellow by the name of Val Simpson was assistant ranger on the Teton–he was a good man. I put him in charge, and I asked for volunteers. I told them, "Before you volunteer you should know you are going to have a tough hike going down and you are going to have a lot tougher hike coming back out. I would prefer younger men, and I prefer men that are in good physical condition." I think there was around thirty, thirty-two men that offered to go, so I looked them over and picked what I figured was the best twenty-five of them. They started down, and I thought going downhill they'd go faster, but if you walk down a steep slope very long it gets tiresome. I sat on this rock and looked, and along in the afternoon about four o'clock I began thinking, "What's happened? They should have made it by now." (They had also dropped in a radio so that I'd have contact with the crew once they arrived on the sandbar.)

Well, I was beginning to think I would have to send a plane down and do some scouting–a little plane like a Piper Cub could have flown down in there. I counted these orange dots a hundred times, I guess. Finally I counted them and I didn't get my ten spots. And I counted again and sure enough, there was one missing. Pretty soon then they started to disappear, and it wasn't too long until all the orange dots were gone. So I knew then that they were down there. I hurried back to camp to see if we had any communication from them on the radio. No word. And all the rest of that afternoon and evening and the next morning, no word! So I called the dispatcher and asked him, "Did you drop a radio?"

He said, "Yes."

"Well," I said, "something has happened. It's not working." And I asked him if he would fly another one in. (I found out afterwards [laughter] some dumbbell in their warehouse . . . . Sleeping bags were dropped free-fall. They tied them real tight and free-falled them in, and they hit the ground and would bounce thirty feet in the air. And he put the radio inside one of those bundles! Of course, it just demolished it.) The next morning they flew a radio in and dropped it on a chute, and then we had communication.

They were a good crew and this Val Simpson was a darned good man to have down there. They did a wonderful job. It was only a couple of days until he had that bottom pretty well caught up, and as soon as they got that bottom stopped, then the fire was over.

One of the things about the Hells Canyon fire that will stay in my mind is that this was the closest I ever came to burning up a bunch of men. Even yet, I get goose bumps when I really get to thinking of it. One morning I had taken a twenty-five-man crew, and I wanted them to work on fires down below where the chutes came up. About three-fourths of the way down the slope there was a fire in one of these chutes or draws, and there was quite a lot of timber there. I thought if we could get that, then we probably got her as far as this part of the fire was concerned. There was a trickle of smoke coming up, but it didn't amount to much, and I thought if I put enough men on it they could knock it down in just a short period of time and get out. We gathered on a ledge where I could look down, and when they got ready to go they had to come off this ledge and go up the ridge a ways before they could drop down into the chute. I was watching this, and they were down just a little bit below me, and all of a sudden I just didn't like the looks of that thing. There was still white smoke coming up, but I could see that it was becoming more active; there was more smoke boiling up. So I thought, "I'm not just going to send those men down there." I yelled down to them and signaled for them to come back up. And boy, they didn't like it! They stood there and talked among themselves for quite a while, and the foreman yelled something up to me; but I couldn't hear him, and I signaled for them to come back out again. Finally they turned around, and they weren't too happy when they got back out. They came on around up to where I was. By this time the smoke had started turning black, and in just a little bit after they got there, here she come! That fire just literally exploded. She came a'roaring up that chute. If they'd got down there a hundred yards more before I turned them around, I don't believe they'd have made it back.

I'd been around fires enough, and looking at the smoke I suspected the worst. Not only the color of the smoke but the way it was rising suggested serious fire. If it just comes up lazy there isn't an awful lot of heat underneath it, but it was coming up fast. In a column like that, even when the smoke is white, there has to be heat underneath it; and when it turns black, you know that it's moving–it isn't burning in a dead log, it's moving! Pretty soon she really started crowning, and up she came, and she had enough heat and slope and wind behind her that when she hit the sagebrush she zipped through it just like that! Those men would never have made it.

On that Hells Canyon fire–now, this is all hearsay, but this is the way it was told to me–a crew had gone down into one of these chutes, and the fire started up and the foreman was trying to get them out. He was moving them just as fast as he could, but one guy kept lagging back; he couldn't keep up with them. The fire was getting practically on top of them, and finally it caught up to this fellow and flashed over him and set him afire. They were able to get down to him, but most of his clothes were gone and he was badly burned. They tried to walk him out, and a fellow who was in the crew told me that his lymph fluids were running out of his body where he was burned, just like water. He got so completely dehydrated that he couldn't go any further. So then they left him underneath a tree, and they left him alone, which they never should have done. Then they went out for a stretcher. Surely they must have had jackets; if they didn't have jackets, they at least had shirts. They could have turned the sleeves inside out and made a very simple stretcher, and they could have put him on that and taken the clothes they could spare, and wrapped him in them and packed him out. I understand he later died.

I guess there were three things that made the Hells Canyon fire so special for me: One was that I was very fortunate that I had exceptional overhead. (Of course, the overhead were all Forest Service people, and maybe that is what made it that way.) I was also fortunate that I had excellent men in the crew. And it was interesting that although I was assigned to be a division boss, I was pure and simple a fire boss, because I had no contact with the fire camp at all. I was totally responsible, and this was a pretty difficult fire.

Hot shots   When I was in the Forest Service, a ranger was fully responsible for fires on a district. If he needed help, he started yelling at the fire dispatcher, who got help to him. If it became a bigger fire yet, the Forest would contact the regional office and start help coming in from various districts and Forests over the region. But this help that came in wasn't organized. They were just rangers and sometimes men out of the regional offices, and once in a while the assistant Forest supervisor. There was no organization, and when they came you had to look your men over and talk to them to find out which ones you wanted to make straw bosses, and which ones you wanted to make division bosses, and all the different positions that you'd want to fill on a fire. One thing that helped in the 1950s, after I was over to Ely for a while, was that the regional office issued each of us a card that said you were qualified as a division boss or a sector boss or a fire boss or a camp boss, or whatever your qualifications were.

Today they have these overhead teams fully organized, as I understand it. They have teams that have worked together, and everybody has an exact position in the organization. And if they have a big fire like the Acorn fire, they just contact the regional office for an overhead team. If it's a great big fire, they might even want to expand on that, and these overhead teams are all organized. When they want additional manpower, they call for a hot-shot crew. A lot of those are Indians. They may come from clear up in Montana or New Mexico or Arizona, but they're all organized. They are, I believe, reservation groups that have been organized, and that's all they do—they fight one fire, and then they move to another one all during the summer. And they're good men; they're well-trained, and they're well-built, husky men.

The first hot-shot crews that I knew of was when we had the Dog Valley fire. There were local hot-shot crews, and they had prison hot-shot crews that I believe had to be back to the prison every night. There were also two Indian crews on that fire—one was Crow, and one was either Sioux or Blackfoot, and they came from eastern Montana. The two groups didn't get along very good together. Crow Indians, as I understand it, never did get along too well with surrounding Indians. They were a different bunch, and the mountain men way back preferred to winter with the Crow more than they did with any other band. From what I could gather.... Well, our youngest son, John, worked on the Dog Valley fire, and he worked quite a bit with the Indians. He said the Crow kind of felt they were a little superior to the other Indian tribes. Whether that was necessarily true or not, I don't know.

As well as I can compare, our summer CCC crews years ago were similar to today's hot-shot crews. They were awful good fire crews. Every man knew his position, and they had their own foremen and their own straw bosses, everything. And twenty-five-man CCC crews could do more work than fifty pick-up men. Today's Indian hot-shot crews could probably put out as much work in a day as the CCC boys, but the recuperative power of the CCC boys.... You could work their tails into the ground, but you give them plenty to eat and a night's sleep, and they were ready to go the next day. I never saw a *tougher* bunch of men.... Of course, they were young, and that makes a difference. They weren't great big kids, but they were well built, and they had more endurance and recuperative power than any bunch of fire fighters that I have ever run across—and I'll call them men because they *were* men, even though they were boys.

But I was very satisfied with the effort that the hot-shot crews put out at Dog Valley and later. I retired in 1965, so I only observed them for five years, but from my impression and from what I have since been told, the Indian hot-shot crews are some of the best. And with the Indians, if they don't cut the mustard, they're weeded out and replaced by the other Indian hot shots.

Let some fires burn? There's been a debate for some time over whether or not fire prevention should even be practiced. This debate was just starting shortly before I retired, but all through my Forest Service career the main point was to prevent as many fires as you possibly could, and put out the fires that you did have as quickly as possible. You did that regardless of where the fire was located. Once in a great while–like, say, on that Loon Creek district–we'd have a lightning strike way back up in the rocks some place where you knew it couldn't go anywhere, and you'd leave it and let it go. But we would *never* deliberately let a fire burn. I don't care whether it was out in the sagebrush or a grass fire or a timber fire, or what it was.

Of course, other rangers and I would sometimes talk back and forth about the effect on a forest's health of putting out every fire–we understood that fire was a natural part of a forest's evolution. But it's kind of interesting that all these prime mature and over-mature stands of timber that we have, that you can walk through just like walking through a park (how pretty they are!) are the result of either fire or insect invasion or disease . . . or clear-cutting and logging. Something in the past has wiped that area out, so you had an even-aged stand of timber started, which matured to the same size and was fairly evenly spaced. (Now, you take clear-cut logging, where they were going back and planting trees, and those trees grow up even-aged: when those trees are, say, two hundred years old, they're going to be just as nice a stand as these stands that we have now that they're trying so hard to protect; the result will be exactly the same.)

Personally, I have conflicting feelings–especially now in retirement, looking back at the things that happened–as to whether maybe we should have let some fires burn. Probably a lot of my feelings are controlled by my education and the years of experience I had in the Forest Service. Even though I'm retired, I'm still a part of the Forest Service . . . but suppose I had never been involved with the Forest Service, and I was looking at these fires from the outside, and had no interest in the cutting or sale of timber—all I could see was this pretty stand of trees there. Then I'd think of nature. I would think that before the white man came to this country trees grew up, fire burned them off or insects killed them, disease killed them, old age killed them; they fell down and died and rotted away. Then looking ahead, what is that stand of timber going to be like, say, in two hundred years? Maybe if I was looking at it from the outside I'd say, "Let her burn. Let's get rid of all that old timber. Let a new, fresh stand of timber come up." This process has been going on for maybe millions of years. So why should man step in now and try to change it? Why not let nature take its course?

But that isn't the thought that I have looking at it as a forestry-educated man. I'm all in favor of going in and putting these fires out, and then going in wherever it's possible and replanting. I'm also in favor of clear-cutting timber (a lot of people object to this), and then going in and planting trees that'll grow best on the site and will produce the most lumber—and then you have an even stand coming up. When it's two hundred years old, it'll be exactly like the even stands of timber now—I don't care whether in Nevada or Washington or Oregon or California—that people admire so much. *Those same stands* in the distant past were created when fire, disease or insect infestation completely removed a stand of timber, and you had an even stand coming up. If that were not true, you'd have an uneven stand of trees of mixed species.

16

An Evolving Mission: Training and Policy

WHEN I STARTED with the Forest Service, we followed as closely as we could the principles set down by Gifford Pinchot.* We believed that our main job was to manage the Forest to furnish the greatest good to the greatest number of people in the long run. Since then, especially right now, I feel that most of Gifford Pinchot's ideas have been thrown out the window. Due to interference by outside interests the Forest Service is having a very difficult time even meeting the expectations of its chief and its regional foresters.

---

* Born in Connecticut in 1865, Gifford Pinchot received a degree from the French National Forestry School in 1890. He was appointed the first chief of the U.S. Forest Service, established in 1905 when the national forest reserves of the Department of Interior were transferred to Agriculture. Pinchot's credo was "to make the forest produce the largest amount of whatever crop or service will be most useful, and keep on producing it for generation after generation of men and trees."

The Buckskin Bibles

When the Forest Service was created there were no forestry schools in the West. The men who applied for jobs were cowpunchers, miners, loggers . . . I doubt that some of them even had a high school education, and they didn't need one. They just had to take this old Ranger Exam that pretty much amounted to showing that they could saddle a horse, shoe a horse, pack a horse and pack it right; and that they could cook for themselves when they were out and not starve to death. The Forest supervisor sent them out, and often they'd be gone for weeks or months at a time. Since there were no telephones then, no radios, and absolutely no contact, they had to be able to take care of themselves . . . but they weren't required to know much else.

Early rangers didn't have to know all that much because they had two manuals that told them everything. One manual was maybe two inches thick, and the other one was smaller, and they were done up in a heavy, buckskin-colored canvas with leather reinforcing on the corners. Rangers called these two volumes the Buckskin Bibles and they were it: you had all your regulations; you had all your laws; everything you needed to run your district was in those two books. The smaller one, you carried with you all the time in your saddle bag. It contained primarily regulations regarding timber trespass, range trespass . . . everything pertaining to administration. The other volume was fiscal regulations. The Buckskin Bibles were still in use when I went to the Lincoln district on the Helena Forest in 1931, and when the first supplement came out I remember the rangers saying, "What am I going to do with this?" Then, of course, it wasn't long until more supplements came out, and now the ranger has just shelves and shelves and shelves of different kinds of manuals, and the staff has all kinds of manuals. It's just unbelievable, the amount of instructions rangers have to have now to run a district.

Up into the 1930s, if a ranger had taken the Ranger Exam and could read and write, he could pretty much get the knowledge that he needed out of those two books. Of course, he had to be an industrious and ambitious individual, because he wasn't under close supervision. The only things his supervisor was interested in were direct results, and that the ranger got along with the people that he associated with. A ranger then was a different sort of man from the kind that began coming in once it was required that you pass the civil service exam. They were down-to-earth, practical, experienced men, and those that survived were pretty tough individuals. A lot of my supervisors and instructors and rangers were old Ranger Exam men, and you didn't want to cross them or try to put anything over on them.

A forestry education Our forest management prof at Montana had taken forestry in Michigan, then he'd come west and passed the old Ranger Exam . . . where if you could pack a horse and build a fire and cook sourdough hotcakes, you had a job! [laughter] He worked for the Forest Service for a number of years, and then he got a job as a prof in the forestry school, where his approach to teaching was more practical than theoretical. Courses that we took in forestry school in the 1920s and 1930s were very basic, and little of what we took–I don't care whether it was forestry or range–could not be applied when we got out on a job. Now, that was different from what was being taught in the 1950s and 1960s, if I can believe what I was told by some of the profs and by students coming out of forestry schools before I retired. A lot of the education that students were receiving by then was too scientific to be applied out on the ground–it wasn't applicable to their jobs. They were being taught more theoretical forms of range management as determined by research done in schools and experiment stations. What they learned may have worked well under a certain set of controlled conditions; however, when you get out on our Forest ranges, be they cattle or sheep, there is very little that is controlled other than numbers and season of use. Your management of the range has to adjust to range conditions, and due to the type of education they were receiving it was hard sometimes for young men just out of college to make this adjustment.

We received more basic training in range or forestry when I was in college, and there were a number of range students who went directly to managing large ranches after graduation. I doubt if a student graduating now, majoring in range, would be able to do this. Then, too, when I went to forestry school there were a lot of boys that had worked in the logging industry before they took forestry, and there were a lot of them who were ranchers' sons, and had worked on the range before they took range management courses. Some of them had even been riders on cattle allotments, and they had fair experience before they started in the Forest Service, which makes an awful lot of difference.

There used to be long periods when the ranger was the only man on a district. During fire season he had lookouts, patrolmen, guards and fire crews, but the rest of the time he was the only man on the district. He was the engineer; he was the fireman; he was a timber man; he was a recreation man; he was a wildlife man; he did his own typing; he did his own filing ... he was pretty much a jack of all trades, and the forestry schools, as much as they could, tried to teach their men to handle situations like that. Besides our regular range subjects we also took surveying–regular and topographic mapping–and in my case I was able to take forest management and a number of other forestry subjects. When you graduated you were able to handle most of the jobs on a ranger district.

Blasting: training and regulations

The big thing the powderman on the Kootenai taught me was to respect dynamite–not to be afraid of it, but to respect it. And I've never had a problem with dynamite: I've never gotten hurt myself, and I've never had anybody with me that got hurt, although for a long time I didn't have a blaster's certificate. It wasn't required. Later on a certificate was required, and you had to take a test before they issued you one. On the Challis Forest (and I imagine it was by request of the regional office) they started requiring dynamiters to have a blaster's certificate along about 1938, and then you had to have it renewed every so often. There, the road foreman and I were the only permanent personnel who had certificates.

Of course, back in those days, we used caps and fuses. We didn't have electric caps then, and we used what they call two-minute fuse, which means that a foot of fuse will burn in two minutes. In the morning the powderman would always check his fuse to see how it would burn, because sometimes it'd absorb a little moisture overnight, and whenever he got a new roll of fuse he'd always check it. (Learning to check fuse probably saved my life when I was on the Rapid River district. At the Seafoam ranger station they didn't have a powder house, but stored their dynamite and rolls of fuse in the loft of the barn. During the summer it got pretty hot in the loft, and the tar on the outside of a fuse would sometimes melt and turn hard. When you would unroll a roll of hardened fuse it would sometimes crack and let a little bit of air in. Once I got this roll of fuse, shiny and just like brand-new, and unrolled a section of it to load a hole, and something just clicked. I thought, "Uh-oh, I'd better check that." So I took a foot of it and lit it, and it was cracked in several places, and the fire ran through there in a matter of three or four seconds. If I'd loaded a hole with that fuse, she'd have gone off too soon and I wouldn't have had a ghost of a chance of getting away from the blast.)

At the time that I learned about dynamite there wasn't any sort of formal education in blasting for people in the Forest Service. You were supposed to pick up these skills by working with experienced powdermen. I was really an apprentice; I think the old man considered me to be his helper, but looking back on it, I was really an apprentice. (Others perhaps didn't have as good a teacher, and suffered the consequences. Years afterward I knew a guy on the Challis Forest who got his leg broken by loading a little too short–the charge went off before he got far enough away from it.) Later on the Forest Service put out quite a lot of instructions. Then, when they outlawed the cap-and-fuse and went to the electric, there was all kinds of literature. But in the 1920s and 1930s there were no pamphlets about blasting, no instructions, nothing standardized–it was just whatever got passed down by word of mouth and experience working with somebody else.

I was still using blasting caps and a fuse as recently as 1957 or 1958. The electric caps and blasting machines came out a long time before that, but I'd been so indoctrinated in the fuse and caps that I preferred them. But I imagine . . . as far as the Forest Service was concerned, I might have been blasting illegally. [laughter] Probably I should have been using a blasting machine. Oh, they were pretty strict there at one time about using the blasting machine instead of caps and fuse.

I doubt that the Forest Service still does as much dynamiting as it did in the 1930s and 1940s, but, oh, we used to use dynamite a lot! On trail work you were always blasting rocks and blasting out beaver dams. If you had a long dam, you'd put your charges on the upside of it; you'd tie your dynamite at the end of a stick, and bring the fuse up and tie it at the top, and lay it on the in (upstream) side of the dam. You'd put charges about five feet apart, and to really take a dam out you had to have the charges go off at the same time. Even in Ely I used quite a bit of dynamite, blasting beaver dams and blasting out post holes and things like that, so I was very fortunate that I was able to be the powder monkey for that old guy on the Kootenai, because what he taught me really paid off.

Property transfer   When I left the Kamas district the property hadn't been completely checked. The tremendous amount of snow made it difficult for Alonzo Briggs, my replacement, to get out to check property with me—in fact, the property at the CCC camp and the guard station and the WPA camp had *never* been checked. It was a mistake I wish I hadn't made. Practically a full year from the time that I left the Kamas district I got a letter from the Wasatch National Forest informing me that Briggs had finally completed a check of the property on the Kamas district and that I was short a lot. (There were a number of sheets of paper listing all this property that I was short.) The letter went on that inasmuch as this property had been used, they had discounted the value. But I was charged, as I remember, close to three thousand dollars. Whatever the precise sum, I know it was a lot more than I was making in a year at that time. I was about ready to quit, but Jane talked me out of it.

The next day when I went to work I was feeling pretty low. Merle Markle, the Challis Forest fire chief, and I sat side by side at two different desks, and he could see something was wrong. He asked me, "Arch, just what's bothering you?" I decided to tell him, and he says, "Well, let's go in and talk to Mac."

I says, "No, this doesn't have anything to do with the Challis Forest. It's my problem, and I'm going to have to work it out myself." I told him definitely not to talk to the supervisor, but he ignored me; later he went and talked to Mac. When Jane was just a little girl living in Pinedale, Wyoming, Mac McKee had been stationed there. Jane's folks and the McKees had been good friends, so she knew Mac, and she saw him uptown, and unbeknownst to me she also mentioned this property bill to him. So Mac already knew a good deal about my problem when he called me in and said, "Arch, I want the whole story. Every bit of it, as close as you can tell me." I told him, and he said, "What do you think happened to that property? Was it stolen?"

I said no. What happened was when I was on leave and they moved the CCC and WPA camps out, they probably took the property with them. Apparently, nobody had ever made a record of that transfer. If that was so, it was also the fault of whoever received that property. The policy was that when property would transfer when you weren't there, the receiving ranger or supervisor's office would execute a property transfer, and you'd get your copy back. That was never done in this case.

Mac told me that Ben Rice, who was stationed in the regional office, was in Salmon City visiting the Salmon Forest. He says, "He's coming through here on his way back to Ogden. I'm going to tell Ben this whole story, and he'll probably want you to come in and talk to him." It wasn't too long afterwards Ben showed up, and Mac called me in. I told Ben the whole story, and he asked me what I thought happened to the property. I told him there was no way the property could have been stolen, because the road was snowed in all winter. Most of the property had been small items–stoves, tools and other things you would find in a CCC camp. But the funny part of it was they told me that a big rock crusher that had been used to crush rock to surface the roads–thing must have weighed forty or fifty tons when it was put together–was missing. They assumed that a piece of equipment as large as that probably wasn't stolen; that it must have been moved out and a transfer had never been made. So they didn't charge me for that, thank goodness.

Ben Rice told me, "Don't you pay any attention to the Wasatch Forest. You completely ignore them. It don't make any difference what they write you, don't even answer them." So I didn't. Time went on; they wrote to me a few times and wanted to know why I wasn't paying up, and I didn't answer them. It must have been late the next summer when I got a letter from the Wasatch National Forest notifying me that they had rechecked my property and I was short about sixteen dollars and something. So I immediately wrote them a check for the sixteen dollars, and that more or less ended it.

Years later, after I'd transferred to Ely, I was to a ranger meeting and a bunch of us rangers were sitting around shooting the bull. We got to talking about different experiences we'd had in our Forest Service career, and I mentioned my experience with missing property on the Wasatch. This ranger from the Wasatch, a man probably in his late fifties, turned and looked at me. He said, "So you're the son-of-a-gun that made us check all our property right in the middle of summer!" What Ben had done was he'd gone back to the Wasatch Forest and ordered them to make a complete property check of the entire Forest. This was right in the middle of summer during the busy part of the season. "That," said this ranger, "was rough!"

On a ranger district at that time you really had to account for property. Even if it wore out, you had to have it condemned before you could dispose of it. You take an axe that was too dull and worn to use, you couldn't throw it away; you had to have it condemned first. But once in a while on a fire you'd gain a little property–maybe you'd gain a few axes or a few Pulaskis and shovels or small items like that. Maybe in the transfer they would send too many or maybe they slipped through without a transfer or something–a *lot* of things can happen on a fire. When you'd make out your property return you'd account for the ones that you were supposed to have, but you might have a few axes or shovels or this and that extra. So the next year if you were short you could put those in. Well, Ben Rice required those rangers on the Wasatch to report *everything*, and he told them that there was going to be a pretty stiff penalty if anybody fudged on their property returns. So those rangers had lost all that buffer equipment that they had. We had a big laugh over it, but he didn't like it too much. He said, "That was tough when you had to drop everything to check property." But we finally got the property on the Kamas district straightened out, and I was completely exonerated.

"Rangers had to furnish everything"

In the 1930s and 1940s rangers had to purchase their horses with their own money, and we were supposed to keep them shod. We also furnished our own saddle equipment. The Forest Service did let us use a government pack saddle on our pack animal, and they paid for whatever grain we had to buy during the summer for our animals. Of course, the Forest Service took care of our horses during the winter, or they'd put them out on feed lots where they fed them hay. In the spring, the government also stood the cost of any work on their teeth, and they would worm them.

I don't know why they made rangers provide their own horses, but I do know what their thinking was when they finally said we didn't have to do it any longer: All along the Forest Service had been buying horses, but they were dual saddle/packhorses, and while most of them were good packhorses, some weren't all that hot as a saddle horse. We rangers generally bought good saddle horses, however, because when you spent a week, ten days or longer in the saddle you wanted a halfway decent horse. So when inspectors from the regional office or supervisor's office would come out, they'd want to ride one of the ranger's horses. Most of us rangers were giving our horses about all the riding that they needed, and we kind of resented the inspectors leaving Forest Service horses there in the corral and riding our horses. With some rangers this resentment got pretty strong, so the Service decided that they'd furnish the horses. After that they bought us good saddle horses, and after I moved to Ely in 1947 I didn't have to buy my own horses anymore. (I continued to furnish my own saddle, chaps, blanket, pads, halter and bridle.) Over there at Ely they had some darned good saddle horses.

Traditionally, rangers had also had to provide their own bedrolls or sleeping bags. And when we were away from the station they fed us, but you didn't collect any per diem. Along about 1938 they changed that, and they said, "You buy your own grub, and we'll pay you so much a day per diem." (I think it started out first at something like two dollars a day.) I liked going on per diem, because with them providing your grub there had been a lot of waste. Maybe you'd have something that was open, and when you brought it back to the station you couldn't use it. And maybe the next time you started out again it'd be spoiled, and you'd have to go get another can of lard or whatever it might be. With per diem, anything left over you could bring into your house and use.

Rangers had to furnish *everything* at the ranger stations. When you arrived, the place was bare. Anything that was moveable.... You had to furnish your own curtains and drapes and rugs and everything you needed to live in the ranger station. It wasn't until I moved to Ely that I even had a government pickup truck. Before that, if I had to move a crew or go in to Challis for some men, I hauled them in my own car. And if I couldn't haul everybody, somebody else brought them out. You got mileage, but it wasn't very much, and it sure didn't come anywhere near paying for the damage to your car. You hauled everything a pack mule could have hauled in your car. Either in it or on it! [laughter] Things have changed a lot since then, but that was the job and you never thought you were being hurt or not treated right.

"Better not be any squawking" Back in those days when you took a job you agreed to what that job required, and if you weren't satisfied with it, you didn't take the job, period. After you signed up and took a job, you'd better adhere to the requirements. Better not be any squawking or back-biting or anything else, because that was it. Today rangers do a lot of complaining that.... Man, if we'd done that back then, we'd have been fired! That's right!

They were a lot stricter then, but there was very little turnover, either due to dissatisfaction or to being fired. A man pretty-well knew what he was going into, and if he didn't like it he didn't apply. When we were in forestry school the dean and most of our professors were ex-Forest Service employees, and they knew what the job consisted of. We were told time and again what to expect if we went out on a district, so none of this was new, at least as far as I was concerned, and I was happy as a lark on those back-country districts. In fact, the best district I ever had in the Forest Service was the Loon Creek district.

Administrative authority   Laws and rules and regulations were handed down directly from the Washington office. Then the regional office could add further regulations or restrictions, and the Forest might come out with some special restrictions of its own. But out on the district you could do pretty much what you wanted to do, as long as you held within the limits of these rules and regulations. If you felt the situation required it, you could be a lot more restrictive than the rules and regulations, providing you could show justification for what you wanted to do. If you *were* restrictive, and a permittee or operator appealed your decision to the Forest supervisor, you had to be ready to back it up. You could get yourself in trouble if you could not.

Starting way back in the late 1930s or early 1940s, there began a gradual change in the authority of rangers, and this change has continued clear up to the present. Things that a ranger wanted to do back then, he could go ahead and do without any approval from anybody. Then as time went on it got so that if you wanted to build a trail, for example, instead of just going out and building it, it had to be approved by the supervisor's office. And I don't mean just for the money to finance it, but the actual location of where the trail went. Same way with a telephone line or the location of a lookout. As time went on, it got so that a certain thing not only had to be approved by the supervisor's office, but it had to be approved by the regional office. You take when I got over on the Toiyabe Forest, where we did so much reseeding–I would approve the reseeding job from the Forest standpoint, but it still had to go into the regional office and be approved. I imagine that now there are probably very few things a ranger plans that do not have to be approved by both the supervisor's office and the regional office.

I was fortunate during my career to work with some excellent Forest supervisors, starting with John Templer, the supervisor of the Helena in 1931-1933. Then when I was on the Challis there was E. E. "Mac" McKee; and on the Nevada Forest, there were Johnny Herbert and Louie Dremolski. When I came to the Toiyabe, Ivan Sack was the supervisor. These men were all excellent Forest supervisors. They told you your job, or you learned your job, and then they left you alone. They gave you all the leeway that you wanted to do the job as long as you stayed within bounds that they set and within the rules and regulations of the Forest Service. You never had any trouble with these men looking down the back of your neck all the time, so you could go out and analyze your project or problem and go ahead and solve it. Another administrative strength of those men was that if you ran into a problem that you needed some help or advice on, you could sit down with them and tell them your problem, and when you came out of their office you had an answer; you had a decision.

Without mentioning the names of supervisors for whom I had less respect (some of them are still alive), I can identify what I believed to be their major weaknesses: you couldn't get a decision out of them, and you couldn't get their support. If you made a decision and it looked like trouble was going to come, they'd never support you. Maybe that was an effective managing art, I don't know, because some of those men went up in the Forest Service: one of them became a regional forester, and others worked their way up into the regional office. Now, you take Johnny Herbert or Louie Dremolski or Mac McKee–if you made a decision and they reviewed that decision, they'd support you clear to the Washington office if necessary! You never worried about lack of support from those men, and that gave you confidence. Decisiveness and support were what distinguished the good supervisors from the bad.

"You had to live a Forest Service life"

When I first started out the Forest supervisors were involved in rangers' lives far more than they are now. For one thing, we had to have our hair cut a certain way, and we weren't allowed to wear a mustache; we weren't allowed to wear sideburns. (I used to wear my sideburns down a little way–not long, but there were times I was questioned about my sideburns.) Of course we had our uniforms, and anytime we were in the office we were supposed to be in full uniform, period. Our dress was always up for inspection, but out of the office we would wear our forest-green pants and green shirt. And our dwellings, whether they were Forest Service cabins or privately rented, had to be kept up neat and clean. You got called on it if you had a dirty house, even if it was your own property . . . because, say some permittee comes in and sees your house in a mess–that's the impression that he's going to get of the Forest Service.

You had to live a Forest Service life night and day, whether you were on work or off. There was a good reason for that, because at that time you'd get probably more calls from ranchers and timber people at your home than you would in the office. (Much of the time you'd be gone out of the office during the day.) If a rancher or a timber man wanted to see you, your home was the only place he could contact you, regardless of the time of day or night. If a permittee wanted some information, you could be coming to the door in your pajamas, but you were still representing the Forest Service.

Now, I know a case where a ranger got low of funds, and he couldn't meet his bill at the store. Whether the storekeeper went to the supervisor or what, I don't know, but the supervisor checked and found out that it was because of illness in the family that the ranger was having problems paying his bills, and the supervisor himself offered to loan him money to get his bills paid up. I know of another case of a ranger who was kind of a renegade–he gambled a lot and he drank a lot. He had a wife and a small child, but he was spending his check about as fast as he was getting it. Boy, the supervisor took him on the carpet! He tried to straighten him out, but eventually . . . . He had a signed resignation for him on his desk, and he told him, "If you fall off the wagon one more time, you're signing that and I'm sending it in," and that's just exactly what he did.

That kind of interest in a ranger's personal life was not so common when I retired. A lot of the restrictions that we had then had gone by the board quite a while before, and by then they didn't care whether you wore a mustache or sideburns. I imagine if you had your hair clear down to your shoulders, maybe somebody'd have said something, but a lot of those restrictions and the restrictions on wearing the uniform, I think, had lessened.

Self-supporting districts

Quite a ways back the Forest Service reached a point, due to higher wages, costs of administration, and various other expenses, that many districts and Forests ceased to pay for themselves. Now, Forests in Oregon and Washington, where they cut a tremendous amount of timber, are probably still paying for themselves, but for most districts I would guess the break-even point would have been passed shortly after the New Deal went out, because that's when the cost of everything started skyrocketing.

Districts that were used primarily for grazing were probably never as self-supporting as those where a lot of timber was cut. Of course, I didn't keep records (the Forest supervisor probably kept track of what revenues came in and what was spent), but I have a feeling that the Ely district wasn't paying for itself when I left. On the other hand, I'm sure some districts where they could graze practically every foot of land were self-supporting, and maybe some still are, because they've raised grazing fees considerably from what they were.

Grazing fees were set in the Washington office, and they were uniformly applied nationwide; they didn't vary from one district to another. Some land is much richer than other land, but to fill a cow's belly it don't make any difference whether it's on good range or poor range—you still pay for that cow-month of feed. If a cow can exist on one acre on a certain area for a month, that's a cow-month; and if over in this other area, it takes her forty acres to get a cow-month, the fee is the same. A cow-month is a cow-month, whether it's on an acre of meadow or out on forty acres of sagebrush.

Armed rangers

I understand that in Region Five, where the public is growing so much marijuana back in the Forest, some of the rangers carry sidearms. Years ago rangers always went armed, but it wasn't for dealing with crime. I always carried a gun with me on my saddle or in my pickup in case a horse broke a leg and I had to shoot it, or for one reason or another. My favorite gun was a .44 Colt revolver on a .45 frame that I bought in 1929. The advantage of a .45 frame for a .44 is that it's a heavier gun, and it don't rear back on you like a smaller-frame unit. Even that .44 on a .45 frame, it'll jump on you. In fact, if you are going to shoot it fast you better sit so you can hold the thing down, because it'll climb with you.

I packed that Colt until we had a little incident on the Rapid River district, where the patrolmen all carried sidearms, mainly on account of rattlesnakes. I don't know really what happened: I heard one story from the camper, and I heard another story from the patrolman himself. This young patrolman down on the river was kind of hotheaded, and he supposedly pulled his gun on a fellow who wouldn't clean up his camp. The camper gave the patrolman some lip, and he decided he'd *see* that the camper cleaned his camp up . . . which he shouldn't have done. I've been in the same position a number of times when I'd have sure liked to have done the same thing, but he shouldn't have pulled a gun.

The camper lodged a complaint to the supervisor's office, so the supervisor wrote a memo to all the districts that there'd be no more sidearms packed by the rangers or anyone else on the districts. I talked to the supervisor and took exception, believing that the packer should carry a gun of some kind—either a saddle gun or a sidearm—to deal with injured horses. So he said, "OK, your packers can carry a gun." So all the time I was on either district, Rapid River or Loon Creek, my packers always carried a gun.

After I got to the Ely district I carried a .22 rifle in my pickup, and sometimes a .30-'06. I never went out in my pickup without a rifle, and I also had a rifle scabbard on my saddle. Sometimes I'd shove the .22 in the scabbard, but during the deer season I generally had my .30-'06 in it, because you never knew when you were going to run on to a crippled deer. (My .30-'06 was one that was checked out to me by the Army to hunt mustangs with. It was a surplus Springfield Armory weapon with iron sights.)

The game warden was never too happy with me for shooting crippled deer. One time a hunter told me as he came in that he had seen a wounded deer that had its whole lower jaw shot off—it was so far gone it was just staggering. So I took my .30-'06 and walked up the road from our camper wagon, our headquarters for checking deer out, but I couldn't find the deer. It probably had just crossed the road and laid down. As I was coming back in, packing this rifle, the game warden pulled into the camp. He knew I'd already taken my permitted deer for the season, and he wanted to know what I was doing with a rifle. I told him, and he said, "It's a good thing you didn't find it. If I had caught you I'd have arrested you; I'd have taken you in."

I didn't find that deer, but I shot a number of deer that had been wounded. I shot an elk once that had been wounded so long that the meat wouldn't have been worth eating, anyway. Hemorrhaging and seepage of stomach contents from a gut-shot animal ruin the flavor of its meat. I considered dressing it out and bringing it in, but I thought, "Well, the meat's probably already spoiled, and it wouldn't be fit to eat, anyway. So why bring it in? Why not leave it for the buzzards and the eagles and the coyotes, and let them clean it up?"

Contracting out

We were still doing our own road work when I retired on the Toiyabe in 1965, but they were contracting out a lot of trail work. Now, at least over on the Ely district, I think they're contracting out all the road work too. One reason for the Forest Service reducing its road and trail work–especially road work–was probably the lack of good foremen and employees; probably, too, the cost of the wages that they'd have to pay those people. Road and trail work also tied up a lot of money in purchasing and maintaining equipment, and then the equipment stood idle for long periods of time. And on big Forests, like the Humboldt and Toiyabe, they had to move their equipment long distances to work their roads.

I haven't been over too many of the Toiyabe trails since retiring, but when I go deer hunting I go over a lot of the Humboldt roads, and the contractors are not doing near the job of maintenance that the Forest Service could have done. In fact, they're doing a darn poor job on a lot of their roads. If I was a ranger, those contractors sure wouldn't get by with some of the maintenance that they're doing. The Forest Service is just throwing money down a rat hole on some of that maintenance, and on trails it's probably the same thing: it's wages and getting the right kind of men to be trail foremen. It's easier to just contract out and let somebody else build the trails. That's about all.

Summer homes on Forests

After World War II the Forest Service, nationwide, wasn't too happy with all the summer homes that had been built on Forest land with special use permits, and in places in Region Four (and probably elsewhere) they canceled permits on some lakes. This was mainly because there was competition between permit holders and fishermen and other users of the streams and lakes. The Forest Service didn't feel that one group should have the monopoly of sitting there right on a lake . . . and lots of times it was difficult to even get around a lake because of these summer homes.

When summer home permits were first issued the Forest Service had actually encouraged applications because this was considered a higher use of Forest land. Back then, there was not the heavy recreational use of

Forests that there is now, and we didn't have all of these environmentalists that don't want *any* improvement on a stream or around a lake. (Some of them, I doubt would even have been in favor of issuing a permit to Utah Power to put those dams in on the Kamas district. They want lakes just to be left as natural as possibly could be.)

Summer home permits were issued on a first-come-first-serve basis, but applicants had to submit a plan, and it had to be a building that would fit into the natural environment. (Most of the early cabins were built out of logs.) If the plan was approved by the regional office, then the applicant was given a permit for twenty years. After the twenty years were up, then they'd reevaluate the special use, and maybe grant it again and maybe they wouldn't. Say a father and a mother took out a special use permit, and then they got older and one or both of them died—it could be transferred to one of their children. But, as I remember, if they wanted to sell the cabin, then whoever they sold it to would have to take out another permit himself. The cabin could be sold, but the permit couldn't.

The amount of land encompassed by a permit varied. We didn't want the summer homes packed together like in town, but the acreage depended both on what the permittees wanted and what the Forest supervisor and ranger decided was adequate for their use. Sometimes it was an acre, and sometimes it might be a little bit smaller than that; and in a few cases, I think there were summer home permits where they kept horses—they wouldn't have grazing room for them, but they'd have a place for a corral and a barn. Homes could not be built right down to the water, each summer home was inspected, and things were regulated even to the point where the cabin would sit. It was all well controlled, and as long as there wasn't other heavy use of the streams and the lakes, it was a very good use of a Forest resource.

I remember the policy beginning to change back in the 1940s on the Challis, where they weren't too happy with some of their summer homes. See, these summer homes weren't always taken up on a lake; they could be taken up any place. In one case on the Wildhorse Ranger district this fellow supposedly (according to him) struck a vein of ore, and he wanted to build a cabin there. Well, if he had a valid mining claim he *could* build a dwelling on it, even without a permit from the Forest Service; but it turned out that he hadn't struck a vein, and he was just using the cabin as a summer home, pure and simple. We suspicioned that, and we asked the mining inspector to come out from the regional office. He came out and checked the point of alleged discovery, and it wasn't a real claim of any kind. The miner then admitted that he just wanted to use it for a summer home. We told him no, and he eventually moved everything out, and in the late fall, when it was permissible, he set the cabin afire. If he hadn't, *we* would have burned it down, because the cabin would invite other people to come in and use it without a permit.

Regional differences: trespassing, fire and budgeting

There were some administrative differences–in a way, I'd say glaring differences–that I could see when I first came down to Region Four from Region One. For example, that trespass case we had with the Frenchman, Chevalier, on the Helena Forest in 1932: The year before we had trespassed him and just charged him for the grazing fees. In 1932 he trespassed again, and the ranger trespassed him a second time, but this time the Forest supervisor added five hundred dollars in *punitive* damages. Now, that was acceptable, apparently, to the regional office; and I presume that at the time that was policy, or he could never have done it. But when I got down to Region Four, there was no way that we could charge punitive damages in any shape or form. Even if we had information from a fisherman or somebody up there camping, or evidence in the form of manure that was left on the ground by cattle or sheep (you can go on that and tell fairly close to how long they have been there) you could only trespass for the *period that you actually had observed the livestock in trespass*. So a lot of these permittees that we trespassed said, just like Chevalier, that they were getting far more value out of the feed their livestock took in trespass than it ever cost them in trespass fees.

Possibly the reason Region Four wouldn't let us levy punitive damages was that it had a lot more grazing than Region One, and it had probably run into many more appeals. In fact, when I left Region One I didn't know there was such a thing as an appeal; I thought that when the Forest Service said a trespass had happened, that was it. When I got to Region Four, I found that wasn't the case. I found out that any time a permittee was disgruntled with anything the ranger or the supervisor did, he started writing to his congressman. Since I couldn't charge punitive damages on the Ely district, I tried another approach, and nobody ever said that it was wrong or that I couldn't do it: like when Bert Robison trespassed some sheep, I computed the animal-months that he was in trespass, and then at the end of his grazing season I deducted that from his permitted animal-months for one band of sheep. He paid for the trespass in the end, in that he had to take his sheep off early. But with a fellow like Dan Clark . . . I don't know; it didn't seem like there was anything that you could do to keep him from trespassing. The problem was that he'd trespass after the season was over (in other words, he'd leave his sheep on longer than he was supposed to), so you couldn't cut his period of use; you couldn't cut his animal-months. The only thing you could do was to trespass him for the amount of time that you knew he was on the Forest. Of course, with Dan Clark I eventually stopped this by taking a 10 percent reduction on his permit.

If we notified a rancher that his stock were in trespass, and he failed to get them off, we could remove them. But there was a hazard in that. It must have been about 1962 or 1963 when the ranger over to Austin found these cattle in trespass in the spring when they were just dropping their calves. He notified the permittee that his cattle were in trespass and asked him to get them off. Well, the permittee didn't do it, so the ranger took one of his men with him and went and rounded them up and drove them off. They drove them a good ways off the Forest, which they probably shouldn't have done, and they met the owner coming up. Three or four cows were bawling their heads off, and the owner got really mad. What had happened was the cows had just had their calves, and they'd hidden their calves, and when the ranger had rounded up the cows their calves had been left back up the mountain. The only thing they could do was turn the cows loose and let them go back and find their calves. When they'd mothered up the next morning, the permittee rounded them up and took them off.

Well, that got the Forest Service into quite a lot of trouble, and it got the ranger into a *lot* of trouble. I had to kind of chew him out for doing it, but it was partly my fault, too, in that he'd asked me once what the procedure was if the stock weren't removed from their allotment on time. I told him that if it was not too many head, round them up and drive them off. So he was following my instructions but not at the right time. [laughter] He should have known that he had some wet cows there when he rounded them up. If you take a wet cow and she don't have a calf by her side, you know the calf is out there lying down someplace. He should never have brought them off.

When I first came to Region Four, another way that I felt there was a considerable difference between Region One and Region Four was in their approach to fires. Of course, Region One is a fire region; in fact, it's the *top* fire region, I guess, in the entire Forest Service, and fire has extremely high priority. In 1933 I had a fair-sized fire, and the minute I called for manpower and for help from the supervisor's office, I got it. There was no time lost, and I had all the manpower I needed. But when I got down to Region Four, on the Kamas district, I had problems trying to get manpower on a fire. It was hard for me at that time to understand the difference between the two regions. But looking back on it now, I'm sure it wasn't a regional policy, because we had received a memo not very long before stating that all going fires would be manned, period. And that meant that they'd be manned 100 percent, regardless of where you had to bring the men from. At that time the Forest had five or six CCC camps, so there was no reason why that fire of mine couldn't have been manned. I kind of think it was a personnel problem within the supervisor's office.

For a while I wasn't too happy with Region Four's approach to fires. However, when I got to the Challis Forest and was on two heavy fire districts, I couldn't have had better support from the fire dispatcher or the fire chief or supervisor in getting supplies and manpower to all of my fires. So I don't want to imply that it was a regional policy. I think that the reluctance to commit resources to fighting fire was strictly confined to the Wasatch National Forest.

Another thing that seemed different between Region One and Region Four was that in the use of funds Region One seemed to adhere closer to fiscal discipline than Region Four did. Of course, I was only in Region One for a very short time, but I did work with the ranger in controlling the expenditure of funds; I really had at least three summers where I did have some knowledge of the expenditure of funds. There, if you were allotted money for different jobs or functions, you adhered right to that allotment, and you didn't dare to spend a penny over it. They wanted you to spend up to it too–you'd get chewed out if you didn't spend all your funds. But down to Region Four, especially when I got to the Ely district, there wasn't that control of the expenditure of funds, and there were rangers that periodically went over their allotments. We would have to dig down into what funds we were saving for trail work or recreation guards, and help bail a ranger out, which I didn't appreciate too much.

Later on, when I got to the Toiyabe, rangers would have various projects–fence jobs or water developments or reseeding–for which they would apply for funds, and we had a regular application form to be filled out. The forms would come in and I'd look them over, and on the fences and water developments I generally pretty much left the figures the way they were; but none of the rangers had much experience on reseedings or spray jobs, so I'd go over their estimates on these, and most of the time I'd increase them some. They'd go in to the supervisor's office and be approved and sent in to the regional office, and generally we would get the money we had requested. Well, when the funds would come back to the supervisor's office, the fiscal officer would start creaming money off the top to pay his staff there in the office. That left the ranger with insufficient funds to complete his project, so next spring he'd have to apply for additional funds to finish it. As soon as *those* forms went into the regional office, they'd fire a letter back and say, "We gave you money for that job. Why didn't you finish it last year?"

It finally came to the point that I told the administrative assistant, "Now, if you feel that the impact of these jobs is great enough on your staff that you need additional money to pay them, fine. But when these forms come in, you add what you want to the amount that the ranger shows, or that I show, so that when the money comes back, you're not taking any money away from the ranger."

This one fiscal officer says, "Oh, we can't do that, because Range Management won't like it if we're taking project funds to pay administrative people." I'm not saying it wasn't legal, but sometimes I wondered whether he was stretching things a little. But as time went on, he started budgeting for the additional personnel he needed to support a project (processing vouchers and payrolls, et cetera), and from then on we didn't have any trouble.

"I was never directed to preserve anything"

It would be safe to say that in the 1930s Forest environmental impact decisions were regularly made by the individual rangers (although the term "environmental impact" had not been coined), and then as time went by, as with other policies, they became more and more decisions that were made at higher levels. Back when I was on the Loon Creek district and we decided to locate a trail, we picked the shortest route from here to there with the least amount of work, and environment played no part in it. We never really thought,

"Well, if we go around this side of a ridge and cut a side hill trail, that's going to cause a lot of erosion." That never entered into our heads. Now you wouldn't dare to put a trail like that through.

As regards what are now called cultural resource issues, I was never directed to protect or preserve anything, but if we had run onto, say, Indian dwellings, we'd have done our best to preserve them. An Indian grave that was up on the ridge above the Loon Creek ranger station had been left just as the Indians had left it. However, horses going up and down the ridge had knocked all the rocks off of it, and so we built a fence around the grave to protect it from the horses. But that was a decision made right there on the spot; there was no policy concerning that sort of thing. I'm sure that down in Arizona and places where they had a lot of Indian dwellings and Indian artifacts, the Forest . . . I would *hope* that they had some kind of a policy protecting them. I can hardly believe that a Forest supervisor would let one of his rangers go out and destroy any of those. But I have read that there was a case down in Arizona where they actually did that on a logging operation.

Sometimes, obviously, there were conflicts between using the Forest and preserving certain parts of it. There was a little meadow and a spring over on the Schell Creek Range on the Ely district that apparently at one time must have been quite a camping spot for Indians. It was what they call a semi-dry meadow–part of the year it was dry, and part of the year it was wet. We were doing a reseeding over there, and they'd bring their Cats down there and park them at night. Where they turned the Cats around, they tore some of the sod up on the dry part of the meadow, and they uncovered *a lot* of Indian arrowheads–a lot of them. That apparently had been an Indian campground used pretty heavy at one time, but there was nothing on the surface to indicate it. The only thing that I had to go by was the abundance of broken arrowheads–all of them were broken, so they'd apparently camped there and taken their broken arrowheads off and put new ones on.

I wound up with a pretty good arrowhead collection myself. When I was a ranger, there was no regulation whatever against collecting artifacts. You could pick up all kinds of arrowheads or Indian hammers, spear points, awls–anything you ran onto. I had a whole bunch of them lying on my desk there at Ely. It was perfectly legal, and there was no Forest Service policy to contact some anthropologist. Anybody that found one, it was theirs.

Public relations, public service

The Ely district included Ely, McGill, Ruth, Lund, Preston, and Baker, and these were all comparatively small towns. Ely was the biggest of the bunch, but it still would be considered a fairly small town, and there wasn't a lot of recreational diversion around the area, except gambling and a movie theater in town. As a consequence, there was a high percentage of the population that spent its free time, at least during the summers, out in the mountains and the streams, fishing and camping. Most of the streams, even though they were very small, contained a lot of fish, and at one time they had two fish hatcheries—one up Duck Creek and one over out of Baker—that kept the streams pretty well stocked. The fact that these people spent a lot of time outdoors meant that most of them knew the country pretty well. For instance, they could recognize where a band of sheep might have been trailing too heavy in an area and had done some damage. So, anything that you had to tell them about the Forest and the wilderness, they were interested in. I think it's for that reason that I was frequently called on to give talks on Forest activities while I was on the Ely district. It didn't make too much difference whether it was fire or recreation or grazing or plants—anything that pertained to the Forest, they were interested in it. No matter whether it was the Lions Club or the Rotary or Chamber of Commerce or the Professional Women's Club—they were always interested in what you had to say. It was a pleasure to get up and talk before a lot of those groups, and for that reason I spent an awful lot of my own time, evenings and weekends, giving lectures and slide presentations. Also, by then I was a good movie projector operator, and sometimes a group that had a film they wanted to see would ask me to bring the Forest Service projector, and come and show it for them. It was very rare that I ever turned anybody down, even though it took a lot of my own personal time.

All this gave you a good relationship with the people, and it paid off in many ways. If I was having problems with a grazing permittee, those people knew about it; it didn't take long before that went around the whole community. There were people who weren't tied in with grazing in any way who would stop me and ask me about it, and some of them would voice their opinion. When I had to take a reduction on Dan Clark, some said it was a good idea, and a few of them questioned it and wanted me to give a good explanation as to why it was done. But I've never been on a district or worked anyplace in the Forest Service where I had as good a working relationship with the public as I had there.

An Evolving Mission

At times, I think I was tied in too much with the public, because I was spending way too much time with them and not enough time with my family. While we were there, I guess you'd say my wife practically raised the children, because there were a lot of Saturdays and Sundays that I was tied up with some community activity and I just wasn't at home. I promised my wife when we left there that I'd never get that involved with the public again, and I never did. In fact I told her I wasn't going to belong to a single thing when I retired. I finally joined this Conservation Forest and Woodland Committee, but that's the only committee or group that I really belong to. Well, I belong to the Old-Timers on the Forest. But I don't even go to the Lions Club or Masons meetings anymore.

I joined the Masonic Lodge when I was at Challis, Idaho, and I got my fifty-year pin in the Masons last year. I think probably years ago Masonry was more prominent in the Forest Service than it is now. There were quite a few . . . not so much in Utah, [laughter] but I've been on Forests where there was quite a few Masons. Like the Challis–the fire chief was a Mason; the forest supervisor was a Mason; I was a Mason. I was brought in by Merle Markle, the fire chief. When I was on the Ely district I belonged to the Lions Club, the Chamber of Commerce, the Boy Scout Committee and Girl Scout Committee, the Rodeo Committee, and there's probably others that I don't remember right now. I was just too much involved, but, at that time from the regional office on down, the Forest Service encouraged everybody–especially rangers–to get involved in the community. I wouldn't say by any means that I did it *because* of pressure from the supervisor or the regional office; it was just that I enjoyed working with the people, and I saw it as a medium through which I could put across a lot of our ideas, a lot of our plans, and also get their views on things, which was important. When we were having such a problem with deer overpopulation, I gave a lot of slide presentations to various groups, showing the consequences of having too many deer on the Forest. I am sure these presentations did a lot in turning around the feeling of the Sportsmen and the general public toward the Forest Service.

When I came to the Toiyabe Forest, I was in a different position, and was no longer expected to have such close contact with the public. I was a staff officer in charge of range, wildlife and watershed at first, and then I was trespass officer and safety officer, and a number of other things. But rangers were always expected to develop good relations with the public.

We were encouraged to join these organizations. If I had not, when the personnel officer from the regional office had a personnel inspection, I'm sure he would have asked me how come I didn't belong to any of these public clubs or civic clubs or associations.

Vandalism Through most of my career, Forest Service people were respected, but I felt when I came to the Toiyabe that things had started to change. Perhaps it was just the area, because you had a larger population and not only some influential ranchers, but other people that were more influential; or maybe people in the Reno area disagreed more with our Forest Service policies. By the time I left Ely we were getting–other than from a few individuals–full support from the community in the things that we were trying to do. That includes the businesses and Kennecott and others. As soon as I got over to the Toiyabe I met resistance. I also began to see more vandalism on the Forest.

Back in 1929, the Forest Service never locked a building out in the back country. And miners or trappers would never lock a cabin–their cabins were always left open. My partner and I could go into a miner's cabin and cook us a meal and re-fill his wood box with firewood, and that would be it. And on occasion when we were gone from our camp–we didn't really have a tent camp; we had a tarp–somebody going through would stop and cook themselves a meal and go on. Of course, it was always mandatory that you did the dishes and left a pile of firewood; if you didn't do that you were in trouble! But as time went by we began locking ranger stations, and when I got to Kamas even the guard stations were kept locked.

Locked buildings or not, we used to have very little vandalism: even at Ely where we built a number of stopover cabins, I never had any. However, we did have some vandalism in our picnic areas, campgrounds. Most of that was when maybe a bunch of guys was on a drunken party and they were too lazy to get firewood, and they'd break up tables. Tables were log tables then, and they'd break them up and burn them. We had one case where they tore down a toilet and used it for firewood, and that was the first seriously destructive vandalism that I encountered. (Of course, in the back country, like at Loon Creek, people would sometimes break a window or two in a lookout just to get in and get warm or see if they could find something to eat.)

While I was on the Ely district we built a number of stopover cabins. They were about sixteen feet square, and each had one of these sheetmetal sheepherder's stoves in it. There were dishes and a cupboard, the door of which let down for a table, and there were two steel cots. In each cabin we always kept a certain amount of grub that wouldn't perish by freezing, and the cabins were kept locked with a Forest Service lock. Nobody ever broke into any of them back then, but I've been told comparatively recently that some of these cabins have been practically destroyed; that they can't keep them locked because the doors and windows will be broken out, and if they have anything in them it's stolen. Vagrants move in and they leave nothing but a mess. Times have changed.

Glossary

Abney level: A small hand-held instrument used by rangers to measure tree height or slope of the land. It has a vernier scale on the side calibrated in either percent or degree, and incorporates a bubble from which true level is determined.

Animal month: A period of use of national forest rangeland equal to one animal grazing for one month.

Baching: From *bachelor*. A man or men living together, not in the company of women. They do their own cooking, washing and housekeeping.

Baseline: An accurate line run with a chain and compass on timber surveys. The parallel lines surveyed through each section were run from the baseline and perpendicular to it.

Batwing chaps: Chaps with sides that extend out and back ten inches or more from the leg. They provide increased protection to the wearer in dense brush.

Bedlog: A foundation log. The term has many applications, among them the short logs resting on the ground to support the framework of log chutes.

Berm: The level space between the edge of a trench, ditch or other such feature and the earth excavated from it.

Biltmore stick: A graduated scale used by timber estimators to determine tree diameters. The stick was a flat piece of hardwood with a T-shaped metal head. When held against a tree or log of known length, the log's estimated volume could be read directly from the stick.

Blaze a tree or stump: To mark a trail or individual tree by chipping off a piece of bark.

Board feet: A unit of quantity for lumber equal to the volume of a board 12 x 12 x 1 inches.

Boom the ties: Ties are driven down a stream to a larger river, where they are held by a circle of steel cable and logs or ties to keep them from going down the river. This restraint is called a boom.

Break bed ground: About daylight, when sheep get up and start leaving the area where they bedded down for the night, they are said to be breaking the bed ground.

Bridge plank: Heavy three- or four-inch planks, usually twelve inches wide, that surface a bridge.

Broad axe: A heavy axe used in hewing ties that may be ten or twelve inches wide. It has a flat face on one side and is sharpened on the other side. The flat face lies next to the tie being hewed.

Brush-hop: To fly low over trees with an airplane.

Cartwheels: Silver dollars.

Cat skinner: One who operated a Caterpillar-type tractor.

Catch horse: An alert horse with good action that works with the cutting horse. Once a steer is driven to the edge of the herd by the cutting horse, the catch horse gets around the steer and drives it away from the herd.

CCC: *See* Civilian Conservation Corps.

Chain: A piece of equipment used in early Forest Service surveys to measure distance. It was sixty-six feet long and consisted of one hundred links. A mile was eighty chains.

Chalk a line: To use a string or cord covered with chalk, usually colored, to establish a straight line. The cord was stretched tight and let loose or snapped, leaving a straight line marked on the ground.

Check dams: Short dams built across contour trenches. The top of the lower side of the contour trench was always higher than the top of the check dam. These dams were placed from fifty to one hundred feet apart, and their purpose was to keep all the water from draining out of the trench if the side should break.

Check scale: In a survey, to check the work of a log scaler or a timber cruiser for accuracy.

Chink: To fill cracks, as in a log cabin, with material such as moss or mud to make it weather-tight.

Cinch: A braided band on a saddle that is four or more inches wide. It is made of hair or cord and has a ring at either end to which the latigo is tied on a saddle.

Civilian Conservation Corps (CCC): Established by Congress in 1933 as a New Deal relief and recovery measure, the CCC annually provided work for as many as five hundred thousand unemployed males between the ages of eighteen and twenty-five. Under the direction of army officers, work camps were established and projects undertaken in reforestation, road construction, the prevention of soil erosion, flood control, et cetera.

Clear cutting: On timber sales, a forested area may consist entirely of merchantable trees. When this is the case, the entire stand is cut and removed: hence, to clear cut. The clear cut area is then generally replanted with a desirable tree species.

Close herd: To confine sheep to a small grazing area on the range through action by a shepherd and his dog.

Closing order: An order closing a Forest or district to use by horses. Issued preliminary to a wild horse or mustang hunt.

Contour trenching: The building of deep parallel trenches on a level grade across the faces of steep slopes. Their purpose was to store water that may run off the slope during a heavy rain or snow-melt.

Cordwood: Firewood sold by the cord.

Count off of water: To count sheep when they start going back out to graze after they have had water.

C ration: Combat ration, a canned field ration of the U. S. Army.

Creosote: A brownish oily liquid consisting of aromatic hydrocarbons obtained by distilling coal tar. Creosote is used primarily as a wood preservative.

Cruise: To cruise is to estimate the volume of merchantable timber (or the extent of pine beetle infestation) in a given area by walking the area and personally observing, measuring, counting and recording the pertinent data.

Cruise sheet: A printed sheet on which all data is recorded from the sample plots that are taken when timberland is surveyed.

Cutting horse: A cutting horse must be quick and intelligent. When working cattle, the rider takes the cutting horse into a herd, picks out a specific animal and works it to where it can be separated from the herd.

Cyclone seeder: A piece of equipment used to sow grass seed. It can be either hand powered or motor powered.

Dally: Temporary twisting of the rope around the saddle horn when roping an animal.

Deadfall: A fallen tree that has toppled from natural causes, usually severe weather or disease or beetle infestation.

Deck of ties: A large pile of railroad ties, ready to be driven down a river.

Deferred grazing: A term used when a grazing allotment has been divided into units, and one unit is taken out of use each year.

D8 Cat: A large bulldozer made by Caterpillar for moving dirt. (D designates dozer, 8 designates size.)

District: Forest Service lands are divided into regions, forests and districts. A ranger district is the smallest subdivision.

Drift pins: Pins made of round iron 3/8", 1/2", or 3/4" in diameter and whatever length necessary. They are hand-made and are sharpened at one end in order to drive them through two or more logs to hold them in place. They are used in the construction of cabins, bridges, et cetera.

Drill seed: Using a seed drill, to place seed in the ground for germination.

Duff: Litter composed of needles, leaves, and other organic material that accumulates on the ground under trees.

Fire plow: A heavy plow designed to plow a fire line or trench around a fire.

Flash point: The temperature at which a fuel or oil will ignite.

Flyblow: A condition that occurs when eggs are deposited by the blowfly in flesh wounds in livestock. When the eggs hatch, the larvae (maggots) eat into the exposed flesh, and the animal is said to be flyblown.

Flying boxcar: C-119 military cargo plane manufactured by Fairchild. The entire rear of the plane could be opened for parachute jumps or cargo dumping.

Forb: An herb other than grass; a broadleaf weed.

Forest: A national forest, the largest subdivision of a U.S. Forest Service region.

Forest Service land: Public and private lands placed into the National Forest System either by presidential proclamation, congressional action, purchase or donation. These lands are administered for the public by the United States Forest Service, an agency of the U. S. Department of Agriculture. There are currently 156 national forests across the country, encompassing more than 191 million acres.

Grain drill: A machine designed to drill or place seed in the ground at specified intervals and depths.

Grub: Food.

Guard: Temporary employee hired by the Forest Service as a fire guard, station guard or recreation guard.

Gyppo: An individual or a small group of men or loggers who work together and contract their logs or ties to a larger company.

Halogeton (*Halogeton glomeratus*): A coarse annual noxious weed that is especially poisonous to sheep.

High lined: When deer (or other animals) have grazed all the branches and twigs off a tree or bush as high as they can reach, causing the trees to have an umbrella-like appearance, the trees are said to be high lined.

Indian ricegrass: A tufted perennial that is Nevada's official state grass. Its grain was used by Great Basin Indians to make bread, and it is also known as mountain ricegrass, Indian millet, sandgrass, sandrice and silkygrass.

Iron mine telephone: A phone invented for use in mines. The box of the phone was made of heavy cast iron. The Forest Service used these phones in the 1920s and 1930s because they were waterproof and could be set up outside.

Iron sight: An open sighting device used for aiming a firearm. The sight consists most often of a small beaded blade placed on top of the muzzle end of the barrel, which must be aligned by the eye with a notched projection at the chamber end.

Jack: Male donkey; jackass.

Jungle bum: A term used to describe unemployed men during the Depression who traveled from place to place on freight trains. They set up camps called hobo jungles near the railroad; hence, jungle bum.

Larkspur (*Delphinium*): A poisonous plant of the buttercup family. Some species are poisonous to cattle, others to sheep.

Lash rope: Generally a long rope usually 1/2" in diameter tied to a cinch at one end. The cinch has a ring at one end and a hook at the other, and the lash rope is tied to the ring end. A pack animal is loaded by tying the pack to the saddle and animal with the lash rope. It is the rope that the diamond hitch is tied with.

Latigos: Narrow strips of leather that attach both ends of a cinch to a riding or pack saddle.

LDS: Church of the Latter Day Saints, popularly known as the Mormons.

Locate a trail: To survey where a trail is to be built, including examining any grades, switchbacks or other topographic features.

Marker: A marker can be a sheep with a bell around its neck, a black sheep, or any off-colored sheep. To count these markers is a good way for a herder to count his sheep. Most herders like to have at least one marker for every one hundred sheep.

Mine prop: A pole or timber used in mines to support the roof of a tunnel. (Term is singular or plural.)

Mop up: A term used in the suppression of fire. After the fire has been controlled, the fire crews go inside the fire line and begin putting out all adjacent fires. This may also include the isolation of areas of unburned fuel. The purpose is to decrease the chance of a fire spreading beyond the established fire line.

Mustanger: One who rounds up wild horses on the open range and sells them, especially for horse meat.

Non-use: An arrangement whereby a rancher does not use part or all of his national forest grazing permit for a year or more.

Nubbins: A little bit of a branch that is sticking out from a tree after deer or cattle have chewed it off.

Off stirrup: The stirrup on the right side, or offside, of the saddle. The rider mounts from the left side.

On water: When sheep come in to water around noon and shade up until the early afternoon, they are said to be on water.

Once-over grazing: When sheep are allowed to graze over the range only once.

Open herd: When sheep are allowed to spread out over the range, rather than be closely controlled in tights herds, they are said to be open herded.

Pacific Marine fire pump: A water pump with a four-cylinder, two-cycle water-cooled engine used in the 1930s and 1940s to fight fires. Later a two-cylinder engine was developed. This high-speed motor had a vertical lift of around five hundred feet.

Palouser lantern: A lantern made from a tapered lard bucket. An X was cut in its bottom side and a candle inserted. It was carried with a bail that went through the top and bottom rims.

Park the ties: To pile the ties in readiness for hauling along a sleigh road.

Peavey: A hooked lever used by lumbermen to handle logs. It had a heavy hardwood handle with a pointed metal tip and a hinged hook near the end.

Permittee: An individual with a permit to graze livestock on national forest land.

Pickaroon: A tool made by the tie hacks to assist them in moving ties. It looked like an axe, but instead of a blade it had a sharp point that could be driven into a tie.

Pigtail: A 1/4" or 3/8" rope that has been doubled, and a loop tied or braided in the middle. The loop extends out behind the pack saddle for a foot or eighteen inches, and the ends are tied to the big ring on either side of the pack saddle that the latigos are tied to. The halter rope of the pack animal behind the first animal is tied to the pigtail with a slipknot.

Pile or **Pile up:** When sheep are bunched too tightly, some will try to climb on the backs of other sheep. This sometimes happens in corrals with dogs bunching them too tightly, or when they are being stalked by a predator such as a bear or a cougar.

Pitch tube: When the pine beetle would bore through the bark of a tree and into the cambium layer, the tree would exude pitch through the hole, and this would harden into a tube-like projection.

Pitcher pump: A small hand-operated suction pump that must be primed with water before it can draw water from its source.

Pole charge: A dynamite charge tied to the end of a pole, with the fuse run up the pole and tied. It is usually used when a charge is set off in deep water, such as when beaver dams are blown up.

Powder house: Where dynamite is stored.

Powderman: One who is experienced in the use of dynamite.

Powder monkey: Usually an apprentice who assists an experienced powderman in setting his dynamite charges.

Preference: The permitted number of livestock that a rancher is allowed to graze on national forest land, and the season of use.

Preference, reduction of: To reduce the animal month use on national forest land.

Project fire: A fire that burned uncontrolled after 10:00 a.m. of the day *after* it started.

Pubescent wheat: Perennial wheatgrass with hairy "pubescent" flowering parts.

Pulaski: An axe-like tool with a head on it consisting of an axe blade on one side and a grub hoe on the other. It was used in fire suppression.

R-1 fire ration: Sometime before the 1930s Region One (R-1) developed its own emergency fire ration. Each ration was done up in a cloth sack and was to last a man working on a fire for a full day.

Range count: When livestock are counted out on the open range.

Ranger exam: A proficiency examination required for employment as a ranger in the U.S. Forest Service. Instituted by Forest Service chief Gifford Pinchot, the exam entailed shooting, riding a horse, using an axe, taking a written test, and throwing a diamond hitch.

Glossary

Reefer: A horse in a pack string that will stop for no reason. He will brace his feet and refuse to move, and usually a halter or halter rope will break as a result. A reefer is also known as a balky driving horse.

Region: A U.S. Forest Service administrative area that may encompass one or more national forests.

Reseeding job: Sagebrush land is plowed and then grass seed is planted to provide improved range for grazing animals.

Rotation grazing: A system of grazing in which the period of use is rotated. For example, a unit may be grazed spring-summer-fall one year and then summer-fall-spring the next.

Rough locks: Chains wrapped around sleigh runners to slow the sleigh down on its way downhill, usually when it held a load of ties.

Round timber: Logs left in their natural shape.

Sacrifice area: A term used to describe an area that has been overgrazed in order to force livestock to use less accessible range.

Salt ground: Ground where salt blocks are left for cattle, or salt is scattered for sheep.

Sapwood: Living wood surrounding the heartwood in a tree.

Saw timber: Trees large enough to be sawed into lumber.

Scale ties: To count ties and stamp them with a U. S. Forest Service stamp after they are hewed.

Scaler: One who scales ties or logs.

SCS: *See* Soil Conservation Service.

Section: A piece of land one square mile or 640 acres in area forming one of the 36 subdivisions of a township in a U.S. public-land survey.

Short erosion: When soil is washed off a slope in the absence of gullies.

Sheepherder's stove: A small sheet-metal stove with an oven that sheepherders used in their camp wagons. The Forest Service also used them in their lookouts and cabins.

Shelterbelt: A barrier of trees and shrubs that protects crops from the wind and lessens erosion.

Shevelen [phonetic spelling]: A type of siding used on Forest Service buildings constructed during the 1930s by the CCC. It had a rounded surface to imitate a log.

Side hill trail: A trail built on a contour across a slope.

Sill log: When building a cabin, the sill log is the first log that sits on a foundation; on a log bridge, it is the log at either end that the bridge rests on.

Single-footer: A horse that has mastered the single-foot gait, like a pace.

Slash: Debris such as limbs and bark left from a logging or tie operation.

Sling rope: The rope used to hang the packs on the sides of a pack saddle.

Slough hay: Wild hay that grows in muddy or wet areas. It is coarse hay but is good winter feed and pasture.

Smokechaser outfit: The equipment a smokechaser used to suppress a fire. The equipment was done up in a pack sack to be carried on his back or on a horse. The smokechaser's tools were carried in his hands.

Smokechaser fire: A small fire, usually caused by a lightning strike. As a rule, only one man went to smokechaser fires to extinguish them.

Smooth brome: A perennial grass used for pasture and forage crop plantings that has a smooth, hairless leaf and outer flower parts. It is commonly found in fields, meadows, and moist waste places such as irrigation ditches. It is also known as common brome or Hungarian brome.

Snag: A dead tree, often with the top broken off; also a fallen tree whose broken limbs can tear the flesh of passing sheep.

Snatch tongs: Tongs large enough to grip a thirty-inch log. They had a handle at the end of each arm so they could be opened and closed by hand. When they were used to move logs with horses, a rope from the team went through one handle and fastened to the other. When the team pulled, the tongs closed.

Snow course: Pre-determined sites established by the Soil Conservation Service where measurements were taken to determine the water content of the snow.

Soil Conservation Service (SCS): Established by the Soil Conservation Act of 1935, the SCS provides for the protection of land resources against soil erosion and for other related purposes. SCS routinely enlists the cooperation of other federal agencies in the discharge of this mission.

Sorrel top: When the pine beetle attacked a lodgepole pine, the cambium of the tree was destroyed and the tree died, maintaining the reddish-brown needles for some time. This sorrel top was indicative of a beetle attack.

Spike camp: Jobs some distance from a main CCC camp required that those working on them establish temporary tent camps, called spike camps. These had their own cooking facilities.

Spitter: A short length of fuse nicked to the powder core at about three-quarter-inch intervals so that when the fuse is ignited, the nicks successively spit fire and are used to light the fuses of a round of loaded holes.

Splash-down: When a stream had an insufficient flow of water to float ties, a splash dam was built. When the dam filled with water, the gates were opened and the ties were floated (or splashed) down on the high water.

Spray job: When herbicide is sprayed on sagebrush to eradicate it in preparation for a reseeding.

Spud: A tool used by tie hacks to peel the bark from trees or ties.

Stonebridge lantern: A collapsible lantern used in the early days by the Forest Service. It was about four inches wide and had a bail by which to carry it. Its windows were covered by translucent isinglass mica, and beef tallow or wax candles were used in it.

Swedish gang saw: A saw that was set in a frame and worked up and down. It was often operated by water power and could be moved sideways so that one-inch, two-inch, or other widths of board could be cut.

Swell-butted: A term used to describe trees that maintain an even diameter until a few feet from the ground, when they swell out and may double in diameter.

Switch tie: An extra long and large tie used at a switch where two tracks came together. They were long enough to go under both tracks at the switch.

Take up (ties, telephone poles, mine prop, cordwood): The act of scaling or determining the number of cords, board feet, et cetera.

1080: A deadly poison (sodium fluoroacetate) used on coyotes and other animals by predatory animal control men.

Tie hauler: One who hauls ties on a truck or sleigh.

Tie hack: One who hews or makes railroad ties.

Top a colt off: A young horse that may buck a rider off is first saddled and ridden in a corral (topped off) until he settles down. Then he can be taken out and ridden.

Trespass: 1) A violation of regulations regarding the use of Forest Service land. Such a violation may be termed a trespass, and the violator (or his animals) said to be in trespass. 2) To trespass may be either the act of trespassing or the act of *charging* a grazing permittee or other Forest user with trespass.

Tup-line or **Tump-line:** A piece of equipment used as an aid in carrying a backpack. A wide strap went around the forehead, down either side of the pack, and buckled around the bottom. Properly adjusted, it enabled a hiker to carry part or all of the weight of his pack on his head.

2,4-D: a white crystalline herbicide used as a weed killer.

Type line: A line drawn on a map between different tree species or trees of different ages, such as mature trees and over-mature trees.

Undercut: A large notch cut at the base of a tree on the side to which you wish the tree to fall.

Vernier scale: As used on an Abney level, a scale graduated in degrees or percent. There was a movable arm from which one could read the difference of line of sight in degrees or percent from a true level.

Warbles: A variety of dipterous flies of the family Oestridae lay eggs on the feet and legs of grazing animals and then bite them. When the animal licks the irritated area of the bite, it ingests the eggs, which hatch internally. The larvae burrow through the tissues to the skin beneath the animal's back, where they pupate and cause lumps and open wounds called warbles.

Wattle: A long piece of skin that has been cut loose and let hang to help identify ranchers' livestock in the wintertime. Wattles are cut because brands and earmarks cannot be seen due to the thickness of fur on the livestock. Each rancher wattles his cattle in different places, such as the hip, shoulder, nose, et cetera.

Wet-mantle flood: A flood that occurs generally in the spring when there is still a heavy snow cover and a warm rain comes and melts the snow in a very short time.

Whippoorwill: A long pole used in chute logging. It was fastened at an angle across a chute to deflect the logs from their downward journey so they would roll out onto a landing.

Widowmaker: The burning top of a dead tree that sometimes falls off and crashes to the ground, making very little noise and giving little or no warning to nearby fire fighters.

Windfall: Trees blown down by the wind.

Wolf tree: A tree generally standing alone that is limbed to the ground and is a good producer of seed.

Works Progress Administration (WPA): Established in 1935 under the Emergency Relief Appropriations Act, the WPA was the major agency of this effort. It was designed to provide work for the jobless. By 1936 over three million persons were employed by the WPA in a wide variety of projects, including road building, bridge construction, and other public works, at the prevailing hourly wage.

Wow: A bend, as in a tree with a bend in it.

WPA: *See* Works Progress Administration.

INDEX

A

Adams, Mr., 232
Aldus, Doc, 221-222, 227
Anaconda Copper Company, 10-12, 37

B

Baring, Walter, 244
Bible, Alan, 244
Bishop, Merlin, 349
Black, Jim, 125-126
Blackfoot Protective Association, 38, 39
Blasting, 23-24, 231-232, 362-363
BLM. *See* Bureau of Land Management
Boulter, Reuben, 61, 62
Boyle, Joe, 145, 160
Boyle's Dude Ranch, 144, 156, 158, 160
Brado, Glenn E. (Bull), 127
Briggs, Alonzo E., 123, 209-210, 364
Brown, Bill, 21-22, 271, 285
Bureau of Land Management, 180-182, 215, 217, 232, 236, 237, 238, 257, 293, 298, 303, 310, 315-316

C

Cannon, Howard, 244
Casey, John, 259-260, 261-262
Cattle, 173-182, 184-187, 219-220, 236, 254-256, 259-261, 315-318, 330, 376
Cazier, Samuel E. (Ed), 59, 128, 131
CCC. *See* Civilian Conservation Corps

Chapin, Bill, 45, 46
Chevalier, Mr., 42, 43, 375
Church, James E., 187, 323
Civilian Conservation Corps, 47, 61-62, 70, 110-113, 117-118, 120, 127, 152, 155, 159, 168, 180, 185, 200, 294, 318, 321, 338, 344-345, 355, 364, 365, 378
Clark, Dan, 213-216, 219, 376, 380
Clark, Mrs. Dan, 214, 215
Clear cutting. See Timber, clear cutting of
Conyers, Shorty, 271, 306
Coonrod, Melvin A., 124
Copper Basin Cattle Association, 174-179, 191
Cougar, 229, 277-283
Croft, A. Russell (Bus), 321
Cruising for bugs. See Pine beetles, infestation and eradication of
Cruising timber. See Timber, surveying of
Cusick, Arthur M., 169-170
Cusick, Cassie, 170

D

Dalley, Robert S., 59, 60, 62
Deer, 221-222, 225-227, 280-282, 372
Dog Valley, 248-254, 329, 355-356
Dremolski, Louis A., 242, 369
Dressler, Fred, 202, 254-255
Dutch Charlie, 134-140

E

Elridge, George, 290, 298-300
Erosion, causes of, 248, 251, 254-255, 256, 316-317, 318, 321
Erosion, control of, 251-254, 256, 317, 320, 321-322
Estinson, Martin, 6

F

Falconberry, Len, 144
Falconberry ranch, 144, 156, 166
Favre, O. E., 53, 59
Finnegan, Mr., 20, 21
Fire, fighting, 14-16, 19-24, 27-28, 70, 111, 116-120, 262, 333-357, 376-377
Fish and Game Department, Nevada state, 221
Fish and Wildlife Service, U.S., 221-222, 225, 227, 229-230, 252, 281, 293-294
Flesher Pass, 44
Freece, Herbert J., 163-165, 166

G

Geiger Lake fire, 22-24, 27
Gissel, Harvey W., 261, 262
Granite Creek fire, 15, 19-21
Great Lakes Timber Company, 108, 109-110
Great Northern railroad, 2-3
Green River, 79, 80

Index

H

Hahn, Jay B., 71, 100, 101, 133, 200
Halverson, Hank, 164, 165, 167, 278, 279-280
Hells Canyon fire, 27, 349-354
Herbert, John M., 207-208, 209, 213, 214, 218, 231, 238, 370
Hinkley, Mr. (Ranger), 132
Horse Creek, 80, 124-127

I

Industrial Workers of the World, 37-40
IWW. *See* Industrial Workers of the World

J

Johansen, Pete, 236-237
Johnston, Velma, 301
Justice, Sterling R., 173, 174, 176

K

Kennecott Copper Corporation, 225, 228, 242, 244, 245, 323, 324
Ketchie, Henry L., 128
Kimball, Thatcher, 283
Kinnear, John, 245
Koch, Lewis B., 275
Kootenai River, 16
Koziol, Felix C. (Kozy), 118, 119

L

Langer, Charlie, 133, 169
Lincoln, Montana, 38
Lovell, Sam, 272

M

Mains, Guy B., 154
Markle, Merle G., 140, 150, 154, 163-164, 167, 364, 381
Martz, Mr., 46
Matthews, Tom, 53, 54, 59, 288
Maw, Edward C., 263
McConkie, Andrew R., 128, 131
McKee, Ernest E. (Mac), 132, 173, 189-191, 207, 208, 364, 365, 369
McKnight, Lyle P. (Mac), 38, 42-43, 44-45, 47, 52
McKnight, Mrs. Lyle P., 47, 52
Melville, Bob, 158
Miles, Clark, 70, 287-288
Mormon, Pop, 202
Murchie, Carol Ann, 197, 203-204
Murchie, Jack, 4
Murchie, James, 2, 3-8, 46, 299
Murchie, Jane (nee Boulter), 62, 64, 71, 98-102, 109, 123, 124, 132, 155-157, 167, 194-198, 199, 202-205, 207-208, 281-282, 364
Murchie, Jean, 194, 205
Murchie, Jimmie, 98-101, 109, 132, 193, 198, 199-200, 203-205
Murchie, John (Archie's uncle), 2

Murchie, Johnny (Archie's son), 191, 194, 197, 203, 205, 281-282, 355
Murchie, Laura (nee Peterson), 4, 7
Murchie, Mamie, 2
Murchie, Mary, 2, 5
Murchie, Pete (Archie's brother), 5, 6
Murchie, Peter (Archie's grandfather), 1-3, 5
Murchie, Susie, 2
Murchie, Will, 2
Mustang(s), 232-236, 288-290, 297-303

N

National Park Service, 243, 244
Newton, Doc, 190
North Dakota Agricultural College (Fargo), 9
Northern Pacific railroad, 37

O

Olsen, Chester J. (Chet), 103-104, 107, 108, 112, 113, 116, 120
Olson, Foyer, 209-210, 234

P

Pack, Bob, 107, 108
Palmer, Snap, 229
Parke, Morgan, 104-105, 118, 122
Parker, John W., 102, 227

Parsons, Dale, 179
Permits, special use, 106-107, 134, 138-139, 373-375
Permittees, grazing, 41, 42-43, 179-182, 184, 208-210, 213-216, 217-219, 220-221, 259-260, 261, 311-312, 315-316, 375-376
Peterson, Lars, 4, 5
Peterson, Olive, 4, 5
Phillabaum, Elva, 8
Pinchot, Gifford, 359
Pine beetles, infestation and eradication of, 55-56, 59-60, 63, 69, 111
Poison and its use, 229-230, 252-253, 257, 285, 292-294, 295, 296-297
Poisonous vegetation, 41-42, 185-186, 291, 294-295
Predatory animal control, 229-230, 276-277, 280, 291-294
Pulaski, Edward, 337

R

Range, overgrazing of, 148, 226, 232-233, 255-256, 309, 320
Range, reseeding of, 256-257, 320, 327-330
Range, salting of, 148, 149, 184, 307, 316-318
Rice, Ben, 365, 366
Roads, construction, maintenance and clearing of, 182-183, 373
Robinette, Les, 221, 227, 280
Robison, Bert, 208-211, 217-220, 375-376
Robison, Shirley, 225

Rosendo, 210-213
Rushton, Stephen M., 255

S

Sack, Ivan, 247-248, 253, 263, 329, 369
Sarles, North Dakota, 3, 6, 7, 8
Scaling ties. *See* Ties, scaling of
SCS. *See* Soil Conservation Service
Shank, Dan, 1, 2
Shank, Henry M., 153-154
Sheep, 41-43, 121, 141, 175-176, 211-212, 215-221, 226, 276, 283, 288-289, 305-315
Sheepherders, 210-213, 216, 276, 306-308, 310
Simes, Suzy, 225
Simpson, Val, 351, 352
Skarra, Perry, 59, 61-62, 66, 69
Smith's Fork, 79, 80
Snow, measuring water content of, 187-189, 323-327
Soil Conservation Service, 132, 187, 188, 323, 324
Standard Timber Company, 70, 71, 72, 73, 79, 82, 88, 90, 93-97, 102, 127, 128
Stemple Pass, 38, 44

T

Taynton, Roger (Shag), 259-260, 327
Templer, John, 51-52, 369
Thompson, Roy, 39-40
Tie hacks, 73-90, 94-98, 100-101
Ties, driving, 126-127
Ties, hewing of, 72-79, 80, 87-88, 94, 128-129
Ties, scaling of, 74, 90-98, 102
Timber, clear cutting of, 81-82, 108-109, 356-357
Timber, cutting of, 73-74, 81-83, 106-109, 356-357
Timber, sales of, 72, 81, 106, 357
Timber, surveying of, 59, 64-69
Trails, locating and maintaining of, 13-14, 142-144, 339, 368, 373
Trespass (range), 41-43, 121-122, 139, 179-182, 220, 257-261, 375-376
Trespass (timber), 106-110
Turner guard station, 14, 15

U

Uhalde, Alfred, 209
University of Montana Forestry School (Missoula), 9, 35, 36

V

Van Meter, Thomas H. (Van), 61

W

Watersheds, 248, 251, 254, 309, 320-321
West, Dick, 285-286
White Pine County Sportsmen's Association, 225-227

Wild Horse Annie. *See*
    Johnston, Velma
Wilson (Meyers Cove rancher),
    272
Winn, Field, 205
Winn, Roma, 205
Woggenson, Woggie, 349, 350
Woods, C. N., 153-155
Woods, Lowell G. (Woody) 64,
    66, 69, 124
Works Progress
    Administration, 113-116,
    117, 123, 364
WPA. *See* Works Progress
    Administration

Photo Credits

Dust jacket	Courtesy Archie Murchie
Frontispiece	Photo by R. T. King
Page 17	Pick Mountain lookout before improvement, Weiser National Forest. Courtesy USDA Forest Service, Region Four
Page 25	Courtesy USDA Forest Service, Region Four
Page 33	Map 1 by Kris Pizarro
Page 57	Lyle Springs crew completing the job of oiling infested trees for burning on the Targhee-Teton National Forest mountain pine beetle control project. Courtesy USDA Forest Service, Region Four
Page 58	Map 2 by Kris Pizarro
Page 77	*Top:* Courtesy Archie Murchie. *Bottom:* Forest ranger counting ties packed for hauling on the

	Wasatch National Forest. Courtesy USDA Forest Service, Region Four
Page 78	Building a splash dam on Bear Creek, Bridger National Forest. Courtesy USDA Forest Service, Region Four
Page 91	Forest service scaler on Thorn Creek sale, Boise National Forest. Courtesy USDA Forest Service, Region Four
Page 135	Courtesy Archie Murchie
Page 136	Map 3 by Kris Pizarro
Page 195	Loon Creek ranger station, Challis National Forest. Courtesy USDA Forest Service, Region Four
Page 196	Courtesy Archie Murchie
Page 223	Courtesy USDA Forest Service, Region Four
Page 224	Map 4 by Kris Pizarro
Page 249	Map 5 by Kris Pizarro
Page 267	Courtesy Archie Murchie
Page 313	Cavanaugh sheepherder's camp on private land in Stanley basin, Challis National Forest. Courtesy USDA Forest Service, Region Four
Page 314	Counting sheep on the Humboldt National Forest in 1938. Courtesy USDA Forest Service, Region Four
Page 325	*Top:* Snow survey studies in 1930 at the Trinity ranger station, Boise National Forest. Courtesy USDA Forest Service, Region four. *Bottom:* Ash Canyon rehabilitation project in 1969 on the

Photo Credits

Toiyabe National Forest. Courtesy USDA Forest Service, Region Four

Page 341   Spring telephone maintenance in 1923, Weiser National Forest. Courtesy USDA Forest Service, Region Four

THE FREE LIFE OF A RANGER

Text and photo mechanicals designed by Helen M. Blue
Camera-ready master composed and printed
at the University of Nevada Oral History Program
in Goudy Old Style using WordPerfect 5.1,
Bitstream Facelift and a Hewlett-Packard Laserjet III

Printed and bound by Braun-Brumfield, Inc.